Dietary Bioactive Compounds: Implications for Oxidative Stress and Inflammation

Dietary Bioactive Compounds: Implications for Oxidative Stress and Inflammation

Guest Editors

Maria Digiacomo
Doretta Cuffaro

MDPI

Basel • Beijing • Wuhan • Barcelona • Belgrade • Novi Sad • Cluj • Manchester

Guest Editors
Maria Digiacomo
Department of Pharmacy
University of Pisa
Pisa
Italy

Doretta Cuffaro
Department of Pharmacy
University of Pisa
Pisa
Italy

Editorial Office
MDPI AG
Grosspeteranlage 5
4052 Basel, Switzerland

This is a reprint of the Special Issue, published open access by the journal *Nutrients* (ISSN 2072-6643), freely accessible at: https://www.mdpi.com/journal/nutrients/special_issues/bioactive_oxidative.

For citation purposes, cite each article independently as indicated on the article page online and as indicated below:

Lastname, A.A.; Lastname, B.B. Article Title. *Journal Name* **Year**, *Volume Number*, Page Range.

ISBN 978-3-7258-3417-4 (Hbk)
ISBN 978-3-7258-3418-1 (PDF)
https://doi.org/10.3390/books978-3-7258-3418-1

Contents

Preface

The following reprint of the Special Issue (SI) "Dietary Bioactive Compounds: Implications for Oxidative Stress and Inflammation" describes different aspects of the nutraceutical potential of dietary bioactives present in food and food by-products. The SI includes original papers and reviews on the implications of dietary bioactive compounds on different molecular pathways, including novel in vitro and in vivo studies and discussion of the crucial step of bioavaibility fundamental to the bioefficacy of every dietary element.

Maria Digiacomo and Doretta Cuffaro
Guest Editors

nutrients MDPI

Editorial

Dietary Bioactive Compounds: Implications for Oxidative Stress and Inflammation

Doretta Cuffaro [1,2], Maria Digiacomo [1,2,*] and Marco Macchia [1,2]

1 Department of Pharmacy, University of Pisa, 56126 Pisa, Italy; doretta.cuffaro@unipi.it (D.C.); marco.macchia@unipi.it (M.M.)
2 Interdepartmental Research Center "Nutraceuticals and Food for Health", University of Pisa, 56100 Pisa, Italy
* Correspondence: maria.digiacomo@unipi.it

Citation: Cuffaro, D.; Digiacomo, M.; Macchia, M. Dietary Bioactive Compounds: Implications for Oxidative Stress and Inflammation. *Nutrients* **2023**, *15*, 4966. https://doi.org/10.3390/nu15234966

Academic Editor: Maria Dolores del Castillo

Received: 25 October 2023
Accepted: 20 November 2023
Published: 30 November 2023

Nowadays, it has been amply demonstrated how an appropriate diet and lifestyle are essential for preserving wellbeing and preventing illnesses [1,2]. Dietary bioactive substances are natural constituents present in food that provide health benefits. These compounds can be defined as substances capable of modulating biological activities and important physiological functions [3–5]. It is known that the consumption of dietary bioactive compounds, especially phytochemicals, contribute to decreasing the risk of chronic disorders, such as cancer and cardiovascular and neurodegenerative diseases [6–10].

In the last few years, the study of the efficacy of these natural bioactive compounds has attracted the increasing interest of researchers. In this regard, this Special Issue in *Nutrients* entitled "Dietary Bioactive Compounds: Implications for Oxidative Stress and Inflammation", which includes a review and 12 original articles, provides an insight on current knowledge about the nutraceutical effects of different bioactive compounds present in foods or their related byproducts on some pathologies, with a special focus on their antioxidant and anti-inflammatory properties.

In particular, the role of various bioactive-rich extracts on gastrointestinal disorders [11] have been further discussed, as documented by three papers that deal with the effects of a number of natural extracts in pathologies affecting the digestive system.

Lo et al. [12] studied the potential treatment of ulcerative colitis (UC) with the oral administration of ethanol extracts from rice bran and whole-grain adlay seeds, in mice with dinitrobenzene sulfonic acid (DNBS)-induced UC. Their results indicated that rice bran ethanol extract reduces UC-induced damage in the colon along with inflammation and oxidative stress in DNBS-induced UC mice. Moreover, whole-grain adlay seed ethanol extract is able to modulate colonic inflammation and clinical symptoms in UC mice. Additionally, both of the extracts reversed DNBS-induced alterations in T-helper-cell-associated cytokines and glutathione in the colon. Unfortunately, in this paper, the composition of the ethanol extracts was not defined; thus, the main bioactive compound(s) in rice bran and whole-grain adlay seeds responsible for their activities was not identified.

Chen et al. [13] explored the effect of Shibi tea (EST) on liver injury in an in vivo mouse study. EST is a non-Camellia tea prepared by the infusion of dried *Adinandra nitida* leaves, which are rich in flavonoids and especially in Camellianin A (CA) [14]. In this work, for the first time, the benefit of EST and CA in liver injury was investigated, with the authors exploiting the hepatoprotective effects of EST and CA extracts in a carbon tetrachloride (CCl_4)-induced acute-liver-injury mouse model. Additionally, the anti-inflammatory, anti-apoptosis, and antioxidative effects of EST and CA in repairing acute liver injury were explored, with the authors analyzing the regulation of the oxidative stress signaling pathways and the expression of inflammatory cytokines and phosphorylated nuclear factors. The results highlight that EST and CA display anti-inflammatory, anti-apoptosis, and antioxidative properties, and could be promising agents in the prevention of liver injuries.

Liu et al. [15] evaluated the role of C-phycocyanin (CPC) and *Lycium barbarum* polysaccharides (LBP) on aspirin-induced gastric damage in rat gastric mucosal (RGM-1) cells.

The primary active compounds, CPC and LBP, found in Spirulina platensis and wolfberry, respectively, possess antioxidative, anti-inflammatory, and immunoregulatory properties [15,16]. Therefore, the aim of this study was to use these food ingredients to contrast the gastric damage caused by the (i.e., acetyl salicylic acid) chronic administration of aspirin, for anti-inflammatory or cardiovascular purposes [17]. The evaluation of CPC and LBP at a high dose of 500 μg/mL demonstrated a promising ability to attenuate aspirin-gastric injury in gastric RGM-1 cells. CPC and LBP affected the activation of ERK and JNK signaling pathways, increasing the expression of the anti-inflammatory interleukin 10 (IL-10) and modulating proinflammatory markers (NF-κB, caspase 3, Bax protein). In particular, CPC and/or LBP exhibited anti-inflammatory effects by inhibiting activation of the ERK signaling pathway, while LBP reduced apoptosis by decreasing activation of the JNK signaling pathway in gastric RGM-1 cells with aspirin-induced epithelial damage.

In other diseases, such as skin disorders, the health benefits of bioactive compounds contained in natural extracts were exploited [18–21]. In fact, many studies discussed the role of natural extracts in skin disorders and in particular in atopic dermatitis (AD) [22–24]. In this Special Issue, Bae et al. [25] exploited the effect of *Daphnopsis costaricensis* extract on an in vivo AD lesion model. This paper outlines the impact of the *D. costaricensis* EtOH extract (DCE) on AD-like lesions in a mouse model induced with oxazolone (OX). The findings indicate that DCE significantly improved AD-like pathology by reducing ear epidermal thickness and mast cell infiltration. Additionally, eleven compounds, with a flavonoid-like structure, were isolated and identified in extract, through the use of 1D and 2D NMR and HR-MS data. These compounds have been evaluated for their anti-inflammatory and anti-allergic activities, demonstrating that 7,8-dimethoxyflavone and 7,2′-dimethoxyflavone were able to inhibit IL-4 overproduction and mast cell degranulation in vitro (in a rat basophilic leukemia cell line, RBL-2H3 cells). Therefore, DCE could be useful in the treatment of AD.

Beyond the nutraceutical properties widely demonstrated of a lot of foods, recently the study of agri-food wastes, as a source of bioactive compounds, has also received growing interest [26–29]. Moreover, these wastes are related to an authentic environmental drawback, and consequently waste disposal becomes a challenge in the agri-food industrial economy [30–33]. Different residues, such as pomace, leaves, wastewaters, peels, stems or flowers are discarded and a real effort is being placed on the valorization of food chain byproducts [34]. The re-use of agrifood wastes as an a renewable, abundant and low-cost source for the production of high value products, is presently being exploited. In this Special Issue, seven papers explored the important valorization of food byproducts, giving a wide point of views on this in-trend topic.

Two papers investigated the valorization of two main byproducts of extra virgin olive oil (EVOO) production: olive mill wastewater (OMWW) and olive leaves. The nutraceutical proprieties of EVOO are already well known and confer a leading role to EVOO in the health benefits of the Mediterranean diet [35]. Despite dietary benefits, EVOO production has a great impact in terms of sustainability due to the difficult management of its related wastes such as olive leaves and OMWW [36–40]. These discards contain phenolic compounds endowed with nutraceutical properties and they are a promising font of bioactive compounds. Therefore, correct waste management, which considers all potential paths for a circular economy of the olive oil supply chain, is crucial.

Cuffaro et al. [41] reported the study of improved nutraceutical properties of EVOO extract by adding different percentages of olive leaf extracts (OLE). In this study, a promising EVOO extract enriched by 8% OLE was selected. This extract contained important nutraceutical polyphenols such as oleocanthal, oleacein and oleuropein in similar quantities, as reported by HPLC analysis. The 8% OLE-EVOO extract showed increased antioxidant and antiradical properties (evaluated by in vitro assays (DPPH, ABTS and FRAP)) with respect to the EVOO extract, evidencing a potential effect in reducing oxidative stress. Moreover, its anti-inflammatory activity was evaluated in terms of cyclooxygenase (COX) enzyme inhibition. The 8% OLE-EVOO extract displayed four-fold improved COX-1 inhibition and two-fold COX-2 inhibition (COX-1 IC_{50} = 475 μg/mL; COX-2 IC_{50} = 383 μg/mL) with respect to the EVOO extract (COX-1 IC_{50} = 1.90 mg/mL; COX-2 IC_{50} = 900 μg/mL).

The same research group studied the potential beneficial effect of bioactive polyphenols in OMWW extracts [42]. The aim of this study was to point out a possible nutraceutical valorization of this byproduct, exploiting its biological properties. Similarly to EVOO [43], the composition of OMWW varied depending on the olive cultivar and extraction system. This study assessed multiple samples of three-phase extraction OMWWs obtained from two cultivars of olives, Leccino (CL) and Frantoio (CF), collected in October (CF1 and CL1) and November (CF2 and CL2). The polyphenolic profile (18 polyphenols) of OMWW extracts was defined using quali-quantitative analysis performed by LC-MS/MS, revealing high amount of tyrosol and hydroxytyrosol in all the samples, and oleacein in the October samples, rarely found in OMWW. Furthermore, the antioxidant profile and the anti-inflammatory effect in terms of COX 1 and COX2 inhibition were evaluated, resulting in significantly low values of IC_{50} COX-2 inhibition for oleacein-rich extracts (ranging 0.080 mg/mL).

Similarly to EVOO, the beer industry is environmentally affected by the difficult management of byproducts [44–47], especially brewers' spent grain (BSG). BSG is the most abundant byproduct of the brewing industry, composed mainly of dietary fiber (50%), proteins (30%), and bioactive compounds, such as hydroxycinnamic acids [48–50]. In this Special Issue, Gutierrez-Barrutia [51] investigated how extrusion process, a thermomechanical procedure characterized by high temperature and pressure for a short period of time, influences the bioaccessibility of BSG nutrients (glucose and amino acids) and non-nutrients (phenolic compounds). The study revealed that the extrusion process did not affect glucose bioaccessibility or gluten digestibility, favoring amino acid release during digestion. On the other hand, significantly improved gastrointestinal and colonic bioaccessibility of BSG phenolic compounds was achieved in extruded BSG. Moreover, extruded BSG intestinal digests inhibited glucose transport through increased α-glucosidase and α-amylase inhibition, compared to untreated BSG. Additionally, extruded BSG intestinal digests inhibited intracellular ROS formation and showed anti-inflammatory properties, unlike untreated BSG. Therefore, the extrusion process improved the nutritional value and biological properties of BSG, so the extruded BSG has the potential to be a sustainable and promising ingredient with positive properties that could be useful in both the prevention and treatment of various non-communicable diseases.

The valorization of byproducts affects not only food production, but also fruit collection [52,53] such as cherry, tangerine and chestnut, and edible mushrooms [54,55]. This Special Issue explored all these different fields of applications thanks to the various reported studies.

Nunes et al. [56] discussed the anti-inflammatory and antimicrobial potential of different extracts from the stems, leaves and flowers of *Prunus avium* L., a cultivar of Portuguese cherry from the Fundão region [57,58]. The hydroethanolic extracts of leaves and stems and the aqueous infusion of flowers were effective in reducing inflammation in terms of a decreased level of NO production on a lypopolisacharide (LPS)-stimulated mouse macrophage (RAW 264.7) cell line. Furthermore, the aqueous infusions of all by-products, especially cherry flowers, demonstrated scavenging activity against NO radicals. Moreover, the leaf extracts showed an antimicrobial effect against most of the tested bacteria.

Tangerine peel (*Citrus reticulatae Pericarpium*, CRP) is the main byproduct in the cultivation of tangerine [59–61]. Dried tangerine peel is a traditional Chinese medicine already studied for its nutraceutical properties. Its main components are volatile oils and flavonoids such as hesperidin, naringenin, nobiletin, and tangeretin [61]. Here, Wang et al. [62] discussed the protective effect of Tangerine peel extract in endothelial dysfunction and vascular inflammation related to diabetes on AMPK activation in a rat model. CRP extract was orally administered for 4 weeks at 400 mg/kg/day in high-fat diet/streptozotocin (HFD/STZ)-induced diabetic rats. After treatment, the rats successfully reversed all the diabetic symptoms, resulting in a normalized blood pressure and plasma lipid profile, as well as plasma levels of liver enzymes in diabetic rats. The CRP extract also suppressed vascular inflammatory markers, inducing AMPK activation in the aortas of diabetic rats. Therefore, administering tangerine peel extract chronically can safeguard arteries from vascular inflammation and endothelial dysfunction in diabetic rats.

Flammulina velutipes (FV) is an edible mushroom presenting nutritional and medicinal values. FV mycorrhizae is a poorly studied by-product of FV representing an important source of bioactive compounds [63,64]. In their study, Luo et al. [65] investigated, at first, the composition of FV mycorrhizae, and secondly its effects on lipid metabolism and inflammation in perirenal adipose tissue (PAT), which is still unknown. The composition of FV mycorrhizae comprehend multifarious nutritive components among them polysaccharide, amino acids and derivatives, and organic compounds. HFD supplemented with 4% FV mycorrhizae (HFDFV) caused the attenuation of HFD-induced lipid disorders, reducing HFD-induced oxidative stress and pro inflammatory cytokines, both in the liver and perirenal adipose tissue (PAT) of mice. These results indicated the promising application of FV mycorrhizae as a functional food and herbal medicine in the treatment of obesity.

Piazza et al. [66] described, for the first time, the nutraceutical potential of *Castanea sativa* Mill. leaf extracts in gastritis caused by *Helicobacter pylori* (*H. pylori*) infection. The authors evaluated the polyphenolic profile in *Castanea sativa* Mill. leaf extracts, with a particular focus on ellagitannins [67,68], a nutraceutical polyphenol class with potential gastroprotective properties. Castalagin and vescalagin were identified and quantified in hydroalcoholic leaf extracts using LC–MS. Thus, the anti-inflammatory and antibacterial activity of leaf extracts, in comparison with pure castalagin, were investigated in a model of human gastric epithelium (GES-1) infected by *H. pylori*. The leaf extract and pure ellagitannins inhibited IL-8 release ($IC_{50} \approx 28$ µg/mL and 11 µM, respectively), partially attenuating NF-κB signaling and reducing bacterial growth and cell adhesion. These results were also confirmed by transcriptional studies in which castalagin was able to decrease genes involved in inflammatory pathways (NF-κB and AP-1) and cell migration (Rho GTPase). These observations suggested that *Castanea sativa* Mill. leaves could be adopted to produce sustainable and bioactive extracts.

As seen in the various studies reported in this Special Issue and also in the numerous papers present in the literature, bioactive compounds include a plethora of molecules, such as phenolics, carotenoids, vitamins, minerals, and fibers, ubiquitously expressed in fruits, vegetables, grains and legumes [53]. Usually, the most representative polyphenols in food are phenolic acids, flavonoids and anthocyanins [69,70]. Anthocyanins are a class of flavonoids offering various health benefits, and are responsible for the blue, purple, and red pigments detected in different type of fruits and vegetables [71,72]. These polyphenols exert a wide range of nutraceutical activities, including antioxidant and anti-inflammatory properties, associated with cardioprotection, anti-carcinogenicity or neuroprotection [71]. Among the more than 1000 types of anthocyanins present in the literature, Malvidin is one of the most studied. In this Special Issue, Merecz-Sadowska et al. [73] collected and reviewed the reported studies in the literature investigating the role of malvidin and its related glucosides in different cell, animal and human models. Besides their colorant capacity, malvidin and its related glycosides revealed a widespread range of beneficial properties promoted by antioxidant and anti-inflammatory mechanisms. In addition, these molecules showed the ability to counteract the onset and progression of several diseases whose pathogeneses are linked to oxidative stress. These findings suggest a potential future application for malvidin and its glycosides as ingredients of functional food, able to offer both aesthetic and nutritional advantages.

Despite the extensive studies demonstrating the health effects of bioactive compounds, such as polyphenols, in humans, one of the major drawbacks in the development of single polyphenols or polyphenol-rich natural extracts, as functional ingredients or dietary supplements, is related to their pharmacokinetic profile, compromised by their poor aqueous solubility, intensive metabolism, and low systemic absorption [74–76]. Only a few studies have considered the aspects of bioavailability and metabolism [77–79]. However, most of these have reported in vitro models of pure compounds in which the matrix effects were not considered [69,80,81].

In this context, innovative strategies could be developed to deliver pure polyphenols or natural extracts rich in bioactive compounds [82,83]. One of the alternative delivery methods, able to increase the solubility of active substances by bypassing metabolism, is represented by electrospun nanofibres incorporating bioactive substances [84–86]. In this Special Issue, Paczkowska-Walendowska et al. [87] reported the development of *P. cuspidati* radix extract nanofibers as an innovative approach in solid dispersion for the buccal delivery system. At first, the authors optimized the extraction process in order to obtain the best extract in terms of richness in bioactive compounds, especially stilbenes such as resveratrol and polydantins. The selected *P. cuspidati* radix extract was incorporated in nanofibers based on polyvinylpyrrolidone/cyclodextrin (PVP/HPβCD) using an electrospinning technique, affording nanofibers the six-fold improved solubility of resveratrol and polydantins compared to pure standards. Thus, the electrospun nanofibers may be easily applied within the oral cavity, immediately releasing the incorporated bioactives. These results, along with the intrinsic antioxidant and anti-inflammatory properties of *P. cuspidati* extract, indicate that the buccal delivery system might be an alternative strategy to improve the bioavailability of bioactives.

In conclusion, the set of studies collected in the Special Issue "Dietary Bioactive Compounds: Implications for Oxidative Stress and Inflammation" have explored different aspects of dietary bioactive compounds. The application of different nutraceutical extracts rich in bioactive compounds on gastrointestinal injury and skin disorder have been analyzed. Furthermore, the actual trend topic of agrifood wastes has been widely explored here, with studies analyzing the potentiality of nutraceutical extracts of different byproducts of the food industry (EVOO and beer), fruits (sweet cherry, chestnuts and tangerine) and mushrooms. Moreover, an insight on anthocyanins, an important class of bioactive compound, was proposed. Additionally, aspects related to metabolism and the bioavailability of bioactive compounds have been attended to in proposition of the use of nanofibers.

All these papers highlight that dietary bioactive compounds endowed with antioxidant and anti-inflammatory properties could be beneficial for health and should be further studied to develop functional foods or food supplements.

Author Contributions: Conceptualization, M.D. and D.C.; writing—original draft preparation, M.D. and D.C.; writing—review and editing, M.D., M.M. and D.C.; funding acquisition, M.M. All authors have read and agreed to the published version of the manuscript.

Funding: This research was funded by Ministry of University and Research (MUR) as a part of the PON 2014–2020 "Research and Innovation" resources—Green/Innovation Action—DM MUR 1062/2021—Title of research "Sviluppo di una piattaforma tecnologica per lo studio delle proprietà nutraceutiche di biomolecole e biomateriali presenti negli scarti derivanti dalla filiera dei prodotti alimentari".

Conflicts of Interest: The authors declare no conflict of interest.

References

1. de Ridder, D.; Kroese, F.; Evers, C.; Adriaanse, M.; Gillebaart, M. Healthy Diet: Health Impact, Prevalence, Correlates, and Interventions. *Psychol. Health* **2017**, *32*, 907–941. [CrossRef] [PubMed]
2. Rockström, J.; Edenhofer, O.; Gaertner, J.; DeClerck, F. Planet-Proofing the Global Food System. *Nat. Food* **2020**, *1*, 3–5. [CrossRef]
3. Bessada, S.M.F.; Barreira, J.C.M.; Oliveira, M.B.P.P. Pulses and Food Security: Dietary Protein, Digestibility, Bioactive and Functional Properties. *Trends Food Sci. Technol.* **2019**, *93*, 53–68. [CrossRef]
4. Liu, R.H. Dietary Bioactive Compounds and Their Health Implications. *J. Food Sci.* **2013**, *78*, A18–A25. [CrossRef]
5. Pan, M.-H.; Lai, C.-S.; Dushenkov, S.; Ho, C.-T. Modulation of Inflammatory Genes by Natural Dietary Bioactive Compounds. *J. Agric. Food Chem.* **2009**, *57*, 4467–4477. [CrossRef]
6. Howes, M.-J.R.; Simmonds, M.S.J. The Role of Phytochemicals as Micronutrients in Health and Disease. *Curr. Opin. Clin. Nutr. Metab. Care* **2014**, *17*, 558. [CrossRef]
7. Quero, J.; Mármol, I.; Cerrada, E.; Rodríguez-Yoldi, M.J. Insight into the Potential Application of Polyphenol-Rich Dietary Intervention in Degenerative Disease Management. *Food Funct.* **2020**, *11*, 2805–2825. [CrossRef]

8. Sharifi-Rad, J.; Rodrigues, C.F.; Sharopov, F.; Docea, A.O.; Can Karaca, A.; Sharifi-Rad, M.; Kahveci Karıncaoglu, D.; Gülseren, G.; Şenol, E.; Demircan, E.; et al. Diet, Lifestyle and Cardiovascular Diseases: Linking Pathophysiology to Cardioprotective Effects of Natural Bioactive Compounds. *Int. J. Environ. Res. Public Health* **2020**, *17*, 2326. [CrossRef]
9. Magrone, T.; Perez de Heredia, F.; Jirillo, E.; Morabito, G.; Marcos, A.; Serafini, M. Functional Foods and Nutraceuticals as Therapeutic Tools for the Treatment of Diet-Related Diseases. *Can. J. Physiol. Pharmacol.* **2013**, *91*, 387–396. [CrossRef]
10. Kromhout, D. Diet and Cardiovascular Diseases. *J. Nutr. Health Aging* **2001**, *5*, 144–149.
11. Dryden, G.W.; Song, M.; McClain, C. Polyphenols and Gastrointestinal Diseases. *Curr. Opin. Gastroenterol.* **2006**, *22*, 165–170. [CrossRef] [PubMed]
12. Lo, H.-C.; Chen, Y.-H.; Wu, W.-T. Ethanol Extracts of Rice Bran and Whole Grain Adlay Seeds Mitigate Colonic Inflammation and Damage in Mice with Colitis. *Nutrients* **2022**, *14*, 3877. [CrossRef] [PubMed]
13. Chen, R.; Lian, Y.; Wen, S.; Li, Q.; Sun, L.; Lai, X.; Zhang, Z.; Zhu, J.; Tang, L.; Xuan, J.; et al. Shibi Tea (*Adinandra nitida*) and Camellianin A Alleviate CCl4-Induced Liver Injury in C57BL-6J Mice by Attenuation of Oxidative Stress, Inflammation, and Apoptosis. *Nutrients* **2022**, *14*, 3037. [CrossRef] [PubMed]
14. Zhang, J.; Yang, J.; Duan, J.; Liang, Z.; Zhang, L.; Huo, Y.; Zhang, Y. Quantitative and Qualitative Analysis of Flavonoids in Leaves of *Adinandra nitida* by High Performance Liquid Chromatography with UV and Electrospray Ionization Tandem Mass Spectrometry Detection. *Anal. Chim. Acta* **2005**, *532*, 97–104. [CrossRef]
15. Liu, Y.-C.; Chang, C.-C.; Matsui, H.; Chao, J.C.-J. C-Phycocyanin and *Lycium barbarum* Polysaccharides Protect against Aspirin-Induced Inflammation and Apoptosis in Gastric RGM-1 Cells. *Nutrients* **2022**, *14*, 5113. [CrossRef] [PubMed]
16. Kwok, S.S.; Bu, Y.; Lo, A.C.-Y.; Chan, T.C.-Y.; So, K.F.; Lai, J.S.-M.; Shih, K.C. A Systematic Review of Potential Therapeutic Use of *Lycium barbarum* Polysaccharides in Disease. *BioMed Res. Int.* **2019**, *2019*, e4615745. [CrossRef] [PubMed]
17. Sostres, C.; Lanas, A. Gastrointestinal Effects of Aspirin. *Nat. Rev. Gastroenterol. Hepatol.* **2011**, *8*, 385–394. [CrossRef]
18. Afaq, F.; Katiyar, S.K. Polyphenols: Skin Photoprotection and Inhibition of Photocarcinogenesis. *Mini Rev. Med. Chem.* **2011**, *11*, 1200–1215. [CrossRef]
19. Nichols, J.A.; Katiyar, S.K. Skin Photoprotection by Natural Polyphenols: Anti-Inflammatory, Antioxidant and DNA Repair Mechanisms. *Arch. Dermatol. Res.* **2010**, *302*, 71–83. [CrossRef]
20. Souto, E.B.; Sampaio, A.C.; Campos, J.R.; Martins-Gomes, C.; Aires, A.; Silva, A.M. Chapter 2—Polyphenols for Skin Cancer: Chemical Properties, Structure-Related Mechanisms of Action and New Delivery Systems. In *Studies in Natural Products Chemistry*; Atta-ur-Rahman, Ed.; Bioactive Natural Products; Elsevier: Amsterdam, The Netherlands, 2019; Volume 63, pp. 21–42.
21. Tuong, W.; Walker, L.; Sivamani, R.K. Polyphenols as Novel Treatment Options for Dermatological Diseases: A Systematic Review of Clinical Trials. *J. Dermatol. Treat.* **2015**, *26*, 381–388. [CrossRef]
22. Wu, S.; Pang, Y.; He, Y.; Zhang, X.; Peng, L.; Guo, J.; Zeng, J. A Comprehensive Review of Natural Products against Atopic Dermatitis: Flavonoids, Alkaloids, Terpenes, Glycosides and Other Compounds. *Biomed. Pharmacother.* **2021**, *140*, 111741. [CrossRef] [PubMed]
23. Di Salvo, E.; Gangemi, S.; Genovese, C.; Cicero, N.; Casciaro, M. Polyphenols from Mediterranean Plants: Biological Activities for Skin Photoprotection in Atopic Dermatitis, Psoriasis, and Chronic Urticaria. *Plants* **2023**, *12*, 3579. [CrossRef] [PubMed]
24. Karuppagounder, V.; Arumugam, S.; Thandavarayan, R.A.; Sreedhar, R.; Giridharan, V.V.; Watanabe, K. Molecular Targets of Quercetin with Anti-Inflammatory Properties in Atopic Dermatitis. *Drug Discov. Today* **2016**, *21*, 632–639. [CrossRef] [PubMed]
25. Bae, Y.; Kim, T.; Park, N.; Choi, S.; Yi, D.; Soto, S.; Zamora, N.; Kim, S.; Yang, M. Ameliorative Effects of *Daphnopsis Costaricensis* Extract against Oxazolone-Induced Atopic Dermatitis-like Lesions in BALB/c Mice. *Nutrients* **2022**, *14*, 4521. [CrossRef] [PubMed]
26. Kumar, K.; Yadav, A.N.; Kumar, V.; Vyas, P.; Dhaliwal, H.S. Food Waste: A Potential Bioresource for Extraction of Nutraceuticals and Bioactive Compounds. *Bioresour. Bioprocess.* **2017**, *4*, 18. [CrossRef]
27. Espro, C.; Paone, E.; Mauriello, F.; Gotti, R.; Uliassi, E.; Bolognesi, M.L.; Rodríguez-Padrón, D.; Luque, R. Sustainable Production of Pharmaceutical, Nutraceutical and Bioactive Compounds from Biomass and Waste. *Chem. Soc. Rev.* **2021**, *50*, 11191–11207. [CrossRef]
28. Vilas-Boas, A.A.; Pintado, M.; Oliveira, A.L.S. Natural Bioactive Compounds from Food Waste: Toxicity and Safety Concerns. *Foods* **2021**, *10*, 1564. [CrossRef]
29. Rudra, S.G.; Nishad, J.; Jakhar, N.; Kaur, C. Food Industry Waste: Mine of Nutraceuticals. *Int. J. Sci. Environ. Technol.* **2015**, *4*, 205–229.
30. Iacovidou, E.; Ohandja, D.-G.; Gronow, J.; Voulvoulis, N. The Household Use of Food Waste Disposal Units as a Waste Management Option: A Review. *Crit. Rev. Environ. Sci. Technol.* **2012**, *42*, 1485–1508. [CrossRef]
31. Cecchi, F.; Cavinato, C. Smart Approaches to Food Waste Final Disposal. *Int. J. Environ. Res. Public Health* **2019**, *16*, 2860. [CrossRef]
32. Thyberg, K.L.; Tonjes, D.J.; Gurevitch, J. Quantification of Food Waste Disposal in the United States: A Meta-Analysis. *Environ. Sci. Technol.* **2015**, *49*, 13946–13953. [CrossRef] [PubMed]
33. Paritosh, K.; Kushwaha, S.K.; Yadav, M.; Pareek, N.; Chawade, A.; Vivekanand, V. Food Waste to Energy: An Overview of Sustainable Approaches for Food Waste Management and Nutrient Recycling. *BioMed Res. Int.* **2017**, *2017*, e2370927. [CrossRef] [PubMed]
34. Said, Z.; Sharma, P.; Thi Bich Nhuong, Q.; Bora, B.J.; Lichtfouse, E.; Khalid, H.M.; Luque, R.; Nguyen, X.P.; Hoang, A.T. Intelligent Approaches for Sustainable Management and Valorisation of Food Waste. *Bioresour. Technol.* **2023**, *377*, 128952. [CrossRef] [PubMed]

35. Carluccio, M.A.; Siculella, L.; Ancora, M.A.; Massaro, M.; Scoditti, E.; Storelli, C.; Visioli, F.; Distante, A.; De Caterina, R. Olive Oil and Red Wine Antioxidant Polyphenols Inhibit Endothelial Activation: Antiatherogenic Properties of Mediterranean Diet Phytochemicals. *Arter. Thromb. Vasc. Biol.* **2003**, *23*, 622–629. [CrossRef] [PubMed]
36. Nunes, M.A.; Pimentel, F.B.; Costa, A.S.G.; Alves, R.C.; Oliveira, M.B.P.P. Olive By-Products for Functional and Food Applications: Challenging Opportunities to Face Environmental Constraints. *Innov. Food Sci. Emerg. Technol.* **2016**, *35*, 139–148. [CrossRef]
37. Foti, P.; Romeo, F.V.; Russo, N.; Pino, A.; Vaccalluzzo, A.; Caggia, C.; Randazzo, C.L. Olive Mill Wastewater as Renewable Raw Materials to Generate High Added-Value Ingredients for Agro-Food Industries. *Appl. Sci.* **2021**, *11*, 7511. [CrossRef]
38. Zahra El Hassani, F.; El Karkouri, A.; Errachidi, F.; Merzouki, M.; Benlemlih, M. The Impact of Olive Mill Wastewater Spreading on Soil and Plant in Arid and Semi-Arid Areas. *Environ. Nanotechnol. Monit. Manag.* **2023**, *20*, 100798. [CrossRef]
39. Erbay, Z.; Icier, F. The Importance and Potential Uses of Olive Leaves. *Food Rev. Int.* **2010**, *26*, 319–334. [CrossRef]
40. Araújo, M.; Pimentel, F.B.; Alves, R.C.; Oliveira, M.B.P.P. Phenolic Compounds from Olive Mill Wastes: Health Effects, Analytical Approach and Application as Food Antioxidants. *Trends Food Sci. Technol.* **2015**, *45*, 200–211. [CrossRef]
41. Cuffaro, D.; Bertini, S.; Macchia, M.; Digiacomo, M. Enhanced Nutraceutical Properties of Extra Virgin Olive Oil Extract by Olive Leaf Enrichment. *Nutrients* **2023**, *15*, 1073. [CrossRef]
42. Cuffaro, D.; Bertolini, A.; Bertini, S.; Ricci, C.; Cascone, M.G.; Danti, S.; Saba, A.; Macchia, M.; Digiacomo, M. Olive Mill Wastewater as Source of Polyphenols with Nutraceutical Properties. *Nutrients* **2023**, *15*, 3746. [CrossRef] [PubMed]
43. Esposito Salsano, J.; Digiacomo, M.; Cuffaro, D.; Bertini, S.; Macchia, M. Content Variations in Oleocanthalic Acid and Other Phenolic Compounds in Extra-Virgin Olive Oil during Storage. *Foods* **2022**, *11*, 1354. [CrossRef] [PubMed]
44. Thiago, R.D.S.M.; Pedro, P.M.D.M.; Eliana, F.C.S. Solid Wastes in Brewing Process: A Review. *J. Brew. Distill.* **2014**, *5*, 1–9. [CrossRef]
45. Kerby, C.; Vriesekoop, F. An Overview of the Utilisation of Brewery By-Products as Generated by British Craft Breweries. *Beverages* **2017**, *3*, 24. [CrossRef]
46. Huige, N.J. Brewery By-Products and Effluents. In *Handbook of Brewing*; CRC Press: Boca Raton, FL, USA, 2006; ISBN 978-0-429-11617-9.
47. Kanagachandran, K.; Jayaratne, R. Utilization Potential of Brewery Waste Water Sludge as an Organic Fertilizer. *J. Inst. Brew.* **2006**, *112*, 92–96. [CrossRef]
48. Mussatto, S.I.; Dragone, G.; Roberto, I.C. Brewers' Spent Grain: Generation, Characteristics and Potential Applications. *J. Cereal Sci.* **2006**, *43*, 1–14. [CrossRef]
49. Lynch, K.M.; Steffen, E.J.; Arendt, E.K. Brewers' Spent Grain: A Review with an Emphasis on Food and Health. *J. Inst. Brew.* **2016**, *122*, 553–568. [CrossRef]
50. Mitri, S.; Salameh, S.-J.; Khelfa, A.; Leonard, E.; Maroun, R.G.; Louka, N.; Koubaa, M. Valorization of Brewers' Spent Grains: Pretreatments and Fermentation, a Review. *Fermentation* **2022**, *8*, 50. [CrossRef]
51. Gutierrez-Barrutia, M.B.; Cozzano, S.; Arcia, P.; del Castillo, M.D. In Vitro Digestibility and Bioaccessibility of Nutrients and Non-Nutrients Composing Extruded Brewers' Spent Grain. *Nutrients* **2022**, *14*, 3480. [CrossRef]
52. Fierascu, R.C.; Sieniawska, E.; Ortan, A.; Fierascu, I.; Xiao, J. Fruits By-Products—A Source of Valuable Active Principles. A Short Review. *Front. Bioeng. Biotechnol.* **2020**, *8*, 319. [CrossRef]
53. Shashirekha, M.N.; Mallikarjuna, S.E.; Rajarathnam, S. Status of Bioactive Compounds in Foods, with Focus on Fruits and Vegetables. *Crit. Rev. Food Sci. Nutr.* **2015**, *55*, 1324–1339. [CrossRef] [PubMed]
54. Antunes, F.; Marçal, S.; Taofiq, O.; Morais, A.M.M.B.; Freitas, A.C.; Ferreira, I.C.F.R.; Pintado, M. Valorization of Mushroom By-Products as a Source of Value-Added Compounds and Potential Applications. *Molecules* **2020**, *25*, 2672. [CrossRef] [PubMed]
55. Guo, J.; Zhang, M.; Fang, Z. Valorization of Mushroom By-Products: A Review. *J. Sci. Food Agric.* **2022**, *102*, 5593–5605. [CrossRef] [PubMed]
56. Nunes, A.R.; Flores-Félix, J.D.; Gonçalves, A.C.; Falcão, A.; Alves, G.; Silva, L.R. Anti-Inflammatory and Antimicrobial Activities of Portuguese *Prunus avium* L. (Sweet Cherry) By-Products Extracts. *Nutrients* **2022**, *14*, 4576. [CrossRef] [PubMed]
57. Jesus, F.; Gonçalves, A.C.; Alves, G.; Silva, L.R. Health Benefits of *Prunus avium* Plant Parts: An Unexplored Source Rich in Phenolic Compounds. *Food Rev. Int.* **2022**, *38*, 118–146. [CrossRef]
58. Hussain, S.; Javed, M.; Abid, M.A.; Khan, M.A.; Syed, S.K.; Faizan, M.; Feroz, F. *Prunus avium* L.; Phytochemistry, Nutritional and Pharmacological Review. *Adv. Life Sci.* **2021**, *8*, 307–314.
59. El Barnossi, A.; Moussaid, F.; Iraqi Housseini, A. Tangerine, Banana and Pomegranate Peels Valorisation for Sustainable Environment: A Review. *Biotechnol. Rep.* **2021**, *29*, e00574. [CrossRef]
60. Omar, A.A.; ElSayed, A.I.; Mohamed, A.H. Tangerine (*Citrus reticulata* L.) Wastes: Chemistry, Properties and Applications. In *Mediterranean Fruits Bio-Wastes: Chemistry, Functionality and Technological Applications*; Ramadan, M.F., Farag, M.A., Eds.; Springer International Publishing: Cham, Switzerland, 2022; pp. 287–302, ISBN 978-3-030-84436-3.
61. Singh, B.; Singh, J.P.; Kaur, A.; Singh, N. Phenolic Composition, Antioxidant Potential and Health Benefits of Citrus Peel. *Food Res. Int.* **2020**, *132*, 109114. [CrossRef]
62. Wang, Y.; Zhang, X.; Zhou, C.; Khan, H.; Fu, M.; Cheang, W.S. *Citri Reticulatae Pericarpium* (Chenpi) Protects against Endothelial Dysfunction and Vascular Inflammation in Diabetic Rats. *Nutrients* **2022**, *14*, 5221. [CrossRef]
63. Wang, W.; Yang, S.; Song, S.; Zhang, J.; Jia, F. *Flammulina velutipes* Mycorrhizae Dietary Fiber Improves Lipid Metabolism Disorders in Obese Mice through Activating AMPK Signaling Pathway Mediated by Gut Microbiota. *Food Biosci.* **2021**, *43*, 101246. [CrossRef]

64. Yan, Z.-F.; Liu, N.-X.; Mao, X.-X.; Li, Y.; Li, C.-T. Activation Effects of Polysaccharides of *Flammulina velutipes* Mycorrhizae on the T Lymphocyte Immune Function. *J. Immunol. Res.* **2014**, *2014*, e285421. [CrossRef] [PubMed]
65. Luo, Z.; Gao, Q.; Li, Y.; Bai, Y.; Zhang, J.; Xu, W.; Xu, J. *Flammulina velutipes* Mycorrhizae Attenuate High Fat Diet-Induced Lipid Disorder, Oxidative Stress and Inflammation in the Liver and Perirenal Adipose Tissue of Mice. *Nutrients* **2022**, *14*, 3830. [CrossRef] [PubMed]
66. Piazza, S.; Martinelli, G.; Fumagalli, M.; Pozzoli, C.; Maranta, N.; Giavarini, F.; Colombo, L.; Nicotra, G.; Vicentini, S.F.; Genova, F.; et al. Ellagitannins from *Castanea sativa* Mill. Leaf Extracts Impair H. Pylori Viability and Infection-Induced Inflammation in Human Gastric Epithelial Cells. *Nutrients* **2023**, *15*, 1504. [CrossRef] [PubMed]
67. Landete, J.M. Ellagitannins, Ellagic Acid and Their Derived Metabolites: A Review about Source, Metabolism, Functions and Health. *Food Res. Int.* **2011**, *44*, 1150–1160. [CrossRef]
68. Lipińska, L.; Klewicka, E.; Sójka, M. The Structure, Occurrence and Biological Activity of Ellagitannins: A General Review. *Acta Sci. Pol. Technol. Aliment.* **2014**, *13*, 289–299. [CrossRef] [PubMed]
69. Durazzo, A.; Lucarini, M.; Souto, E.B.; Cicala, C.; Caiazzo, E.; Izzo, A.A.; Novellino, E.; Santini, A. Polyphenols: A Concise Overview on the Chemistry, Occurrence, and Human Health. *Phytother. Res.* **2019**, *33*, 2221–2243. [CrossRef] [PubMed]
70. Teodoro, A.J. Bioactive Compounds of Food: Their Role in the Prevention and Treatment of Diseases. *Oxidative Med. Cell. Longev.* **2019**, *2019*, e3765986. [CrossRef]
71. Alam, M.A.; Islam, P.; Subhan, N.; Rahman, M.M.; Khan, F.; Burrows, G.E.; Nahar, L.; Sarker, S.D. Potential Health Benefits of Anthocyanins in Oxidative Stress Related Disorders. *Phytochem. Rev.* **2021**, *20*, 705–749. [CrossRef]
72. Enaru, B.; Drețcanu, G.; Pop, T.D.; Stănilă, A.; Diaconeasa, Z. Anthocyanins: Factors Affecting Their Stability and Degradation. *Antioxidants* **2021**, *10*, 1967. [CrossRef]
73. Merecz-Sadowska, A.; Sitarek, P.; Kowalczyk, T.; Zajdel, K.; Jęcek, M.; Nowak, P.; Zajdel, R. Food Anthocyanins: Malvidin and Its Glycosides as Promising Antioxidant and Anti-Inflammatory Agents with Potential Health Benefits. *Nutrients* **2023**, *15*, 3016. [CrossRef]
74. Bertelli, A.; Biagi, M.; Corsini, M.; Baini, G.; Cappellucci, G.; Miraldi, E. Polyphenols: From Theory to Practice. *Foods* **2021**, *10*, 2595. [CrossRef] [PubMed]
75. Aatif, M. Current Understanding of Polyphenols to Enhance Bioavailability for Better Therapies. *Biomedicines* **2023**, *11*, 2078. [CrossRef] [PubMed]
76. Wu, X.; Li, M.; Xiao, Z.; Daglia, M.; Dragan, S.; Delmas, D.; Vong, C.T.; Wang, Y.; Zhao, Y.; Shen, J.; et al. Dietary Polyphenols for Managing Cancers: What Have We Ignored? *Trends Food Sci. Technol.* **2020**, *101*, 150–164. [CrossRef]
77. Naeem, A.; Ming, Y.; Pengyi, H.; Jie, K.Y.; Yali, L.; Haiyan, Z.; Shuai, X.; Wenjing, L.; Ling, W.; Xia, Z.M.; et al. The Fate of Flavonoids after Oral Administration: A Comprehensive Overview of Its Bioavailability. *Crit. Rev. Food Sci. Nutr.* **2022**, *62*, 6169–6186. [CrossRef] [PubMed]
78. Fernández-Ochoa, Á.; Cádiz-Gurrea, M.D.; Fernández-Moreno, P.; Rojas-García, A.; Arráez-Román, D.; Segura-Carretero, A. Recent Analytical Approaches for the Study of Bioavailability and Metabolism of Bioactive Phenolic Compounds. *Molecules* **2022**, *27*, 777. [CrossRef]
79. Cuffaro, D.; Pinto, D.; Silva, A.M.; Bertolini, A.; Bertini, S.; Saba, A.; Macchia, M.; Rodrigues, F.; Digiacomo, M. Insights into the Antioxidant/Antiradical Effects and In Vitro Intestinal Permeation of Oleocanthal and Its Metabolites Tyrosol and Oleocanthalic Acid. *Molecules* **2023**, *28*, 5150. [CrossRef]
80. Rana, A.; Samtiya, M.; Dhewa, T.; Mishra, V.; Aluko, R.E. Health Benefits of Polyphenols: A Concise Review. *J. Food Biochem.* **2022**, *46*, e14264. [CrossRef]
81. Tangney, C.; Rasmussen, H.E. Polyphenols, Inflammation, and Cardiovascular Disease. *Curr. Atheroscler. Rep.* **2013**, *15*, 324. [CrossRef]
82. Yang, B.; Dong, Y.; Wang, F.; Zhang, Y. Nanoformulations to Enhance the Bioavailability and Physiological Functions of Polyphenols. *Molecules* **2020**, *25*, 4613. [CrossRef]
83. Yin, C.; Cheng, L.; Zhang, X.; Wu, Z. Nanotechnology Improves Delivery Efficiency and Bioavailability of Tea Polyphenols. *J. Food Biochem.* **2020**, *44*, e13380. [CrossRef]
84. Sinha, S.; Ali, M.; Baboota, S.; Ahuja, A.; Kumar, A.; Ali, J. Solid Dispersion as an Approach for Bioavailability Enhancement of Poorly Water-Soluble Drug Ritonavir. *AAPS PharmSciTech* **2010**, *11*, 518–527. [CrossRef] [PubMed]
85. Raja, I.S.; Preeth, D.R.; Vedhanayagam, M.; Hyon, S.-H.; Lim, D.; Kim, B.; Rajalakshmi, S.; Han, D.-W. Polyphenols-Loaded Electrospun Nanofibers in Bone Tissue Engineering and Regeneration. *Biomater. Res.* **2021**, *25*, 29. [CrossRef] [PubMed]
86. Khoshnoudi-Nia, S.; Sharif, N.; Jafari, S.M. Loading of Phenolic Compounds into Electrospun Nanofibers and Electrosprayed Nanoparticles. *Trends Food Sci. Technol.* **2020**, *95*, 59–74. [CrossRef]
87. Paczkowska-Walendowska, M.; Miklaszewski, A.; Cielecka-Piontek, J. Is It Possible to Improve the Bioavailability of Resveratrol and Polydatin Derived from Polygoni Cuspidati Radix as a Result of Preparing Electrospun Nanofibers Based on Polyvinylpyrrolidone/Cyclodextrin? *Nutrients* **2022**, *14*, 3897. [CrossRef]

nutrients
MDPI

Article

Consumption of Feed Supplemented with Oat Beta-Glucan as a Chemopreventive Agent against Colon Cancerogenesis in Rats

Joanna Harasym [1,*], Katarzyna Dziendzikowska [2], Łukasz Kopiasz [2,*], Jacek Wilczak [3], Rafał Sapierzyński [4] and Joanna Gromadzka-Ostrowska [2]

1 Department of Biotechnology and Food Analysis, Wroclaw University of Economics and Business, 53-345 Wroclaw, Poland
2 Department of Dietetics, Institute of Human Nutrition Sciences, Warsaw University of Life Sciences, 02-776 Warsaw, Poland; katarzyna_dziendzikowska@sggw.edu.pl (K.D.); joanna_gromadzka-ostrowska@sggw.edu.pl (J.G.-O.)
3 Department of Physiological Sciences, Institute of Veterinary Medicine, Warsaw University of Life Sciences, 02-776 Warsaw, Poland; jacek_wilczak@sggw.edu.pl
4 Department of Pathology and Veterinary Diagnostic, Institute of Veterinary Medicine, Warsaw University of Life Sciences, 02-776 Warsaw, Poland; rafal_sapierzynski@sggw.edu.pl
* Correspondence: joanna.harasym@ue.wroc.pl (J.H.); lukasz_kopiasz@sggw.edu.pl (Ł.K.)

Abstract: Colorectal cancer (CRC) accounts for 30% of all cancer cases worldwide and is the second leading cause of cancer-related deaths. CRC develops over a long period of time, and in the early stages, pathological changes can be mitigated through nutritional interventions using bioactive plant compounds. Our study aims to determine the effect of highly purified oat beta-glucan on an animal CRC model. The study was performed on forty-five male Sprague–Dawley rats with azoxymethane-induced early-stage CRC, which consumed feed containing 1% or 3% low molar mass oat beta-glucan (OBG) for 8 weeks. In the large intestine, morphological changes, CRC signaling pathway genes (RT-PCR), and proteins (Western blot, immunohistochemistry) expression were analyzed. Whole blood hematology and blood redox status were also performed. Results indicated that the histologically confirmed CRC condition led to a downregulation of the WNT/β-catenin pathway, along with alterations in oncogenic and tumor suppressor gene expression. However, OBG significantly modulated these effects, with the 3% OBG showing a more pronounced impact. Furthermore, CRC rats exhibited elevated levels of oxidative stress and antioxidant enzyme activity in the blood, along with decreased white blood cell and lymphocyte counts. Consumption of OBG at any dose normalized these parameters. The minimal effect of OBG in the physiological intestine and the high activity in the pathological condition suggest that OBG is both safe and effective in early-stage CRC.

Keywords: azoxymethane; bioactive compounds; colorectal cancer; oat beta-glucan; oxidative stress; rats; CRC signaling pathway

Citation: Harasym, J.; Dziendzikowska, K.; Kopiasz, Ł.; Wilczak, J.; Sapierzyński, R.; Gromadzka-Ostrowska, J. Consumption of Feed Supplemented with Oat Beta-Glucan as a Chemopreventive Agent against Colon Cancerogenesis in Rats. *Nutrients* **2024**, *16*, 1125. https://doi.org/10.3390/nu16081125

Academic Editors: Maria Digiacomo and Doretta Cuffaro

Received: 1 March 2024
Revised: 5 April 2024
Accepted: 10 April 2024
Published: 11 April 2024

1. Introduction

Beta-glucans, a group of polysaccharides, are found in microorganisms like bacteria and yeast, as well as in cereals such as barley and oats. The structural complexity of beta-glucan varies significantly based on its source and the method of isolation, primarily differing in the distribution and length of the side chains. Factors such as primary structure, solubility, degree of branching, and molecular weight play a crucial role in determining the biological activity of beta-glucan [1]. Our previous study showed that dietary administration of beta-glucan isolated from oats exhibited anti-inflammatory, antioxidant, and prebiotic effects in artificially induced enteritis [2]. Furthermore, our results have shown that these effects are related to the molar mass of beta-glucan [2,3], with low molar mass fractions showing more pronounced effects. Additionally, in our recent studies focusing on the effects of oat beta-glucan supplementation on large intestine inflammation (in the absence of early signs of carcinogenesis), we observed that low molar mass oat beta-glucan significantly downregulated the expression of genes associated with pro-inflammatory proteins, as well as reduced the colonic levels of these proteins [3].

It is important to emphasize that in the case of gastrointestinal disorders, especially inflammatory bowel disease or gastrointestinal cancers, multi-component, unpurified oat grain preparations, such as oatmeal or fiber, should not be used as a food ingredient. These preparations contain a variety of active ingredients, including a fraction of insoluble dietary fiber that irritates the intestinal mucosa, as well as proteins that may cause hypersensitivity and/or intolerance. As a more suitable alternative, the use of high purity, low molecular weight 1-3, 1-4-D-beta-glucan as a dietary supplement is recommended. This formulation is devoid of other compounds, including soluble proteins. We intend to use this specific formulation in our *in vivo* CRC animal model. Based on previous research of its anti-tumor activity in *in vitro* studies on skin tumor and lung epithelial cancer cell lines, we anticipate similar effects in this model [4,5].

Colorectal cancer (*colorectal carcinoma*—CRC) is a common malignancy of the gastrointestinal tract, and its incidence is steadily increasing year by year. Epidemiological statistics forecast a rise in the number of new cancer cases and associated deaths in Europe, the United States, and other countries. As a result, colorectal cancer represents a significant global health challenge and ranks as the second most dangerous type of cancer, affecting women and men equally. It is estimated that there were more than 1.9 million new cases of colorectal cancer worldwide in 2020, resulting in approximately 930,000 deaths. The highest occurrence of this cancer type was noted in Australia/New Zealand and various parts of Europe [6]. CRC is also one of the most common cancers in the Polish population. In 2018, it was the third most common malignant neoplasm in men and the second in women. In that year, almost 12,500 deaths due to CRC were recorded in Poland. This type of cancer was also the second most common cause of death from malignant neoplasms in men and the third in women. Despite improvements over the years, 5-year survival rates for colorectal cancer patients in Poland are still among the lowest in Europe [7].

Colorectal cancer primarily originates from the superficial glandular epithelial cells lining the colon or rectum. Aberrant crypt foci (ACF), recognized as preneoplastic lesions of the colon, represent the earliest identifiable intermediate precancerous lesions in CRC. This phenomenon is observed in both laboratory rodents [8,9] and humans [10] and can be identified microscopically on the surface of the entire colonic mucosa, serving as a valuable histological marker in the development of CRC. Furthermore, ACF are also considered as a unique biomarker to evaluate the potential preventive efficacy of various factors or compounds against CRC in the early stages of cancer development [10]. Ezuka et al. (2015) [11], in their comprehensive review, confirmed the potential utility of ACF as a surrogate biomarker for CRC. Additionally, ACF have been used as an endpoint to identify and evaluate the preventive or therapeutic roles of natural and pharmacological compounds, including dietary factors, in CRC development [12–15]. It is important to note that ACF are a critical juncture at which cancer or tumors may develop. They are readily induced by colon-specific carcinogenesis in rodents, offering a valuable model for studying the process of colorectal cancer development.

Rodents, such as rats, typically exhibit a very low incidence of spontaneous cancer. Therefore, to study colon carcinoma, they are often administered a carcinogen. Azoxymethane (AOM), a chemical mutagen and potent genotoxic agent, is frequently used in these studies. Rodent models of AOM-induced CRC have proven to be highly valuable for investigating the mechanisms underlying carcinogenesis [16]. Additionally, these models are instrumental in studying the preventive effects of various dietary factors [17,18]. The AOM-induced CRC rat model shares many features with human CRC, making it an important model for exploring various aspects of this disease. It has also been found that short-term AOM-induced rat ACF serves as an ideal early biomarker for identifying a range of preventive factors and compounds against CRC [14,19].

Naturally occurring substances of plant origin, particularly polymers, used in animal model experiments are very often unpurified and contain other biologically active compounds, such as polyphenols, protein residues, or other dietary fiber fractions. This raises concerns about the unambiguous assessment of the bioactivity of the tested formulations. Therefore, in our study, we used a chemically pure preparation of oat beta-glucan as an integral component of the rat feed. This preparation was devoid of protein and peptide residues, as well as other components with significant biological potential, such as polyphenols. The main objective of the current study is to assess the effect of a chemically pure fraction of low molar mass oat beta-glucan (OBG), administered *per os* at two doses (1% or 3% *w*/*w*) as a dietary supplement, on the early stages of AOM-induced CRC development in rats. An additional objective is to monitor changes in selected peripheral blood parameters.

2. Materials and Methods

2.1. Isolation and Characterization of Low Molar Mass Oat Beta-Glucan

A low molar mass oat 1-3, 1-4, beta-D-glucan isolate was prepared from oat bran (Bestpharma, Warszawa, Poland) through a unique, patented process outlined in previous works [20,21]. Initially, the oat bran underwent a freezing process, followed by significant milling while still frozen, repeated several times to effectively diminish particle size. Subsequently, beta-glucan was extracted via alkaline water (pH = 8.5; NaOH). The residual fraction was then separated by centrifugation at 11,000× *g* and discarded. The supernatant was deproteinized at the isoelectric point (pH = 4.5), and the resulting protein precipitate was also removed by centrifugation. A combination of proteolytic, peptidolytic, and amylolytic enzymes was utilized for additional purification, along with enzymatic precipitation of proteins at previously specified isoelectric points [21]. The resulting beta-glucan preparations exhibited 99.3% purity as verified by enzymatic methodology (employing AACC Method 32-23.01, AOAC Method 995.16, AOAC Method 992.28, CODEX Method Type II, EBC Method 3.10.1, ICC Standard No. 166, and RACI Standard Method). These preparations also displayed a molar mass of $5.2 \times 10^4 \pm 0.6 \times 10^4$ g/mol (52 ± 6 kDa), as confirmed by size exclusion HPLC as described elsewhere [20]. The isolates were analyzed for total polyphenol content using Folin–Ciocalteu reagent [22] and antioxidant activity vs. DPPH, as described previously [23]. Protein content (Lowry method) and total nitrogen (Kjeldahl) were assessed as well.

2.2. Animals and In Vivo Experimental Design

The *in vivo* experiment was performed on male outbred Sprague–Dawley rats (n = 45) purchased from Charles River Laboratories (Sulzfeld, Germany). Experimental details were described in our previous publication [24]. Briefly, rats were divided into two main groups: (1) the treatment group (CRC group), in which the early stage of colorectal cancer was chemically induced by peritoneal injection of azoxymethane (AOM) (Sigma-Aldrich, Saint Louis, MO, USA), and (2) the control group, consisting of animals administered the equal volume of 0.9% NaCl solution in the same way as the treatment group. After the last administration of AOM/NaCl, both the CRC and control groups were further divided into 3 dietary subgroups fed during 8 weeks with pellet feed (ZooLab, Sędziszów, Poland) supplemented with different levels of low molar mass oat beta-glucan (OBG), as shown in Table 1. All feeds were formulated based on AIN-93M pellets, which were designed for rats in accordance with the species-specific nutritional recommendations of Reeves et al. (1993) [25]. Body weight was recorded on a weekly basis, and feed intake was monitored every day. Clinical status indicators (including appetite loss, body weight loss of more than 10%, loose and bloody stools, diarrhea, anus edema, and rectal bleeding) were monitored twice daily throughout the experiment.

All experimental procedures were conducted with the approval of the 2nd Local Ethical Committee in Warsaw, Poland (resolution No. WAW2/040/2019, 15 March 2019) in accordance with the UE Directive (2010/63/UE), Polish law and the principles of 3R rules (Replacement, Reduction, and Refinement). According to the 3R rules, the number of rats in the study subgroups is the smallest possible from the point of view of statistical verification of the results and was determined based on the scientific literature [26] using available tools for estimating the number of animals in the experimental groups (http://biomath.info/power/ttest.htm; accessed on 3 March 2019).

Table 1. Experimental groups.

	Control Groups (n = 21)			CRC Groups (n = 24)		
	OBG_0	**OBG_1**	**OBG_3**	**OBG_0**	**OBG_1**	**OBG_3**
Feed	Without OBG	+1% OBG	+3% OBG	Without OBG	+1% OBG	+3% OBG

Symbols indicate CRC—rats after AOM injection; OBG—low molar mass oat beta-glucan.

2.3. Blood and Tissues Sampling

After 8 weeks of feeding, the rats were bled by cardiac puncture under Izoflurane (Aerrane Isoflurane USP, Baxter, Poland) general anesthesia. Subsequently, the entire large intestine of each rat was collected. A piece from each of the three parts of the large intestine (cecum, colon, and rectum) was collected and frozen together in liquid nitrogen for Western blot analysis or fixed in 10% buffered formalin, designed for histopathology, histomorphometry, and immunohistochemistry analysis. The other parts were immediately frozen in liquid nitrogen and stored at −80 °C until biochemical analysis.

2.4. Hematological Analysis and Blood Plasma Separation

Peripheral blood collected from the heart was separated into two portions: one for whole blood and the other for plasma. The whole blood sample was subjected to hematological analysis using the Abacus Junior Vet analyzer (Diatron, Budapest, Hungary). This analysis included the determination of the total number of white blood cells (WBC), monocytes and eosinophils (MID), lymphocytes (LYM), granulocytes (GRA), red blood cells (RBC), and platelets (PLT). Additionally, mean corpuscular volume (MCV), red cell distribution width (RDW), mean corpuscular hemoglobin concentration (MCHC), and mean corpuscular hemoglobin (MCH) in cells, as well as hemoglobin concentration (HGB) and hematocrit (HCT) of blood were measured. Blood plasma was obtained by centrifugation of whole blood at 2200× g for 10 min at 4 °C. The obtained plasma samples were stored at −20 °C until biochemical analyses were performed.

2.5. Histopathological and Histomorphometry Evaluation

The collected biological samples of the large intestine specimens were fixed in 10% buffered formaldehyde, subsequently dehydrated through graded ethanol and xylene baths, and then embedded in paraffin wax. Sections of 4 μm thickness were prepared and stained with hematoxylin and eosin (H&E) for examination. The analyses were examined using a standard light microscope Olympus BX41 (Olympus, Tokyo, Japan) according to histological criteria of multistep carcinogenesis as proposed by Perse and Cerar (2011) [14]. The early stage of carcinogenesis was characterized by recognizable histological changes, beginning with aberrant crypt foci (ACF), which were identified by a veterinary histopathologist. Mucosal lesions were classified into various categories: normal mucosa, hyperplastic crypts, dysplastic crypts, adenoma, and carcinoma.

In addition, a histomorphometric examination of colonic crypts, including their length and width, was performed to provide a more detailed characterization of changes in the early stages of carcinogenesis. The samples selected for histomorphometry were those of the highest quality. Histomorphometric analyses were conducted using an Olympus BX41 microscope (Olympus, Tokyo, Japan) coupled to a computer equipped with the CellA® analysis system. For each colon specimen, areas with clearly visible and well-stained crypts with visible histologic elements were selected. In each sample, 10 to 30 crypts were analyzed. The software was used to measure the length of the crypt and its width in the upper, middle, and lower thirds of the length.

2.6. Blood Plasma Biochemical Parameters

The blood cell lysate used to assess antioxidant enzyme activity was prepared according to the protocol described in the instructions provided with the Randox kit (Randox in County Antrim, UK). The blood cell lysate was used for the analysis of total antioxidant status (TAS) and thiobarbituric acid reactive substances concentration (TBARS), as well as glutathione reductase (GR), glutathione peroxidase (GPx), and superoxide dismutase (SOD) activity. TAS, GR, GPx, and SOD were determined using the commercial Randox assay kits (TAS, Glutathione Reductase, RANSEL, and RANSOD kit; Randox Laboratories Ltd., Crumlin, Co. Antrim, UK) according to the manufacturer's protocols. The catalog numbers of the kits were as follows (TAS, cat. no.: NX2332; GR, cat. no.: GR2368; GPx, cat. no.: RS504/505/506; SOD, cat. no.: SD125). Thiobarbituric acid reactive substances (TBARS) were analyzed according to the method described by Aguilar Diaz De Leon and Borges (2020) [27].

2.7. Methods for RNA Isolation, Conversion to cDNA, and Gene Expression Analysis

RNA was extracted from large intestine samples using the RNeasy Lipid Tissue Mini Kit (Qiagen, Hilden, Germany) according to the supplied protocol. The purity and concentration of the extracted RNA were determined using the NanoDrop™ 2000 spectrophotometer (Thermo Fisher Scientific, Waltham, MA, USA). To assess the integrity of the RNA, the Agilent Bioanalyzer 2100 system was used in conjunction with the RNA 6000 Nano LabChip® Kit (Agilent Technologies, Palo Alto, CA, USA) on randomly selected RNA samples. This assessment revealed a low rate of RNA degradation as evidenced by an RNA integrity number (RIN) exceeding 9. The RT2 First Strand Kit (Qiagen, Hilden, Germany) was then used to synthesize complementary DNA. For gene expression analysis focusing on specific pathways, the Custom RT2 Profiler™ PCR array, provided by Qiagen, was applied according to the manufacturer's guidelines. This analysis was performed in two technical replicates for each colon sample to ensure the accuracy and reliability of the results. The RT2 Profiler PCR arrays (Qiagen, Hilden, Germany) contain primers for a curated analysis of genes essential for the mechanisms of carcinogenesis induced by AOM. These include *Ctnnb1* (catalog number PPR57457A), *Wnt1* (catalog number PPR57724A), *Apc* (catalog number PPR52699A), *Dclk1* (catalog number PPR52232A), and *Smad4* (catalog number PPR06574C). Amplification was performed using the AriaMx Real time PCR System (Agilent Technologies, Palo Alto, CA, USA), starting with an initial denaturation at 95 °C for 10 min. This was followed by 40 amplification cycles, each including 15 s of denaturation at 95 °C and 1 min of annealing/extension at 60 °C. The threshold cycle (Ct) method was used to quantify relative gene expression using *B2m* (catalog number PPR42607A) and *Ldha* (catalog number PPR56603B) as reference genes for normalization. Analysis and calculation of results were facilitated by GeneGlobe Qiagen Software (https://geneglobe.qiagen.com/pl, accessed on 3 April 2023). The expression levels of the target genes, normalized to *B2m* and *Ldha*, are presented as fold changes relative to a control group, which is established at a baseline value of 1 to illustrate changes in expression.

2.8. Immunohistochemical Analysis

Immunohistochemical analysis was performed as described by Kopiasz et al. (2024) [24]. Briefly, tissue slices (5 μm thickness) containing three representative pieces of the large intestine from each rat were deparaffinized in xylene and rehydrated in a series of decreasing concentrations of ethanol. After recovering antigens and blocking endogenous enzymes, the samples were incubated with primary antibodies overnight at 4 °C using a rabbit anti-β-Catenin polyclonal antibody (Cat#8480, 1:200, Cell Signaling Technology, Danvers, MA, USA). The samples were then labeled with polymers consisting of anti-rabbit antibodies conjugated to the horseradish peroxidase (HRP) enzyme complex. 3,3′-diaminobenzidine

(DAB) was used to obtain a brown color, and hematoxylin was used to counterstain nuclei. The immunohistochemically stained slides were examined using a NIKON Eclipse Ti2 microscope (Nikon, Melville, NY, USA—funding details in Supplementary Data). On the recorded images, six mucosal areas were marked for each of the three sections of the colorectum. A total of eighteen marked mucosal areas for the colorectum from one rat were analyzed together. Colorimetric saturation (brown colors reflecting antigen expression) and object area were measured using the NIS-Elements BR 5.01 program.

2.9. Western Blot Analysis

Western blot analysis was conducted following the previously described protocol [24]. Briefly, samples from three different parts of the large intestine from each rat were homogenized together in RIPA buffer supplemented with a cocktail of phosphatase and protease inhibitors. The lysates were then centrifuged, and the supernatants were collected. The samples (normalized to a 50 μg protein concentration) were resolved by sodium dodecyl sulfate-polyacrylamide gel electrophoresis and transferred onto a PVDF membrane. Next, the membranes were incubated overnight at 4 °C with primary antibodies, including rabbit anti-β-Catenin monoclonal antibody (Cat#A19657, 1:1250, ABclonal, Woburn, MA, USA) and mouse anti-β-actin (8H10D10) monoclonal antibody (Cat#3700, 1:5000, Cell Signaling Technology, Danvers, MA, USA), under gentle shaking. After that, the membranes were incubated with secondary antibodies conjugated with IR fluorophore (IRDye® 800 CW anti-rabbit antibody) at 1:5000 dilution. The protein expression was analyzed using the ChemiDoc Imaging System (Bio-Rad, Hercules, CA, USA). The integrated optical density (IOD) was quantified using the Image Lab 6.1 Software (Bio-Rad, Hercules, CA, USA). The relative levels of the analyzed proteins were normalized to β-actin.

2.10. Statistical Analysis

The collected data were analyzed using Statistica software (version 13.3 PL; StatSoft, Krakow, Poland). Prior to further analyses, the equality of variance and normality of distribution were determined for all datasets. The results of MID required a square root transformation to achieve normal distribution and equality of variance. Two-way analysis of variance (ANOVA) was used to evaluate the effect of two experimental factors—the early stage of CRC and the type of dietary intervention—as well as their interaction. Tukey's post hoc test was used to determine the significance of outcome differences among the groups. Differences between the control group (control OBG_0) and the other groups were analyzed using Dunnett's post hoc test. Pearson's chi-square test with Cramer's V value was used to analyze the results of histopathological changes. Statistical significance was established at a p-value of less than 0.05. All graphs were generated using GraphPad Prism, version 9.3.1, from GraphPad Software (GraphPad Software Inc., San Diego, CA, USA).

3. Results

3.1. Chemical Characterization of Low Molar Mass Oat Beta-Glucan

The detailed characteristics of the oat beta-glucan isolates used in the present study are comprehensively presented in Table 2. The extensive process of extraction, followed by isoelectric precipitation, effectively removes the majority of proteins. Subsequent dissolution and organic solvent precipitation denatures and hydrolyzes any proteinaceous residues, as well as enzymes used [21]. Exhaustive extraction and purification are performed to remove any potentially bioactive molecules, which could bias the assessment of oat beta-glucan impact. Small molecule antioxidant compounds such as polyphenols can modulate the system response to an administered substance. The tested sample of oat beta-glucan isolate was analyzed for total polyphenols content and antioxidative activity vs. DPPH radical, revealing negligible activity and proving to be free from these two bioactive contaminants. The bioactivity of beta-glucan depends not only on its purity but also on the molar mass of the analyzed fractions. The average molar mass of the isolate was determined using internal standards of oat beta-glucan, resulting in $5.2 \times 10^4 \pm 0.6 \times 10^4$ g/mol.

Table 2. Physicochemical characterization of low molar mass oat beta-glucan used as a feed additive (mean ± SE).

OBG	Molar Mass [g/mol]	Purity [%]	TPC [GAE/ 1 g s.m.]	DPPH [μmol TE/1 g s.m.]	Protein (%/db)	Nitrogen × 5.83 (%/s.m.)
Low molar mass	$5.2 \times 10^4 \pm 0.6 \times 10^4$	99.3 ± 0.23	0.17 ± 0.040	5.94 ± 0.21	0.12 ± 0.04	1.52 ± 0.18

Symbols indicate TPC (Total phenolic compounds); GAE (Gallic acid equivalents); DPPH (2,2-diphenyl-1-picrylhydrazyl radical); TE (*Trolox equivalents*).

3.2. Feeds Nutritional Characteristic, Feed Intake, and Body Weight Gain

In our study, we used a chemically pure beta-glucan from oats as a part of rat feed in early-stage colon cancer models. The beta-glucan was free of proteins, peptides, polyphenols, and other biologically active compounds. The purity of this polysaccharide allowed for an accurate evaluation of the therapeutic effects of beta-glucan without the risk of additional intestinal irritation or hypersensitivity. Our choice of low molar mass beta-glucan was based on its proven anti-inflammatory and indirect antioxidant properties in previous studies [2,3]. OBG was added to the feed at 1% and 3% levels, substituting for equivalent carbohydrate content, maintaining the feed's balance without altering the insoluble fiber fraction. Two doses were used based on dose conversion from human to animal and vice versa, taking into account body surface area, metabolic rate, and physiological processes. Using appropriate formulas [28], the consumption of feed supplemented with the addition of OBG at 1% *w*/*w* in rats is equivalent to consuming approximately 8–10 g OBG for an adult human (BW 60–70 kg), while 3% corresponds to approximately 24–30 g. Furthermore, a dose of 3% was administered to observe any potential dose-dependent effects.

The semi-synthetic purified feeds (pellets) used in the present study were formulated using natural products (corn starch, maltodextrin, sucrose, α-cellulose, casein, and soybean oil) as detailed in Supplementary Table S1. The declared and analyzed compositions of these experimental and control feeds, including proteins, carbohydrates, fat, and fibers, are shown in Supplementary Table S2. The content of 1-3, 1-4, beta-D-glucan in the feeds was determined according to AOAC 995.16 method (K-BGLU, Megazyme, Ireland) based on lichenase digestion. Additionally, the metabolizable energy (ME) of the feeds and the feed growth efficiency (Table 3) were also calculated using the following formulas in accordance with Bielohuby et al. (2010) [29].

$$\text{ME (MJ/kg)} = 0.0223 \times \text{protein (g)} + 0.0341 \times \text{fat (g)} + 0.017 \times \text{starch (g)} + 0.0168 \times \text{sugar (g)} + 0.0074 \times (\text{organic matter} - \text{protein} - \text{fat} - \text{starch} - \text{sugar} - \text{fiber})\ (\text{g}) - 0.0109 \times \text{fiber (g)} \quad (1)$$

$$\text{Feed growth efficiency} = \text{total weight gain (g)}/\text{the calculated amount of energy intake with the feed (MJ)} \quad (2)$$

Table 3. Feed intake, feed growth efficiency, and total body weight gain (mean ± SE).

	Control Groups			CRC Groups		
	OBG_0	OBG_1	OBG_3	OBG_0	OBG_1	OBG_3
Feed intake [g/day]	25.84 ± 0.05	25.29 ± 0.44	25.48 ± 0.15	25.24 ± 0.33	24.98 ± 0.17	25.23 ± 0.33
Total body weight gain [g]	207.6 ± 3.57	196.1 ± 7.5	207.9 ± 10.58	205.1 ± 8.42	212.0 ± 5.82	199.1 ± 6.65
Feed growth efficiency [g/MJ] (mg/kcal)	9.006 ± 0.16 (38.65 ± 0.7)	8.842 ± 0.23 (37.68 ± 1.0)	9.644 ± 0.47 (40.53 ± 2.0)	9.107 ± 0.34 (39.09 ± 1.5)	9.691 ± 0.24 (41.29 ± 1.0)	9.342 ± 0.33 (39.26 ± 1.4)

The differences are not statistically significant.

The nutrient content of the experimental feeds, as declared by the manufacturer, aligns with the recommended levels for adult rats. The analytically determined beta-glucan content in the feeds did not differ from the assumed level in the experiment. Additionally, the growth efficiency of all feeds used in the experiment showed no significant differences This uniformity in feed quality led to consistent feed intake across all rat groups, with no notable disparities in body weight gain. The body weight gain of all groups of rats followed a similar trajectory throughout the study, as shown in Figure 1.

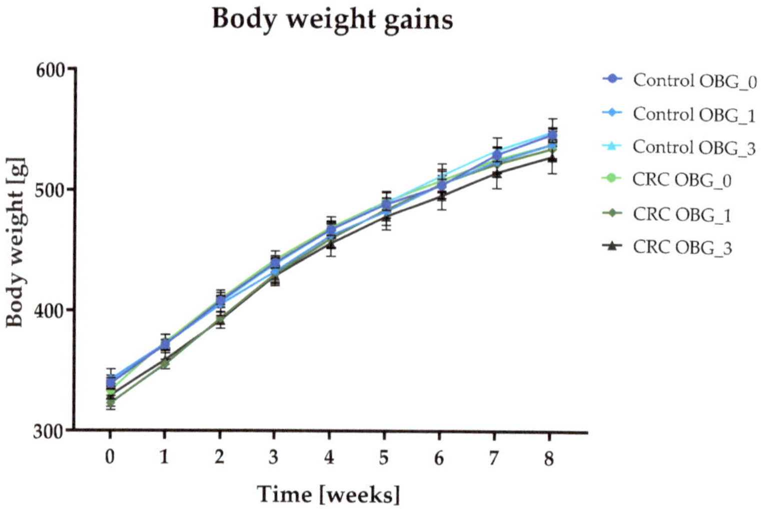

Figure 1. Body weight gain curves during 8 weeks of experiments [mean ± SE].

3.3. Clinical Signs, Necropsy Observations, and Macroscopic Large Intestine Evaluation

During the 8-week period following the AOM peritoneal injection, the rats were monitored twice daily for several health indicators, including appetite, morbidity, stool consistency, and the appearance of the anus. These clinical observations did not reveal any signs indicative of poor health, such as lack of appetite, abnormal appearance of the anus, hair loss, or diarrhea. Similarly, during necropsy, no pathological changes were observed upon visual examination of the large intestine.

3.4. Histopathological Changes in the Large Intestine

Histopathological examination assessed the appearance of intestinal crypts for the presence of aberrant crypt foci (ACF), which may be hyperplasia/dysplasia/adenoma/cancer, as shown in Figure 2. In the Control OBG_0 group and Control OBG_1 group, no morphological changes of crypts were observed; in one rat from the Control OBG_3 group, the presence of single hyperplastic crypts was observed. In the CRC groups, hyperplastic crypts were observed mostly in group CRC_OBG_0 and less commonly in groups CRC OBG_1 and CRC OBG_3. Detailed data on the occurrence of hyperplastic lesions in intestinal crypts of particular groups of rats are presented in Table 4 and Supplementary Table S3.

Table 4. Large intestine epithelium changes.

	Control Groups			CRC Groups			χ^2	p-Value/ V
	OBG_0 (n = 7)	**OBG_1 (n = 7)**	**OBG_3 (n = 7)**	**OBG_0 (n = 8)**	**OBG_1 (n = 8)**	**OBG_3 (n = 8)**		
colon epithelium changes	NC = 7 (100%)	NC = 7 (100%)	NC = 6 (86%) HP = 1 (14%)	NC = 3 (37.5%) HP = 5 (62.5%)	NC = 6 (75%) HP = 2 (25%)	NC = 7 (87.5%) HP = 1 (12.5%)	13.08	0.023/ 0.54

Symbols indicate HP—hyperplasia; NC—large intestine epithelium cells without histopathological changes; χ^2—Pearson's chi-square; V—Cramer's V value.

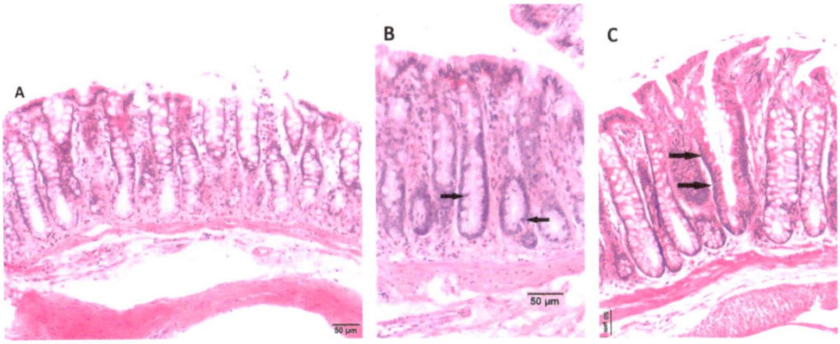

Figure 2. Histopathological changes in the large intestine mucosa. (**A**)—normal mucosa, (**B**)—hyperplastic crypts with an increased number of epithelial cell nuclei (crowded nuclei), lower number of goblet cells, and increased mitotic activity (some mitotic figures indicated by black arrows) restricted to the lower two-thirds of the crypts, (**C**)—one aberrant crypt (black arrows) is present among normal crypts—slightly enlarged crypt diameter and low number of goblet cells are visible. Hematoxylin–eosin staining, magnification: (**A**)—100×, (**B**)—200×, (**C**)—200×.

No significant differences were observed in the measurements of the crypt length and width at the bottom, upper and middle parts of the crypts between the CRC and control groups, as detailed in Table 5.

Table 5. The length and width of intestinal crypts (mean ± SE).

Intestinal Crypts Parameters	Control Groups			CRC Groups		
	OBG_0	OBG_1	OBG_3	OBG_0	OBG_1	OBG_3
Crypt length [µm]	196.8 ± 10.23	196.4 ± 9.96	214.2 ± 9.10	220.5 ± 6.28	207.5 ± 8.20	223.5 ± 9.45
Crypt width bottom [µm]	33.41 ± 0.33	32.63 ± 0.52	32.88 ± 0.62	36.17 ± 1.33	34.21 ± 1.39	35.82 ± 0.69
Crypt width middle part [µm]	34.76 ± 1.28	34.37 ± 0.68	35.58 ± 0.87	34.02 ± 1.36	33.15 ± 0.97	34.88 ± 1.36
Crypt width upper part [µm]	37.39 ± 1.05	38.96 ± 1.32	39.96 ± 0.73	40.60 ± 1.69	38.51 ± 0.90	40.65 ± 1.09

The differences are not statistically significant.

3.5. Red Blood Cells and Platelet Parameters

As shown in Table 6, analysis of variance for RBC, HGB, HCT, MCV, MCH levels, and PLT counts revealed no significant differences between the study groups. However, rats from the CRC group that consumed the feed supplemented with 3% OBG had higher MCHC values compared to the control OBG_3 ($p < 0.05$) and OBG_0 ($p < 0.01$) groups. Additionally, significant differences ($p < 0.05$) in RDWc values were observed between the control group (OBG_0) and the CRC-induced group fed with 3% OGB (OBG_3).

Table 6. Erythrocyte parameters and platelet counts (mean ± SE).

Feature	Control Groups			CRC Groups		
	OBG_0	OBG_1	OBG_3	OBG_0	OBG_1	OBG_3
RBC [10^{12}/L]	8.18 ± 0.29	8.01 ± 0.19	7.85 ± 0.28	8.07 ± 0.15	7.99 ± 0.17	8.09 ± 0.17
HBG [g/dL]	14.42 ± 0.26	14.23 ± 0.23	13.94 ± 0.42	14.45 ± 0.17	14.24 ± 0.32	14.50 ± 0.15
HCT [%]	42.52 ± 0.6	41.24 ± 0.56	41.03 ± 0.92	40.92 ± 0.84	39.92 ± 0.8	41.42 ± 0.59
MCV [fl]	53.00 ± 0.68	51.71 ± 0.87	51.83 ± 0.95	51.25 ± 0.67	50.00 ± 0.46	51.25 ± 0.80
MCH [pg]	17.64 ± 0.3	17.80 ± 0.25	17.77 ± 0.18	17.93 ± 0.21	17.80 ± 0.19	17.98 ± 0.28
MCHC [g/dL]	33.91 ± 0.2	34.49 ± 0.29	33.97 ± 0.43 ^{A}	34.96 ± 0.24 *	35.31 ± 0.19 **	35.30 ± 0.22 $^{**, A}$
RDWc [%]	17.59 ± 0.23	18.29 ± 0.24	17.98 ± 0.14	17.70 ± 0.21	18.04 ± 0.18	18.43 ± 0.19 *
PLT [10^9/L]	718.7 ± 38.3	747.3 ± 32.2	788.9 ± 25.2	710.6 ± 22.7	781.4 ± 45.5	759.0 ± 36.0

OBG—low molar mass oat beta-glucan; RBC—red blood cells; HGB—hemoglobin; HCT—hematocrit; MCV—mean corpuscular volume; MCH—mean corpuscular hemoglobin; MCHC—mean corpuscular hemoglobin concentration; RDWc—red blood cell distribution width; PLT—platelet. The same letters denote significant differences between experimental groups determined by Tukey's post hoc test (^{A} $p < 0.05$). Significant differences from the control group (control OBG_0) were determined by the Dunnett's post hoc test (* $p < 0.05$, ** $p < 0.01$).

3.6. White Blood Cells Parameters

The total white blood cell and lymphocyte counts were lower in rats from the CRC OBG_0 group compared to the control OBG_0 group ($p < 0.01$). However, there was no significant difference in the total number of other types of leukocytes among all groups. Notably, the most significant differences between groups were found in the number of granulocytes. In the control rats, consumption of feed with 3% OBG resulted in a significant reduction in the number of granulocytes ($p < 0.05$). A similar difference was observed in the CRC rats fed with feed containing 3% OBG compared to the control group ($p < 0.001$) (Figure 3).

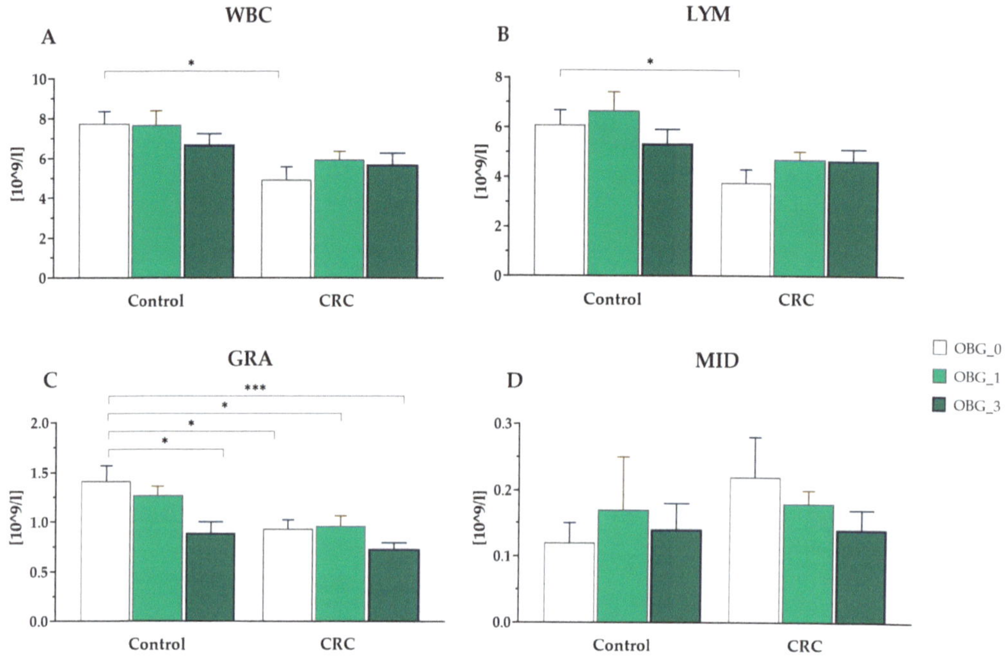

Figure 3. White blood cell parameters (mean ± SE). (**A**)—total white blood cell count; (**B**)—lymphocyte count; (**C**)—granulocyte count; (**D**)—total number of the other types of white blood cells not classified as lymphocytes or granulocytes. Significantly different from the control group (OBG_0) (* $p < 0.05$; *** $p < 0.001$) (Dunnett's post hoc test).

3.7. Peripheral Blood Plasma Redox Status Parameters

Thiobarbituric acid reactive substances (TBARS) are a by-product of lipid peroxidation and are recognized as a marker of accelerated lipid oxidation. In the present study, an increased accumulation of TBARS was detected in the systematic circulation of CRC rats consuming feed without OBG supplementation, with significantly higher values compared to other CRC and control groups ($p < 0.0001$). The eight-week dietary intervention with OBG significantly reduced TBARS levels, with the most significant reduction noted in the OBG_3 group. In the control groups, TBARS levels were higher in the OBG_1 and OBG_3 groups compared to the values of this parameter in the OBG_0 group (Figure 4A). Total antioxidant status (TAS) of peripheral blood was significantly lower in the CRC rats fed with feed without OBG compared to other CRC groups, regardless of the level of OBG supplementation, and to all control groups ($p < 0.0001$). Specifically, in the control rats, a significantly lower TAS was observed in the OBG_1 group compared to the OBG_0 group, as shown in Figure 4B. There was also a significant increase in the activity of antioxidant defense enzymes (GPx and SOD) in CRC animals fed with feed without OGB compared to control animals fed the same feed ($p < 0.0001$). SOD activity in rats from the CRC OBG_3 group was significantly decreased compared to the CRC OBG_0 group and did not differ from the control groups (Figure 4C). In addition, blood GPx activity in the CRC rats from the OBG_0 group was significantly higher than that observed in the OBG_1 and OBG_3 groups ($p < 0.001$ and $p < 0.0001$, respectively). In control animals, GPx activity did not differ among the nutritional subgroups (Figure 4D). In the blood of CRC rats, GR activity was found to be higher in the OBG_0 group compared to the OBG_1 and OBG_3 groups ($p < 0.001$ for all comparisons). Conversely, in the control rats, the highest GR activity was observed in the OBG_1 group ($p < 0.001$) (Figure 4E).

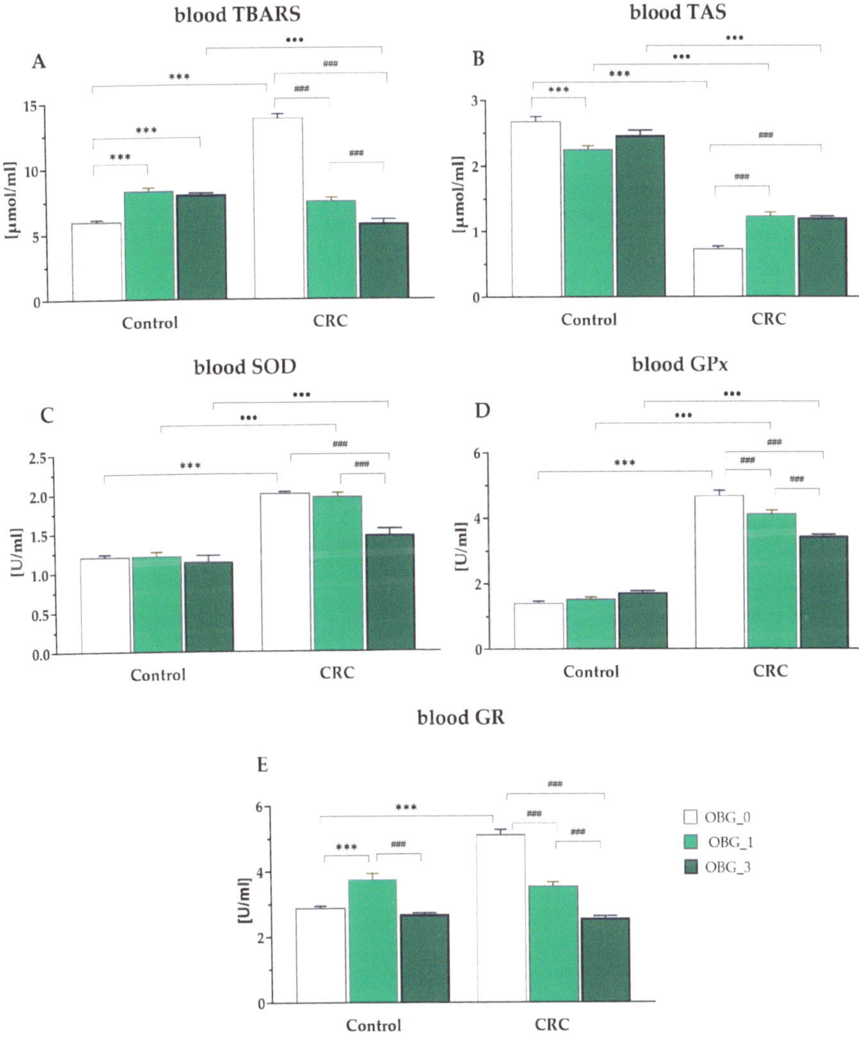

Figure 4. Peripheral blood redox status parameters in CRC and control rats (mean ± SE): (**A**)—thiobarbituric acid-reactive substances (TBARS). (**B**)—total antioxidant status (TAS); (**C**)—superoxide dismutase (SOD); (**D**)—glutathione peroxidase (GPx); (**E**)—glutathione reductase (GR); Significantly different from the control group (OBG_0) (*** $p < 0.001$) (Dunnett's post hoc test). Significant differences within control and CRC groups between dietary subgroups (### $p < 0.001$) (Tukey's post hoc test). Significant differences between the respective control and CRC groups on the same feed (••• $p < 0.001$) (Tukey's post hoc test).

3.8. CRC Signaling Pathways

The gene expression results associated with the WNT/β-catenin signaling pathway, which is implicated in CRC tumorigenesis, showed a significant effect of both OBG and rat health status on the modulation of *Wnt1* expression (ANOVA, $p < 0.01$ and $p < 0.05$, respectively, Figure 5A). *Wnt1* gene expression was downregulated in rats from the CRC groups, as well as in animals fed with feed supplemented with OBG. Post hoc analysis revealed that OBG supplementation decreased *Wnt1* gene expression at the highest OBG concentration (3%) in both control and CRC groups ($p < 0.05$). Additionally, in the CRC OBG_1 group, *Wnt1* gene expression was significantly lower compared to control rats receiving feed without OBG.

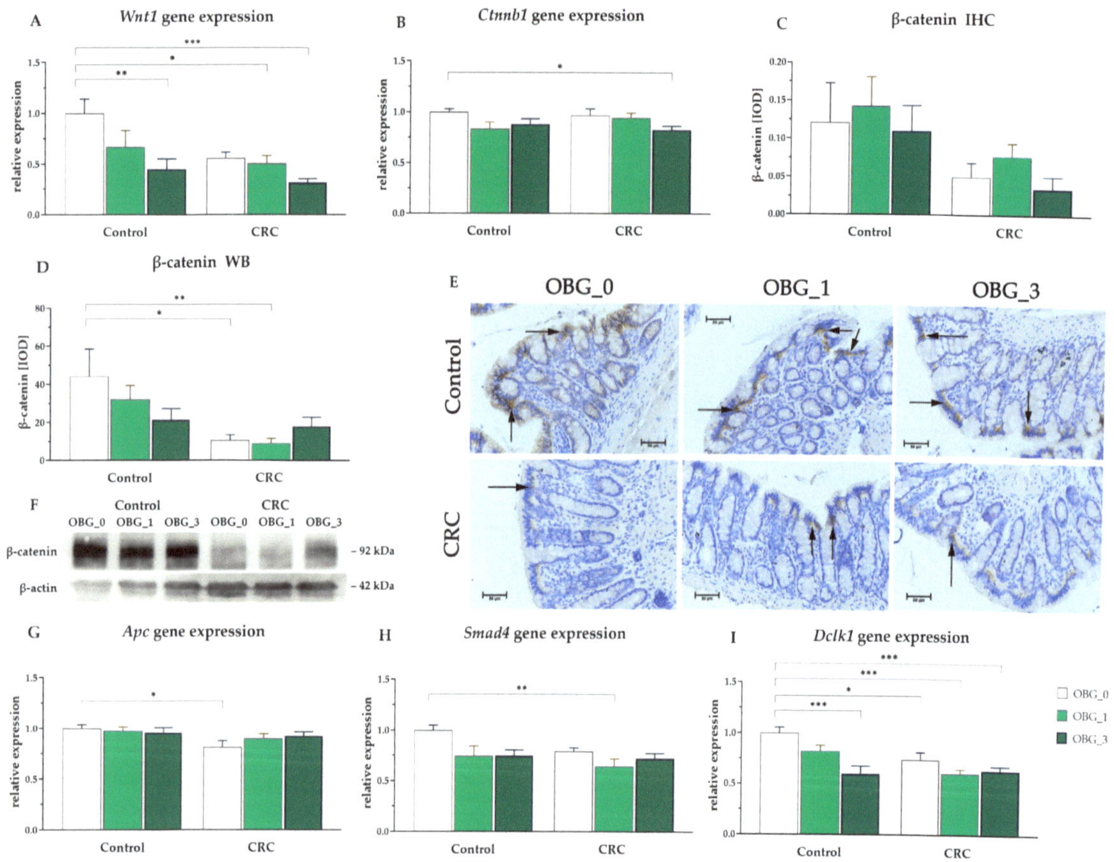

Figure 5. CRC signaling pathways. (**A**,**B**,**G**–**I**)—Changes in relative gene expression in the large intestine of rats: (**A**)—*Wnt1* (Wnt family member 1); (**B**)—*Ctnnb1* (β-catenin); (**G**)—*Apc* (Adenomatous polyposis coli); (**H**)—*Smad4* (Mothers against decapentaplegic homolog 4); (**I**)—*Dclk1* (Doublecortin-like kinase 1). Data are presented in arbitrary units as a ratio of the expression of the target gene to the mean expression of the reference genes (*B2m* and *Ldha*) with the control group calculated as 1. (**C**–**F**)—Changes in β-catenin protein expression; (**C**)—in the large intestine mucosa using immunohistochemistry analysis (IHC) presented as integrated optical density (IOD), (mean ± SE); (**D**)—in the whole large intestine wall presented as IOD, (mean ± SE); (**E**)—Light micrographs imaged (×400 magnification). Black arrows indicate areas with high expression of β-catenin (brown precipitate); (**F**)—Representative immunoblot images. Significantly different from the control group (OBG_0) (* $p < 0.05$, ** $p < 0.01$, *** $p < 0.001$) (Dunnett's post hoc test).

The expression of β-catenin (*Ctnnb1*), a crucial component of the Wnt/β-catenin signaling pathway implicated in cell proliferation and tumorigenesis, was examined at both the gene and protein levels. *Ctnnb1* mRNA expression in colorectal tissue is shown in Figure 5B. A significant effect of OBG supplementation on the expression of *Ctnnb1* was observed (ANOVA, $p < 0.05$). Significant downregulation of *Ctnnb1* was observed in the CRC group supplemented with the highest level of OBG (CRC OBG_3) compared to the control group without OBG supplementation (control OBG_0) ($p < 0.05$). Significant effects on β-catenin protein expression were found only by Western blot analysis (in the entire large intestine wall). Induced early-stage carcinogenesis significantly decreased the expression of this protein (ANOVA, $p < 0.01$), whereas, in the CRC OBG_3 group, β-catenin expression was restored to levels comparable to those observed in the control groups (Figure 5D,F). The expression of this protein in the mucosa did not significantly differ between the study subgroups (Figure 5C,E). However, two-way ANOVA revealed that the early stage of cancerogenesis decreased the expression of this protein (ANOVA, $p < 0.01$).

Two-way ANOVA revealed that the induction of early-stage CRC significantly affected the downregulation of the *Apc* gene (Figure 5G), which is a critical regulator in the WNT signaling pathway and a known tumor suppressor gene, as well as the *Smad4* gene (Figure 5H), also a suppressor gene (ANOVA, $p < 0.05$). Post hoc analysis revealed a significant downregulation of *Apc* expression in the CRC group without OBG supplementation (CRC OBG_0) compared to the control group also without OBG (Control OBG_0) ($p < 0.05$). Interestingly, supplementation with oat beta-glucan mitigated the adverse effects of CRC on *Apc* expression. In the CRC groups receiving OBG, *Apc* expression levels were restored to levels comparable to those observed in the control groups. Furthermore, two-way ANOVA revealed significant effects of both CRC-induced and OBG supplementation on *Smad4* gene expression (ANOVA, $p < 0.05$). In particular, a distinct downregulation of *Smad4* expression was revealed in CRC OBG_1 compared to the control OBG_0 group. The *Smad4* gene was downregulated by OBG. However, the inclusion of 3% OBG in the feed of CRC-affected rats restored *Smad4* expression levels to those comparable to the corresponding control group. The expression of the oncogenic *Dclk1* gene (Figure 5I) was influenced by all experimental factors and their interactions (ANOVA, $p < 0.05$). Post hoc analysis revealed that *Dclk1* gene expression was lower in rats from the CRC groups fed with the control feed compared to the control group without OBG supplementation (control OBG_0) ($p < 0.05$). Additionally, *Dclk1* gene expression was lower in the CRC OBG_1, CRC OBG_3, and control OBG_3 groups compared to the control OBG_0 ($p < 0.05$).

4. Discussion

In our 8-week study, we focused on the impact of OBG supplementation on the rat CRC model with AOM-induced carcinogenesis. Throughout the study period, rats were observed for health indicators, including appetite and physical appearance, with no adverse signs noted in those fed with OBG-supplemented feed. In contrast, CRC rats exhibited early signs of carcinogenesis, such as ACF in the mucosal layer, as well as its erosion and hyperplasia. Blood analysis across different groups showed no major differences in most hematological values, but there were slight MCHC and RDWc changes in CRC rats fed with 3% OBG, alongside notable disparities in white blood cells and granulocytes, indicating an immune response induction. The induction of CRC resulted in a significant increase in the concentration of the oxidative stress marker (TBARS level) and a decrease in antioxidant defense, as evidenced by a significant reduction in total antioxidant status (TAS). This was accompanied by an increase in endogenous antioxidant enzymes, such as superoxide dismutase (SOD) and glutathione peroxidase (GPx). Furthermore, OBG supplementation reduced the TBARS level and enhanced the antioxidant defense, as evidenced by the increased TAS and decreased activities of antioxidant enzymes such as GPx and SOD. This suggests a protective role of OBG against oxidative damage. The analysis of gene and protein expression revealed the effect of OBG supplementation and CRC status on the modulation of the WNT/β-catenin signaling pathway. This pathway is crucial in cell proliferation and tumorigenesis. OBG appeared to mitigate the adverse effects of CRC on this pathway, suggesting its potential as a therapeutic agent for CRC management.

Colorectal cancer (CRC) develops from the superficial glandular epithelial cells lining the colon or rectum through a multistep process known as carcinogenesis [30]. Aberrant crypt foci (ACF) are recognized as pre-neoplastic lesions in the colon and represent the earliest microscopically identifiable intermediate precancerous lesions during colon carcinogenesis in both laboratory rodents [31–33] and human populations [34]. The identification of ACF as pre-neoplastic lesions in the colon underscores their significance in CRC research. Numerous studies have used ACF as a surrogate endpoint for chemoprevention in rats, illustrating their value in early detection and prevention strategies for CRC [10,35]. ACF are recognized as a valuable histological biomarker that indicates the development stage of colorectal carcinogenesis [36]. They have been used to identify and evaluate the preventive or therapeutic role of various natural and pharmacological compounds, including dietary factors, in colorectal cancer development [37]. The size and density of ACF can increase over time, with significantly higher densities noted in the later stages of CRC compared to the early stages, highlighting their potential as indicators of disease progression [36]. In our study, colorectal ACF were present in some rats, indicating very early stages of CRC development, which is also confirmed by other studies [38,39]. Although they were few in number, these ACF signify the initiated process of carcinogenesis.

The classical diagnosis of CRC in humans primarily relies on colonoscopy, followed by examination of biopsied tissue using hematoxylin and eosin staining. While considered the gold standard in the CRC diagnosis, this assessment can be somewhat subjective, relying on the expertise of pathologists, particularly in ambiguous cases. However, the integration of artificial intelligence (AI) and other technological innovations into diagnosis is improving both its accuracy and efficiency [40]. To improve the precision of diagnosis, novel approaches such as histomorphometry are being developed to provide an unbiased evaluation of ACF. This method offers both qualitative and quantitative analysis of tissue structures, providing insights that might be ambiguous to the pathologist, thereby facilitating standardized examination and objective diagnosis. In our study, histological examinations successfully identified the presence of ACF, a key early marker in the development of CRC. However, the histomorphometric analysis did not reveal any significant changes in the crypt parameters. This discrepancy suggests that while histology is sensitive in detecting early morphological alterations associated with ACF, histomorphometry in our study did not provide additional insights into these early lesions. The lack of detectable histomorphometric changes despite the histologic presence of ACF may reflect the subtle nature of these changes during the early stages of CRC development, which are below the detection threshold of histomorphometric techniques.

The AOM-induced colon cancer model in laboratory rodents is frequently used to study the underlying mechanisms of human sporadic colorectal cancer formation. This is because the tumor development mechanism in animals is very similar to that observed in humans. Intraperitoneal administration of AOM causes ACF, leading to the development of malignant adenoma. AOM does not directly interact with DNA and must be activated *in vivo* to promote carcinogenesis. Azoxymethane is metabolized by cytochrome P450, specifically by its CYP2E1 isoform. The first stage of this transformation is the hydroxylation of the AOM methyl group to form methylazoxymethanol (MAM) and its subsequent decomposition into formaldehyde and a highly reactive methyl diazonium. This chemical substance causes DNA mutation, which can initiate the tumorigenesis process by acting on several key genes in intracellular signaling pathways [16].

Disturbances in redox status play a significant role in the actions of AOM and the associated risk of CRC formation and development. This association between oxidative stress and CRC has also been confirmed by various studies in the human population [41–44] as well as by findings from experiments using animal models of chemically induced CRC [45,46]. Redox imbalance primarily results in an increase in oxidative stress, which involves the excessive generation and accumulation of free radicals in cells. Oxidative stress plays a critical role in cellular damage and mutation, contributing to the initiation and progression of CRC [47,48].

Excessive accumulation of free radicals in the cells of the intestinal mucosa causes oxidative stress, which significantly alters the immune response of this part of the gastrointestinal tract. These alterations lead to metabolic changes that are critical for the initiation of cancer formation through multiple pathways [49,50]. These pathways can impair biological membranes by oxidizing their constituent lipids, cause DNA oxidative damage within the nucleus, and oxidize proteins and carbohydrates among the chemical components of the cell. The byproducts of these pathways can act as potentially destructive factors during the initiation and development of CRC [51]. In our current study, we found that the plasma concentration of TBARS was significantly higher in the CRC rats compared to the control animals. Additionally, this alteration was accompanied by an increase in the activity of antioxidant enzymes such as SOD, GPx, and GR, which are natural protective mechanisms that counteract the destructive effects of oxidative stress [47]. The activities of SOD and GPx in peripheral blood are reliable biomarkers of oxidative stress used in CRC patients [52], as well as in *in vivo* studies [53]. Antioxidant enzymes, including SOD, serve as the primary line of endogenous antioxidant defense and are highly susceptible to oxidative damage from carcinogens. Agents that counteract the action of carcinogens suppress the activity of antioxidant enzymes and reduce the overproduction of free radicals, thereby restoring the redox balance [54]. Supplementation with OBG significantly decreased TBARS levels and the activity of GPx, GR, and SOD in animals with CRC. Our previous study showed similar results in animals with TNBS-induced colitis, demonstrating that low molar mass OBG fractions could reduce oxidative stress and inflammation [55]. In the present study, we used OBG, which lacks antioxidant activity. This suggests that its effect on the redox status of CRC rats was indirect. The proposed mechanism of OBG's indirect action may involve the phosphoinositide-3-kinase-protein kinase B/Akt (PI3K-Akt) signal transduction pathway. Furthermore, water-soluble non-starch polysaccharides may also stimulate the insulin receptor α (IRα)-mediated PI3K-Akt signaling pathway to exert antioxidant activity [56].

The results of our study indicate that the hematological parameters, including red blood cell (RBC) counts, hemoglobin (HGB) levels, hematocrit (HCT) percentages, mean corpuscular volume (MCV), mean corpuscular hemoglobin (MCH), and platelet counts, remained consistent among the different experimental groups. This suggests that the induction of early-stage CRC and the dietary intervention with an OBG feed did not significantly alter these parameters. These results are consistent with research on patients with colorectal cancer, where changes in RBC parameters were found in the later stages of this cancer [57]. Based on a large meta-analysis, the authors suggest that RBC parameters, as well as WBC and platelet counts, can be used as predictors for referral to clinical testing for cancer diagnosis in the later stages of CRC [58]. However, as was shown by our results, a notable exception was observed in the MCHC level, which was significantly higher in rats from the CRC group consuming the feed supplemented with 3% OBG, indicating a specific response to dietary intervention at the cellular level that warrants further investigation for its potential implications in CRC pathophysiology.

During tumorigenesis in the early stages of colorectal cancer, slight changes in hematologic parameters may occur. These changes can provide important insight into the body's response to the developing cancer. For example, an increase in RDWc values, associated with higher variability in erythrocyte size, is a condition that can be caused by several factors, including inflammation and nutritional deficiencies, which are common in CRC patients [59,60]. Variations in RDWc are particularly significant as they may reflect the early systemic impact of CRC. Therefore, monitoring these parameters could provide valuable prognostic information in the early stages of colorectal cancer [61].

The study observed a reduction in WBC, GRA, and LYM in rats from the CRC OBG_0 group compared to the control group, indicating a potential immunomodulatory effect of the CRC condition. A decrease in WBC, GRA, and LYM levels in CRC animals could potentially indicate a compromised immune response, which is crucial in the body's defense against cancer progression. Lymphocytes, including T and NK cells, play a crucial role in antitumor immunity. A decrease in lymphocyte count may indicate impaired immune system function, which can promote tumor cell survival and progression [62,63]. Notably, the significant decrease in GRA counts in control rats fed 3% OBG feed compared to the control group highlights the anti-inflammatory potential of the OBG feed. This suggests that dietary OBG may play a role in modulating immune responses.

CRC is a complex disease characterized by the dysregulation of multiple signaling pathways, among which the WNT/β-catenin pathway plays a pivotal role in tumorigenesis. This pathway is crucial for maintaining the balance between cell proliferation and apoptosis, which is essential for intestinal homeostasis and regeneration [64]. The sequence of sporadic events leading to the development of CRC is still not sufficiently well understood. Most evidence suggests that the initiating event of CRC formation is hyperactivation of the WNT signaling pathway, primarily through mutations in β-catenin or the APC gene. Mutations in key components of the WNT/β-catenin pathway, such as APC and β-catenin, are frequently observed in CRC patients, further underscoring the significance of this pathway in CRC development. Abnormalities in the WNT pathway occur in most patients with sporadic CRC. More than 80% of adenomas exhibit APC mutations, and an additional 5–10% have mutations or epigenetic changes in other parts of the WNT signaling pathway, such as β-catenin. It has been demonstrated that azoxymethane causes mutations in β-catenin at codons 33 and 41, leading to the accumulation of β-catenin in the carcinogenesis process [16]. The results of the present study indicate that OBG supplementation modulates the WNT/β-catenin pathway, which is a key pathway in the development and progression of CRC. CRC condition and oat beta-glucan supplementation significantly modulated the expression of *Wnt1* and *Ctnnb1* genes. CRC led to the downregulation of both genes, while OBG supplementation further decreased their expression, particularly at the highest concentration (3%). This illustrates the complex interplay between dietary factors and cancer status in affecting key signaling pathways involved in colorectal tumorigenesis. Additionally, WB and IHC results confirm that the effect of AOM-induced early stage of CRC is the downregulation of β-catenin protein expression in large intestine tissue. These results differ from the conventional understanding based on the literature, where the Wnt/β-catenin pathway is typically upregulated in intestinal cancer [65]. It is interesting to note that dietary factors like OBG might modulate this critical tumorigenic signaling pathway.

Carcinogenesis in the large intestine can lead to the loss of tumor suppressor genes, such as *APC* and *SMAD4*, which normally prevent cancer development [16]. On the other hand, during CRC development, the oncogene *Dclk1* is upregulated, promoting the proliferation of cancer stem cells, increasing tumor growth, and reducing sensitivity to conventional therapies. Therefore, *Dclk1* plays a crucial role in the progression of the disease [66]. In the present study, early-stage CRC significantly downregulates the tumor suppressor genes *Apc* and *Smad4*, reflecting their pivotal roles in cancer suppressive pathways. However, OBG supplementation mitigates this effect against the *Apc* gene, restoring its expression level closer to normal. The study demonstrates the potential of dietary interventions in modulating gene expression associated with cancer suppression. Specifically, the inclusion of 3% OBG in the diet restored Apc expression levels in CRC rats. This finding suggests that OBG supplementation may be a promising approach for cancer prevention. These results contribute valuable insights into the mechanistic pathways by which dietary components, such as beta-glucan, can influence the genetic transformations that occur during the development of colorectal cancer. This offers promising avenues for preventive strategies against tumorigenesis.

When investigating the impacts of dietary supplements on health, it is important to consider their long-term safety. It is worth noting that the consumption of 1-3, 1-4 beta glucan at 1% and 3% levels did not worsen systemic parameters or affect the clinical condition of the animals in the control groups, including fecal consistency. The addition of this polysaccharide to the feed did not affect its sensory qualities or growth performance. This is supported by the lack of differences in feed intake and equalized weight gains in all groups, including both CRC and controls. It is important to note that the OBG isolate has been shown to lack antioxidant activity. Therefore, it can be concluded that its effect on the redox status of CRC rats was indirect. This confirms that 1-3, 1-4 beta-glucan has a safe bioactivity profile. Additionally, its positive effects in pathological conditions, such as the early stages of colon cancer, make beta-glucan a promising candidate for consideration as a nutraceutical.

5. Conclusions

An *in vivo* study was conducted in a CRC model with a nutritional intervention that involved adding low molar mass oat beta-glucan at concentrations of 1% or 3% (*w*/*w*) to animal feed. The study showed the normalizing effects of this cereal polysaccharide. Particularly interesting were its effects at the 3% level on the redox status of peripheral blood and the signaling pathways associated with colon carcinogenesis. The precise extraction and purification process ensured the purity of OBG, which is crucial for accurately assessing its bioactive impact. This is particularly important in modulating genetic and biochemical pathways linked to tumorigenesis. The lack of adverse effects on the healthy intestine, combined with the strong beneficial effects of low molar mass oat beta-glucan in the early stages of carcinogenesis, underscores the safety and potent therapeutic impact of this compound. These observations support the consideration of this oat beta-glucan fraction as a promising nutraceutical. In summary, these findings highlight the potential of OBG as a dietary intervention to restore or maintain the balance of gene expressions crucial for preventing CRC progression.

Supplementary Materials: The following supporting information can be downloaded at: https://www.mdpi.com/article/10.3390/nu16081125/s1, Table S1: Product composition of AIN-93M experimental feeds declared by the producer; Table S2: Declared and analyzed composition of macronutrients (% *w*/*w*) in experimental feeds; Table S3: Results of histopathological examination of intestinal crypts; Supplementary Data: Microscope funding.

Author Contributions: Conceptualization, J.H., K.D. and J.G.-O.; methodology, J.H., K.D. and J.G.-O.; formal analysis, J.H., K.D. and Ł.K.; investigation, J.H., K.D., Ł.K., J.W. and R.S.; resources, J.H., K.D. and Ł.K.; writing—original draft preparation, J.H., K.D. and J.G.-O.; writing—review and editing, J.H., K.D., Ł.K., J.W., R.S. and J.G.-O.; visualization, J.H., K.D. and Ł.K.; supervision, J.G.-O.; project administration, K.D. and J.G.-O.; funding acquisition, J.G.-O. All authors have read and agreed to the published version of the manuscript.

Funding: This research was funded by the National Science Centre, Poland through grant number 2018/29/B/NZ9/01060.

Institutional Review Board Statement: The animal study protocol was approved by the 2nd Local Ethical Committee in Warsaw (Resolution No. WAW2/040/2019, 15 March 2019) in accordance with the EU Directive 2010/63/EU for animal experiments and Polish law and 3R rules (Replacement, Reduction and Refinement).

Informed Consent Statement: Not applicable.

Data Availability Statement: The data that support the findings of this study are available on request from the corresponding author. The data are not publicly available due to privacy.

Acknowledgments: We would like to acknowledge Diana Stopka from the Institute of Veterinary Medicine, Warsaw University of Life Sciences (IVM-WULS), Warsaw, Poland, for assistance with histomorphometric evaluation and *in vivo* experiments.

Conflicts of Interest: The authors declare no conflicts of interest.

References

1. Nakashima, A.; Yamada, K.; Iwata, O.; Sugimoto, R.; Atsuji, K.; Ogawa, T.; Ishibashi-Ohgo, N.; Suzuki, K. β-Glucan in foods and its physiological functions. *J. Nutr. Sci. Vitaminol. (Tokyo)* **2018**, *64*, 8–17. [CrossRef] [PubMed]
2. Wilczak, J.; Błaszczyk, K.; Kamola, D.; Gajewska, M.; Harasym, J.P.; Jałosińska, M.; Gudej, S.; Suchecka, D.; Oczkowski, M.; Gromadzka-Ostrowska, J. The effect of low or high molecular weight oat beta-glucans on the inflammatory and oxidative stress status in the colon of rats with LPS-induced enteritis. *Food Funct.* **2015**, *6*, 590–603. [CrossRef]
3. Żyła, E.; Dziendzikowska, K.; Kamola, D.; Wilczak, J.; Sapierzyński, R.; Harasym, J.; Gromadzka-Ostrowska, J. Anti-inflammatory activity of oat beta-glucans in a crohn's disease model: Time- and molar mass-dependent effects. *Int. J. Mol. Sci.* **2021**, *22*, 4485. [CrossRef]
4. Choromańska, A.; Kulbacka, J.; Harasym, J.; Oledzki, R.; Szewczyk, A.; Saczko, J. High- and low-molecular weight oat beta-glucan reveals antitumor activity in human epithelial lung cancer. *Pathol. Oncol. Res.* **2017**, *24*, 583–592. [CrossRef]
5. Choromańska, A.; Kulbacka, J.; Rembiałkowska, N.; Piłat, J.; Olędzki, R.; Harasym, J.; Saczko, J. Anticancer properties of low molecular weight oat beta-glucan—An in vitro study. *Int. J. Biol. Macromol.* **2015**, *80*, 23–28. [CrossRef]
6. Morgan, E.; Arnold, M.; Gini, A.; Lorenzoni, V.; Cabasag, C.J.; Laversanne, M.; Vignat, J.; Ferlay, J.; Murphy, N.; Bray, F. Global burden of colorectal cancer in 2020 and 2040: Incidence and mortality estimates from GLOBOCAN. *Gut* **2023**, *72*, 338–344. [CrossRef]
7. Jaroszyńska, Z.; Wiśniewska, K. Epidemiology of colorectal cancer (C18-C21) in Poland. *J. Educ. Health Sport* **2021**, *11*, 143–156. [CrossRef]
8. Shen, J.; Li, P.; Liu, S.; Liu, Q.; Li, Y.; Zhang, Z.; Yang, C.; Hu, M.; Sun, Y.; He, C.; et al. The chemopreventive effects of Huangqin-tea against AOM-induced preneoplastic colonic aberrant crypt foci in rats and omics analysis. *Food Funct.* **2020**, *11*, 9634–9650. [CrossRef] [PubMed]
9. Sahebi, Z.; Emtyazjoo, M.; Mostafavi, P.G.; Bonakdar, S. Promising Chemoprevention of Colonic Aberrant Crypt Foci by Portunus segnis Muscle and Shell Extracts in Azoxymethane-Induced Colorectal Cancer in Rats. *Anticancer Agents Med. Chem.* **2020**, *20*, 2041–2052. [CrossRef]
10. Clapper, M.L.; Chang, W.C.L.; Cooper, H.S. Dysplastic Aberrant Crypt Foci: Biomarkers of Early Colorectal Neoplasia and Response to Preventive Intervention. *Cancer Prev. Res.* **2020**, *13*, 229–240. [CrossRef] [PubMed]
11. Ezuka, A.; Sakai, E.; Kawana, K.; Nagase, H.; Kakuta, Y.; Uchiyama, S.; Ohkubo, H.; Higurashi, T.; Nonaka, T.; Endo, H.; et al. Association between factors associated with colorectal cancer and rectal aberrant crypt foci in humans. *Oncol. Lett.* **2015**, *10*, 3689–3695. [CrossRef] [PubMed]
12. Kobaek-Larsen, M.; Nielsen, D.S.; Kot, W.; Krych, Ł.; Christensen, L.P.; Baatrup, G. Effect of the dietary polyacetylenes falcarinol and falcarindiol on the gut microbiota composition in a rat model of colorectal cancer. *BMC Res. Notes* **2018**, *11*, 411. [CrossRef] [PubMed]
13. Madka, V.; Kumar, G.; Pathuri, G.; Zhang, Y.; Lightfoot, S.; Asch, A.S.; Mohammed, A.; Steele, V.E.; Rao, C.V. Bisphosphonates zometa and fosamax synergize with metformin to prevent AOM-induced colon cancer in f344 rat model. *Cancer Prev. Res.* **2020**, *13*, 185–194. [CrossRef] [PubMed]
14. Perše, M.; Cerar, A. Morphological and molecular alterations in 1,2 dimethylhydrazine and azoxymethane induced colon carcinogenesis in rats. *J. Biomed. Biotechnol.* **2011**, *2011*, 473964. [CrossRef] [PubMed]
15. Yang, J.; Wei, H.; Zhou, Y.; Szeto, C.H.; Li, C.; Lin, Y.; Coker, O.O.; Lau, H.C.H.; Chan, A.W.H.; Sung, J.J.Y.; et al. High-Fat Diet Promotes Colorectal Tumorigenesis through Modulating Gut Microbiota and Metabolites. *Gastroenterology* **2022**, *162*, 135–149.e2. [CrossRef]
16. Chen, J.; Huang, X.F. The signal pathways in azoxymethane-induced colon cancer and preventive implications. *Cancer Biol. Ther.* **2009**, *8*, 1313–1317. [CrossRef] [PubMed]
17. Im, S.A.; Kim, J.W.; Kim, H.S.; Park, C.S.; Shin, E.; Do, S.G.; Park, Y.I.; Lee, C.K. Prevention of azoxymethane/dextran sodium sulfate-induced mouse colon carcinogenesis by processed Aloe vera gel. *Int. Immunopharmacol.* **2016**, *40*, 428–435. [CrossRef]
18. Saki, E.; Saiful Yazan, L.; Mohd Ali, R.; Ahmad, Z. Chemopreventive Effects of Germinated Rough Rice Crude Extract in Inhibiting Azoxymethane-Induced Aberrant Crypt Foci Formation in Sprague-Dawley Rats. *Biomed Res. Int.* **2017**, *2017*, 9517287. [CrossRef]
19. Raju, J. Azoxymethane-induced rat aberrant crypt foci: Relevance in studying chemoprevention of colon cancer. *World J. Gastroenterol.* **2008**, *14*, 6632–6635. [CrossRef] [PubMed]
20. Harasym, J.; Suchecka, D.; Gromadzka-Ostrowska, J. Effect of size reduction by freeze-milling on processing properties of beta-glucan oat bran. *J. Cereal Sci.* **2015**, *61*, 119–125. [CrossRef]
21. Harasym, J.; Żyła, E.; Dziendzikowska, K.; Gromadzka-Ostrowska, J. Proteinaceous residue removal from oat β-glucan. *Molecules* **2019**, *24*, 1729. [CrossRef]
22. Harasym, J.; Olędzki, R. The Mutual Correlation of Glucose, Starch, and Beta-Glucan Release During Microwave Heating and Antioxidant Activity of Oat Water Extracts. *Food Bioprocess Technol.* **2018**, *11*, 874–884. [CrossRef]
23. Harasym, J.; Satta, E.; Kaim, U. Ultrasound treatment of buckwheat grains impacts important functional properties of resulting flour. *Molecules* **2020**, *25*, 3012. [CrossRef]
24. Kopiasz, Ł.; Dziendzikowska, K.; Oczkowski, M.; Harasym, J.; Gromadzka-Ostrowska, J. Low-molar-mass oat beta-glucan impacts autophagy and apoptosis in early stages of induced colorectal carcinogenesis in rats. *Int. J. Biol. Macromol.* **2024**, *254*, 127832. [CrossRef]

25. Reeves, P.G.; Nielsen, F.H.; Fahey, G.C. AIN-93 purified diets for laboratory rodents: Final report of the American Institute of Nutrition ad hoc writing committee on the reformulation of the AIN-76A rodent diet. *J. Nutr.* **1993**, *123*, 1939–1951. [CrossRef] [PubMed]
26. Dell, R.B.; Holleran, S.; Ramakrishnan, R. Sample size determination. *ILAR J.* **2002**, *43*, 207–212. [CrossRef]
27. Aguilar Diaz De Leon, J.; Borges, C.R. Evaluation of Oxidative Stress in Biological Samples Using the Thiobarbituric Acid Reactive Substances Assay. *JoVE* **2020**, *159*, e61122.
28. Nair, A.; Jacob, S. A simple practice guide for dose conversion between animals and human. *J. Basic Clin. Pharm.* **2016**, *7*, 27–31. [CrossRef] [PubMed]
29. Bielohuby, M.; Bodendorf, K.; Brandstetter, H.; Bidlingmaier, M.; Kienzle, E. Predicting metabolisable energy in commercial rat diets: Physiological fuel values may be misleading. *Br. J. Nutr.* **2010**, *103*, 1525–1533. [CrossRef] [PubMed]
30. Orlando, F.A.; Tan, D.; Baltodano, J.D.; Khoury, T.; Gibbs, J.F.; Hassid, V.J.; Ahmed, B.H.; Alrawi, S.J. Aberrant crypt foci as precursors in colorectal cancer progression. *J. Surg. Oncol.* **2008**, *98*, 207–213. [CrossRef] [PubMed]
31. Velmurugan, B.; Singh, R.P.; Tyagi, A.; Agarwal, R. Inhibition of azoxymethane-induced colonic aberrant crypt foci formation by silibinin in male Fisher 344 rats. *Cancer Prev. Res.* **2008**, *1*, 376–384. [CrossRef]
32. Suzuki, R.; Kohno, H.; Sugie, S.; Tanaka, T. Sequential observations on the occurrence of preneoplastic and neoplastic lesions in mouse colon treated with azoxymethane and dextran sodium sulfate. *Cancer Sci.* **2004**, *95*, 721–727. [CrossRef]
33. Raju, J.; Swamy, M.V.; Cooma, I.; Patlolla, J.M.R.; Pittman, B.; Reddy, B.S.; Steele, V.E.; Rao, C.V. Low doses of β-carotene and lutein inhibit AOM-induced rat colonic ACF formation but high doses augment ACF incidence. *Int. J. Cancer* **2005**, *113*, 798–802. [CrossRef] [PubMed]
34. Kowalczyk, M.; Orłowski, M.; Klepacki, Ł.; Zinkiewicz, K.; Kurpiewski, W.; Kaczerska, D.; Pesta, W.; Zieliński, E.; Siermontowski, P. Rectal aberrant crypt foci (ACF) as a predictor of benign and malignant neoplastic lesions in the large intestine. *BMC Cancer* **2020**, *20*, 133. [CrossRef] [PubMed]
35. Corpet, D.E.; Taché, S. Most Effective Colon Cancer Chemopreventive Agents in Rats: A Systematic Review of Aberrant Crypt Foci and Tumor Data, Ranked by Potency. *Nutr. Cancer* **2002**, *43*, 1–21. [CrossRef]
36. Cheng, L.; Lai, M. De Aberrant crypt foci as microscopic precursors of colorectal cancer. *World J. Gastroenterol.* **2003**, *9*, 2642–2649. [CrossRef] [PubMed]
37. Macharia, J.M.; Ngure, V.; Emődy, B.; Király, B.; Káposztás, Z.; Rozmann, N.; Erdélyi, A.; Raposa, B. Pharmacotherapeutic Potential of *Aloe secundiflora* against Colorectal Cancer Growth and Proliferation. *Pharmaceutics* **2023**, *15*, 1558. [CrossRef] [PubMed]
38. Gigola, G.; Carriere, P.; Díaz, M.B.N.; Perdigon, G.; Zwenger, A.O.; Gentili, C. Survival effect of probiotics in a rat model of colorectal cancer treated with capecitabine. *World J. Gastrointest. Oncol.* **2021**, *13*, 1518–1531. [CrossRef]
39. Bi, W.; Liu, H.; Shen, J.; Zhang, L.H.; Li, P.; Peng, B.; Cao, L.; Zhang, P.; He, C.; Xiao, P. Chemopreventive effects of Ku-jin tea against AOM-induced precancerous colorectal lesions in rats and metabolomic analysis. *Sci. Rep.* **2017**, *7*, 15893. [CrossRef]
40. Wang, K.S.; Yu, G.; Xu, C.; Meng, X.H.; Zhou, J.; Zheng, C.; Deng, Z.; Shang, L.; Liu, R.; Su, S.; et al. Accurate diagnosis of colorectal cancer based on histopathology images using artificial intelligence. *BMC Med.* **2021**, *19*, 76. [CrossRef]
41. Wu, R.; Feng, J.; Yang, Y.; Dai, C.; Lu, A.; Li, J.; Liao, Y.; Xiang, M.; Huang, Q.; Wang, D.; et al. Significance of Serum Total Oxidant/Antioxidant Status in Patients with Colorectal Cancer. *PLoS ONE* **2017**, *12*, e0170003. [CrossRef]
42. Janion, K.; Szczepańska, E.; Nowakowska-zajdel, E.; Strzelczyk, J.; Copija, A. Selected Oxidative Stress Markers in Colorectal Cancer Patients in Relation to Primary Tumor Location—A Preliminary Research. *Medicina* **2020**, *56*, 47. [CrossRef] [PubMed]
43. Gu, H.; Li, B.; Xiang, L.; Xu, Z.; Tang, Y.; Zhu, Z.; Jiang, Y.; Peng, L.; He, H.; Wang, Y. Association between oxidative stress exposure and colorectal cancer risk in 98,395 participants: Results from a prospective study. *Front. Nutr.* **2023**, *10*, 1284066. [CrossRef] [PubMed]
44. Acevedo-León, D.; Gómez-Abril, S.Á.; Sanz-García, P.; Estañ-Capell, N.; Bañuls, C.; Sáez, G. The role of oxidative stress, tumor and inflammatory markers in colorectal cancer patients: A one-year follow-up study. *Redox Biol.* **2023**, *62*, 102662. [CrossRef] [PubMed]
45. Wang, J.; Ding, K.; Wang, Y.; Yan, T.; Xu, Y.; Deng, Z.; Lin, W.; Zhang, L.; Zhu, W.; Zhao, R.; et al. Wumei Pill Ameliorates AOM/DSS-Induced Colitis-Associated Colon Cancer through Inhibition of Inflammation and Oxidative Stress by Regulating S-Adenosylhomocysteine Hydrolase- (AHCY-) Mediated Hedgehog Signaling in Mice. *Oxid. Med. Cell. Longev.* **2022**, *2022*, 4061713. [CrossRef]
46. Waly, M.I.; Ali, A.; Guizani, N.; Al-Rawahi, A.S.; Farooq, S.A.; Rahman, M.S. Pomegranate (*Punica granatum*) peel extract efficacy as a dietary antioxidant against azoxymethane-induced colon cancer in rat. *Asian Pac. J. Cancer Prev.* **2012**, *13*, 4051–4055. [CrossRef]
47. Bardelčíková, A.; Šoltys, J.; Mojžiš, J. Oxidative Stress, Inflammation and Colorectal Cancer: An Overview. *Antioxidants* **2023**, *12*, 901. [CrossRef] [PubMed]
48. Liu, H.; Liu, X.; Zhang, C.; Zhu, H.; Xu, Q.; Bu, Y.; Lei, Y. Redox Imbalance in the Development of Colorectal Cancer. *J. Cancer* **2017**, *8*, 1586–1597. [CrossRef]
49. Moradi-Marjaneh, R.; Hassanian, S.M.; Mehramiz, M.; Rezayi, M.; Ferns, G.A.; Khazaei, M.; Avan, A. Reactive oxygen species in colorectal cancer: The therapeutic impact and its potential roles in tumor progression via perturbation of cellular and physiological dysregulated pathways. *J. Cell. Physiol.* **2019**, *234*, 10072–10079. [CrossRef] [PubMed]

50. Lin, S.; Li, Y.; Zamyatnin, A.A.; Werner, J.; Bazhin, A.V. Reactive oxygen species and colorectal cancer. *J. Cell. Physiol.* **2018**, *233*, 5119–5132. [CrossRef]
51. Basak, D.; Uddin, M.N.; Hancock, J. The Role of Oxidative Stress and Its Counteractive Utility in Colorectal Cancer (CRC). *Cancers* **2020**, *12*, 3336. [CrossRef] [PubMed]
52. Malinowska, K.; Mik, M.; Dziki, Ł.; Dziki, A.; Majsterek, I. Evaluation of antioxidant defense in patients with colorectal carcinoma. *Pol. Prz. Chir. Pol. J. Surg.* **2015**, *87*, 357–361. [CrossRef]
53. Kumar, V.L.; Verma, S.; Das, P. Artesunate suppresses inflammation and oxidative stress in a rat model of colorectal cancer. *Drug Dev. Res.* **2019**, *80*, 1089–1097. [CrossRef] [PubMed]
54. Qi, X.; Liu, Y. Anti-inflammatory and Antioxidant Effect of Lycoperoside H against the 1,2-Dimethyl Hydrazine (DMH) Induced Colorectal Cancer in Rats. *J. Oleo Sci.* **2022**, *71*, 1021–1029.
55. Kopiasz, Ł.; Dziendzikowska, K.; Gajewska, M.; Wilczak, J.; Harasym, J.; Żyła, E.; Kamola, D.; Oczkowski, M.; Królikowski, T.; Gromadzka-Ostrowska, J. Time-dependent indirect antioxidative effects of oat beta-glucans on peripheral blood parameters in the animal model of colon inflammation. *Antioxidants* **2020**, *9*, 375. [CrossRef] [PubMed]
56. Wu, J.; Chen, M.; Shi, S.; Wang, H.; Li, N.; Su, J.; Liu, R.; Huang, Z.; Jin, H.; Ji, X.; et al. Hypoglycemic effect and mechanism of a pectic polysaccharide with hexenuronic acid from the fruits of *Ficus pumila* L. in C57BL/KsJ db/db mice. *Carbohydr. Polym.* **2017**, *178*, 209–220. [CrossRef] [PubMed]
57. Sala, R.J.; Ery, J.; Cuesta-Peredo, D.; Muedra, V.; Rodilla, V. Complete Blood Count Alterations Prior to the Diagnosis of Colorectal Cancer May Help in the Detection of Synchronous Liver Metastases. *J. Clin. Med.* **2023**, *12*, 6540. [CrossRef]
58. Virdee, P.S.; Marian, I.R.; Mansouri, A.; Elhussein, L.; Kirtley, S.; Holt, T.; Birks, J. The Full Blood Count Blood Test for Colorectal Cancer Detection: A Systematic Review, Meta-Analysis, and Critical Appraisal. *Cancers* **2020**, *12*, 2348. [CrossRef]
59. May, J.E.; Marques, M.B.; Reddy, V.V.B.; Gangaraju, R. Three neglected numbers in the CBC: The RDW, MPV, and NRBC count. *Cleve Clin. J. Med.* **2019**, *86*, 167–172. [CrossRef]
60. Goyal, H.; Lippi, G.; Gjymishka, A.; John, B.; Chhabra, R.; May, E. Prognostic significance of red blood cell distribution width in gastrointestinal disorders. *World J. Gastroenterol.* **2017**, *23*, 4879–4891. [CrossRef]
61. Wen, Z.L.; Zhou, X.; Xiao, D.C. Is red blood cell distribution width a prognostic factor for colorectal cancer? A meta-analysis. *Front. Surg.* **2022**, *9*, 945126. [CrossRef]
62. Waldman, A.D.; Fritz, J.M.; Lenardo, M.J. A guide to cancer immunotherapy: From T cell basic science to clinical practice. *Nat. Rev. Immunol.* **2020**, *20*, 651–668. [CrossRef] [PubMed]
63. Laskowski, T.J.; Biederstädt, A.; Rezvani, K. Natural killer cells in antitumour adoptive cell immunotherapy. *Nat. Rev. Cancer* **2022**, *22*, 557–575. [CrossRef] [PubMed]
64. Cheng, X.; Xu, X.; Chen, D.; Zhao, F.; Wang, W. Therapeutic potential of targeting the Wnt/β-catenin signaling pathway in colorectal cancer. *Biomed. Pharmacother.* **2019**, *110*, 473–481. [CrossRef]
65. Bian, J.; Dannappel, M.; Wan, C.; Firestein, R. Transcriptional Regulation of Wnt/β-Catenin Pathway in Colorectal Cancer. *Cells* **2020**, *9*, 2125. [CrossRef] [PubMed]
66. Kalantari, E.; Razmi, M.; Tajik, F.; Asadi-Lari, M.; Ghods, R.; Madjd, Z. Oncogenic functions and clinical significances of DCLK1 isoforms in colorectal cancer: A systematic review and meta-analysis. *Cancer Cell Int.* **2022**, *22*, 217. [CrossRef] [PubMed]

nutrients

MDPI

Article

Olive Mill Wastewater as Source of Polyphenols with Nutraceutical Properties

Doretta Cuffaro [1,2,†], Andrea Bertolini [3,†], Simone Bertini [1], Claudio Ricci [4], Maria Grazia Cascone [4,5], Serena Danti [4], Alessandro Saba [2,3], Marco Macchia [1,2] and Maria Digiacomo [1,2,*]

1 Department of Pharmacy, University of Pisa, Via Bonanno 6, 56126 Pisa, Italy; doretta.cuffaro@unipi.it (D.C.); simone.bertini@unipi.it (S.B.); marco.macchia@unipi.it (M.M.)
2 Interdepartmental Research Center "Nutraceuticals and Food for Health", University of Pisa, 56100 Pisa, Italy; alessandro.saba@unipi.it
3 Department of Surgery, Medical, Molecular and Critical Area Pathology, University of Pisa, 56126 Pisa, Italy; a.bertolini2@student.unisi.it
4 Department of Civil and Industrial Engineering, University of Pisa, 56122 Pisa, Italy; claudio.ricci@unipi.it (C.R.); maria.grazia.cascone@unipi.it (M.G.C.); serena.danti@unipi.it (S.D.)
5 Department of Translational Research and New Technologies in Medicine and Surgery, University of Pisa, 56126 Pisa, Italy
* Correspondence: maria.digiacomo@unipi.it
† These authors contributed equally to this work.

Abstract: Background: Agrifood waste products are often considered rich sources of bioactive compounds that can be conveniently recovered. Due to these peculiar characteristics, the study of these waste products is attracting great interest in nutraceutical research. Olive mill wastewaters (OMWWs) are generated by extra virgin olive oil (EVOO) production, and they pose environmental challenges due to their disposal. This study aimed to characterize the polyphenolic profile and to evaluate the nutraceutical properties of OMWW extracts from two Tuscan olive cultivars, Leccino (CL) and Frantoio (CF), collected during different time points in EVOO production. Method: After a liquid–liquid extraction, the HPLC and LC–MS/MS analysis of OMWW extracts confirmed the presence of 18 polyphenolic compounds. Results: The polyphenol composition varied between the cultivars and during maturation stages. Notably, oleacein was detected at remarkably high levels in CL1 and CF1 extracts (314.628 ± 19.535 and 227.273 ± 3.974 µg/mg, respectively). All samples demonstrated scavenging effects on free radicals (DPPH and ABTS assays) and an anti-inflammatory potential by inhibiting cyclooxygenase (COX) enzymes. Conclusions: This study highlights the nutraceutical potential of OMWW extracts, emphasizing their antioxidant, antiradical, and anti-inflammatory activities. The results demonstrate the influence of olive cultivar, maturation stage, and extraction process on the polyphenolic composition and the bioactivity of OMWW extracts. These findings support a more profitable reuse of OMWW as an innovative, renewable, and low-cost source of dietary polyphenols with potential applications as functional ingredients in the development of dietary supplements, as well as in the pharmaceutical and cosmetics industries.

Keywords: olive mill wastewater; extra virgin olive oil; nutraceuticals; antioxidants; anti-inflammatory properties

Citation: Cuffaro, D.; Bertolini, A.; Bertini, S.; Ricci, C.; Cascone, M.G.; Danti, S.; Saba, A.; Macchia, M.; Digiacomo, M. Olive Mill Wastewater as Source of Polyphenols with Nutraceutical Properties. *Nutrients* **2023**, *15*, 3746. https://doi.org/10.3390/nu15173746

Academic Editor: Anna Gramza-Michałowska

Received: 23 June 2023
Revised: 19 August 2023
Accepted: 25 August 2023
Published: 26 August 2023

1. Introduction

Extra virgin olive oil (EVOO) contributes to the health benefits of the Mediterranean diet, being the main source of fats and displaying numerous nutraceutical properties [1,2]. EVOO production usually occurs in a specific period, i.e., from September to December, though depending on the area and on the pedoclimatic conditions. It consists of the mechanical extraction of olives and includes several steps: leaf removal and washing, crushing, mixing or malaxation, and, finally, the extraction process. The modern EVOO extraction techniques involve two different processes: two- and three-phase decanter extraction. The two-phase system does not require the further addition of water, resulting in olive oil and a semi-solid olive cake. The three-phase decanter process requires the addition of hot water and results in olive oil, olive mill wastewater (OMWW), and olive pomace. Through the three-phase process, during the extraction of EVOO, a huge volume of OMWW is generated, considering that the usual ratio of olive oil production to OMWW is 1.0:2.5 L and that, in Italy, 1.4 million m^3 of OMWW is produced, as well as 30 million m^3 in the Mediterranean basin [3]. OMWW is characterized by marked acidity and a high content of phenolic components, and due to these peculiarities, it can potentially be toxic to the air, microorganisms, and plants [4]. Therefore, OMWW disposal constitutes an environmental concern with high-costing procedures, and correct management represents a challenge. Nowadays, a great effort is being directed towards the reuse of food-chain wastes in terms of sustainability and revalorization.

Generally, the disposal of OMWWs in waterways or in soil still represents a serious issue due to their phytotoxicity and antimicrobial properties that can affect the balance of ecological systems. In many cases, direct disposal of OMWWs into water streams or in irrigation has resulted in environmental consequences due to the OMWW-high content of phenolic compounds, toxic elements for plants, resulting in deleterious effects on the soil [5–8]. Despite this, polyphenols are widely recognized as natural bioactive molecules with important nutraceutical properties. Therefore, the recovery of phenolic compounds from OMWWs not only provides an economic opportunity but also decreases the environmental impact [5]. In fact, phenolic compounds present in OMWW display well-known biological effects and health benefits, which should be valorized [9].

As has emerged from the literature, the main phenolic compounds present in OMWW are hydroxytyrosol, tyrosol, verbascoside, and cinnamic acids, such as caffeic, gallic, vanillic and syringic acids [10–12]. Recently, researchers aimed to recover polyphenols from OMWW for possible use in the food industry, mainly as nutraceuticals [13,14]. Nowadays, the recovery of valuable bioactive polyphenols from food supply chain by-products is of great interest in terms of sustainability and revalorization. It was demonstrated that phenolic compounds extracted from OMWW exhibit anti-inflammatory action, as they are able to decrease the production of nitric oxide (NO) in LPS-stimulated cells [15]. Moreover, OMWW extracts display in vitro and in vivo antiangiogenic and chemo-preventive capacities [16], inhibiting the proliferation, migration, and invasion of endothelial cells [17]. Due to the interesting properties of OMWW polyphenols, recently, different applications of OMWW extracts in cosmetic fields [18] or in food applications [19] have been successfully described. Moreover, the polyphenol-rich OMWW extracts could be applied as active ingredients in the development of dietary supplements or food products as added value in the formulation, presenting new functionalities such as enhanced nutritional properties.

Considering that OMWW composition primarily depends on the extraction system and olive cultivar, in this work, we studied several samples of OMWWs generated by a three-phases EVOO extraction system. In particular, we analyzed OMWWs produced by two different cultivars of olives: cultivar Leccino (CL) and Frantoio (CF). The collection was conducted in October (CF1 and CL1) and November (CF2 and CL2) in order to investigate the variations of polyphenols content over the time. The OMWW extracts have been obtained via liquid–liquid extraction, allowing for the recovery of important dietary polyphenols. The polyphenolic profile of OMWW extracts was defined by the analytical quantification performed by LC–MS/MS analysis, including the identification and quantification of 18 polyphenols. Furthermore, with the aim to valorize the polyphenolic extract obtained by these waste products and to highlight their nutraceutical properties, the antioxidant profile, and the anti-inflammatory effect in terms of cyclooxygenases (COX 1 and COX2) inhibition were evaluated. This study well-aligns with the purposes of environmental sustainability, exploiting the recovery of valuable bioactive polyphenols from food supply chain by-products such as OMWW.

2. Materials and Methods

2.1. Chemistry

Solvents used for extraction procedures and high-performance liquid chromatography (HPLC) coupled to tandem mass spectrometry (LC–MS/MS) analyses, such as LC–MS-grade and HPLC-grade water, LC–MS- and HPLC-grade methanol (MeOH), LC–MS- and HPLC-grade acetonitrile (ACN), LC–MS-grade 2-propanol, HPLC-grade *n*-hexane, HPLC-grade ethanol (EtOH), HPLC-grade ethyl acetate, HPLC-grade acetic acid (AcOH, 100%), and LC–MS-grade formic acid (FA, ≥8%), were all purchased from Sigma Aldrich-Merck (Merck srl, Milan, Italy). The pure standards of oleocanthal, oleocanthalic acid, and oleacein were obtained via EVOO extraction and purification using the method developed in our previous study [20,21]. The commercial analytical standards of pinoresinol, *p*-hydroxyphenylacetic acid, oleouropein, ibuprofen, and Trolox were purchased from Merck (Merck srl, Milan, Italy); tyrosol, hydroxytyrosol, caffeic acid, vanillic acid, ferulic acid, *p*-coumaric acid, syringic acid, and vanillin were purchased from TCI (Zwijndrecht, Belgium); the commercial analytical standards luteolin-7-glucoside, apigenin-7-glucoside, verbascoside, and rutin were purchased from TRC (Toronto chemicals). 1-acetoxypinoresinol was identified after isolation via semipreparative HPLC and confirmed via LC–MS/MS analysis and quantified using the Pinoresinol calibration curve as already reported [22]. 2,2-Diphenyl-1-picrylhydrazyl (DPPH) and 2,4,6-Tris(2-pyridyl)-s-triazine (TPTZ) and (ABTS) were purchased from Merck (Darmstadt, Germany).

2.2. Extraction of OMWW

OMWW samples belong to two different cultivars: Cultivar Leccino (CL) and Cultivar Frantoio (CF). The samples were collected twice during EVOO production: October 2022 (CL1 and CF1) and November 2022 (CL2 and CF2). After direct collection from the decanter, samples were frozen and stored at −20 °C. Liquid–liquid extraction was carried out following the procedure reported by De Marco et al. [23]. Briefly, 10 mL of OMWW was acidified to pH 2 with HCl 1N and washed with *n*-hexane (15 mL) in order to remove the lipid fraction. The mixture was vigorously shaken and centrifuged for 5 min at 3000 rpm. The phases were separated, and the step was repeated successively two times with *n*-hexane. Extraction of phenolic compounds was then carried out with 10 mL of ethyl acetate. The mixture was vigorously shaken and centrifuged for 5 min at 3000 rpm. The phases were separated, and the extraction was repeated successively three times. The ethyl acetate was evaporated under reduced pressure, obtaining the solvent-free dry residue.

2.3. LC–MS/MS Instrumental Layout and Parameters

The instrument layout consisted in an Agilent (Santa Clara, CA, USA) 1290 UHPLC system, including a binary pump, a column oven set at 40 °C, and a thermostated autosampler, coupled to an AB-Sciex (Concord, Ontario, Canada) QTRAP 6500+ mass spectrometer working as a triple quadrupole, equipped with an IonDrive™ Turbo V source set for electrospray ionization (ESI). Chromatographic separation was achieved by using a 4.6 × 150 mm, 5 µm particle size Agilent Zorbax SB-C18 StableBond Analytical column (Santa Clara, CA, USA), and using (A) MeOH and (B) water (100%), both mixed with 0.025% CH_3COOH, as mobile phases. Gradient elution was performed at 800 µL/min as follows: 0.0–1.0 min (B) 100%; 12.0–14.0 min (B) 5%; 15.0–18.0 min (B) 100%. The injection volume was set at 10 µL. System control and data acquisition were carried out with ABSciex Analyst® version 1.7 software, data processing was carried out with Multiquant™ software version 3.0.3, while data analysis was performed using Microsoft 365® Excel software version 2307 (Albuquerque, NM, USA) and GraphPad (Boston, MA, USA) Prism version 9.0.2.

A mass spectrometry selected reaction monitoring (SRM) method was operated in negative-ion mode. For each compound, after the optimization of declustering potential (DP), collision energy (CE), and collision exit potential (CxP), three transitions were considered in the analysis. One of them was integrated and used as a quantifier (Q), and the other two were used as qualifiers (q), as reported in Table 1. Further operative parameters were gas source 1 (GS1), 40 arbitrary units; gas source 2 (GS2), 40 arbitrary units; ion spray voltage (ISV), −4.0 kV; source temperature (TEM), 450 °C; Curtain gas (CUR), 20 arbitrary units; and collision gas (CAD) N_2—operative pressure with CAD gas on, 2.93 mPa.

Table 1. MS operative parameters. Retention time variations are expressed as mean ± SEM.

Analyte	Retention Time (min)	SRM Transition (Da)	DP (V)	EP (V)	CE (V)	CxP (V)
Tyrosol	7.55 ± 0.01	137.0 →106.0 (Q) 137.0 →107.0 (q) 137.0 →119.0 (q)	−35	−7.0	−20 −21 −20	−5.0 −5.4 −5.9
Oleocanthal	10.32 ± 0.01	303.1 →59.0 (Q) 303.1 →165.0 (q) 303.1 →182.6 (q)	−30	−7.0	−10 −13 −13	−6.0 −8.4 −9.6
Oleocanthalic Acid	10.40 ± 0.01	319.2 →110.8 (q) 319.2 →155.0 (q) 319.2 →199.0 (Q)	−25	−7.0	−23 −25 −19	−5.3 −7.5 −9.8
Hydroxytyrosol	6.49 ± 0.01	152.8 →81.2 (Q) 152.8 →93.3 (q) 152.8 →122.6 (q)	−90	−7.0	−29 −30 −18	−13.0 −11.0 −6.0
Caffeic acid	8.04 ± 0.01	179.0 →79.2 (q) 179.0 →107.4 (q) 179.0 →134.6 (Q)	−70	−10.0	−32 −30 −20	−14.0 −9.0 −7.0
Syringic acid	8.27 ± 0.01	197.0 →78.2 (Q) 197.0 →106.3 (q) 197.0 →123.4 (q)	−70	−10.0	−37 −30 −35	−12.0 −8.7 −6.0
p-Coumaric Acid	8.99 ± 0.01	163.0 →65.2 (q) 163.0 →93.3 (q) 163.0 →119.2 (Q)	−70	−7.5	−52 −40 −20	−10.0 −5.0 −3.5
Ferulic Acid	9.12 ± 0.01	193.0 →89.2 (Q) 193.0 →106.3 (q) 193.0 →149.8 (q)	−60	−10.0	−35 −31 −30	−10.0 −9.0 −8.0
Vanillic Acid	8.12 ± 0.01	167.0 →65.2 (q) 167.0 →91.3 (q) 167.0 →108.4 (Q)	−60	−5.5	−37 −25 −26	−10.0 −7.0 −5.0
Vanillin	8.71 ± 0.01	150.8 →92.3 (Q) 150.8 →108.4 (q) 150.8 →136.5 (q)	−60	−7.0	−27 −30 −20	−7.5 −5.0 −5.0
Luteolin-7-glucoside	9.22 ± 0.01	447.3 →65.2 (q) 447.3 →83.2 (q) 447.3 →132.6 (Q)	−125	−7.0	−130 −90 −82	−7.0 −6.7 −6.3
Verbascoside	8.63 ± 0.01	623.5 →59.0 (q) 623.5 →85.2 (Q) 623.5 →133.6 (q)	−140	−6.0	−116 −92 −110	−7.0 −10.0 −6.0
Pinoresinol	10.07 ± 0.10	357.3 →92.2 (Q) 357.3 →121.6 (q) 357.3 →136.6 (q)	−60	−8.0	−70 −36 −45	−8.0 −6.0 −6.5
1-Acetoxy pinoresinol	10.11 ± 0.10	415.3 →59.1 (q) 415.3 →92.3 (q) 415.3 →136.6 (Q)	−110	−9.5	−71 −70 −38	−6.0 −5.0 −4.7
Oleuropein	9.63 ± 0.20	539.4 →59.1 (Q) 539.4 →89.3 (q) 539.4 →95.3 (q)	−90	−8.5	−85 −50 −50	−6.5 −7 −8

Table 1. *Cont.*

Analyte	Retention Time (min)	SRM Transition (Da)	DP (V)	EP (V)	CE (V)	CxP (V)
Oleacein	9.63 ± 0.01	319.3 →59.1 (Q) 319.3 →69.2 (q) 319.3 →95.4 (q)	−110	−9.5	−18 −38 −36	−6.4 −7.0 −8.0
Apigenin-7-glucoside	9.73 ± 0.01	431.2 →63.2 (Q) 431.2 →83.1 (q) 431.2 →117.3 (q)	−185	−10.0	−101 −80 −70	−9.8 −9.3 −9.0
Rutin	9.33 ± 0.01	609.4 →65.1 (Q) 609.4 →107.6 (q) 609.4 →108.44 (q)	−220	−9.0	−160 −90 −90	−10 −5.0 −5.0
4-Hydroxyphenyl acetic acid (IS)	7.81 ± 0.01	150.8 →79.3 (q) 150.8 →105.3 (q) 150.8 →107.4 (Q)	−70	−7.5	−24 −24 −10	−9.5 −4.0 −4.0

Standard stock solutions and sample preparation: a stock solution of each standard compound, including the internal standard, was prepared in MeOH at a concentration of 100 μg/mL and stored at −20 °C. Calibration curves were freshly prepared in the range of concentrations reported in Table S1. A mixture of all polyphenols at their higher concentration was prepared and then serially diluted in water. Different starting concentrations for each analyte were optimized in order to achieve linearity in the widest range possible. Each calibration point was mixed with the appropriate amount of the internal standard to achieve the same final concentration (500 ng/mL) in both the submitted samples and in the curve. Since standards were not commercially available, and considering the similar behavior and fragmentation patterns of both 1-acetoxypinoresinol and pinoresinol, the quantitation of 1-acetoxypinoresinol was achieved using an external calibration curve built on the common SRM transition 357.3 → 136.6 and 415.3 → 136.6 (Q). Extracted OMWW samples (10–100 mg), stored at −20 °C, were thawed at room temperature (rt) and resuspended in 1000 μL of a H_2O:MeOH mixture (50:50, *v/v*). Then, samples were diluted in water at different ratios—1:10, 1:100—1:1000, and 1:20,000— to allow each analyte to be correctly quantified in its relatively built calibration curve, mixed with the appropriate amount of internal standard, and 10 μL were injected into the LC–MS/MS system. All samples were analyzed in triplicate.

2.4. DPPH• Radical Scavenging Activity

The antioxidant activity of OMWW extracts (CT1, CT2, CF1, and CF2) was evaluated using the DPPH• free radical scavenging assay as described by Cuffaro et al. [24]. Samples were solubilized in MeOH at different concentrations (from 0.1 mg/mL to 1.5 mg/mL) and added to a methanol solution of DPPH• (40.0 μg/mL). After 45 min of incubation at rt and in the dark, the absorbance was read at 517 nm in a SPECTROstarNano (200–1000 nm) UV/Vis spectrophotometer. MeOH and Trolox® were used as blank and standard positive control, respectively, and were treated under the same conditions as the samples. The percent of antioxidant activity (%AA) was calculated according with the following formula:

$$\%AA = (AbsDPPH - Abssample/AbsDPPH) \times 100,$$

where

AbsDPPH = absorbance of the DPPH solution, subtracted of absorbance of MeOH;

Abssample = absorbance of the DPPH solution containing the test compound subtracted of the absorbance of test compound solution without DPPH.

The results were expressed as inhibitory concentrations at 50% (IC_{50}), calculated via linear regression. All experiments were performed in triplicate.

2.5. Ferric Reducing Antioxidant Power Assay (FRAP)

The method described by Borges et al., with some modifications [24,25], was used to assess the antioxidant activity. FRAP assay is based on iron reduction, measuring the ferric-reducing ability of a sample in an acid medium (pH 3.6) through the formation of a specific blue color as the ferric tripyridyltriazine (Fe^{3+}–TPTZ) complex due to the reduction to the ferrous (Fe^{2+}) form. FRAP reagent was prepared by mixing 0.3 M acetate sodium buffer pH = 3.6, 20 mM ferric chloride, and 10 mM TPTZ in 40 mM HCl at a ratio 10:1:1. Amounts of 20 μL of the extracts were mixed with 280 μL of the FRAP solution, and the mixture was incubated at 37 °C for 30 min. The absorbance of the reaction mixture was read at 595 nm in a SPECTROstarNano (200–1000 nm) UV/Vis spectrophotometer. The calibration curve was built using different concentrations of Trolox (0.01–0.2 mg mL^{-1}), and the results are expressed as the mmol of Trolox equivalents per g of the sample. All experiments were conducted in triplicate.

2.6. ABTS Assay

The free radical scavenging activity of samples was determined by the ABTS radical cation decolorization assay using the protocol reported by Pellegrini et al. with some modifications [24,26]. Briefly, ABTS solution was prepared by mixing 7 mM of aqueous solution of ABTS with 2.45 mM potassium persulfate in a 1:1 ratio. The solution was incubated for 12 h in the dark at rt; then, it was diluted with EtOH to an absorbance of 0.7 at 750 nm. Then, 180 μL of the ABTS solution were mixed with 10 μL of samples in EtOH, and the solution was incubated 5 min at rt; then, the final absorbance was read at 734 nm. Calculations were performed by evaluating the percentage of inhibition of the ABTS radical cation as follows:

$$\% \text{ scavenging ability} = (\text{AbsABTS} - \text{Abssample}/\text{AbsABTS}) \times 100,$$

where

AbsABTS = the absorbance of the ABTS solution;

Abssample = the absorbance of the ABTS solution containing the test compound.

The percentage of scavenging ability was calculated against the sample concentration to obtain the inhibitory concentration. All experiments were performed in triplicate.

2.7. Cyclooxygenase Enzyme Inhibitory Assay

The ability of CT1, CT2, CF1, and CF2 extracts to inhibit COX-1 and COX-2 was evaluated using a COX-1 (ovine) and COX-2 (human)-inhibitor screening assay (kit No. 701050 from Cayman Chemical Co., Ann Arbor, MI, USA) following the manufacturer's protocols. COX-1/COX-2 inhibitory evaluation test was performed at 1 mg/mL for all the extracts. Then, increasing concentrations of each sample (0.01–1 mg/mL) were tested. Arachidonic acid at 1.1 mM was the substrate and ibuprofen was used as a control. The peroxidase activity was examined colorimetrically at 590 nm after an incubation of 120 min at rt using a SPECTROstarNano (200–1000 nm) UV/Vis spectrophotometer (BMG Labtec, Germany). All tests were performed three times. The percent (%) inhibition of COX-1 and COX-2 is derived from the following formula:

$$\% \text{ inhibition} = (\text{EAA} - \text{AIA})/\text{EAA} \times 100,$$

where

EAA = Enzyme test activity absorbance;

AIA = Activity inhibition test absorbance.

Results were expressed as percent (%) inhibition or inhibitory concentration at 50% (IC_{50}), calculated via a logarithmic concentration curve. All experiments were performed in triplicate.

2.8. Statistical Analysis

Data were presented as mean ± standard error (SD) of three independent experiments. A two-way analysis of variance (ANOVA) was applied to determine the differences in concentration of all polyphenols between samples, and comparisons of the means were carried out using Šídák's multiple or Tukey's comparisons tests using GraphPad Prism version 9.0.2. Significance was accepted for $p < 0.05$.

3. Results

3.1. LC–MS/MS Analysis and Characterization

To evaluate the methods ability to differentiate the molecules of interest from other possible components and interferents present in samples, specificity was checked via repeated injections of analytes into the system, and their retention times were monitored. The presence of the analytes was also confronted and confirmed using the HPLC-DAD method. The LC–MS/MS method we developed achieved great linearity for all analytes (r > 0.998) and was able to correctly characterize and quantify eighteen different phenolic compounds in the different OMWW extracts, even if not all of the analytes were always detectable. The complete quantification results are reported in Table 2.

Table 2. Quantification results of phenolic compounds in analyzed OMWW extracts. Each sample was analyzed in triplicate. Results are expressed as Mean ± SD.

Analyte	CL1 (μg/mg)	CL2 (μg/mg)	CF1 (μg/mg)	CF2 (μg/mg)
Hydroxytyrosol	55.875 ± 1.511	6.821 ± 0.058	71.919 ± 3.943	25.253 ± 1.283
Tyrosol	17.602 ± 0.792	8.119 ± 0.318	18.808 ± 0.447	12.213 ± 1.157
Oleocanthal	2.271 ± 0.071	0.181 ± 0.006	1.059 ± 0.009	0.140 ± 0.006
Oleocanthalic acid	10.628 ± 0.484	3.908 ± 0.151	2.149 ± 0.091	0.120 ± 0.010
Caffeic acid	2.791 ± 0.101	1.453 ± 0.044	2.234 ± 0.112	3.144 ± 0.142
Syringic acid	N/A	N/A	N/A	N/A
p-Coumaric acid	0.756 ± 0.028	0.513 ± 0.019	0.428 ± 0.032	0.406 ± 0.016
Ferulic acid	0.108 ± 0.020	0.195 ± 0.018	0.078 ± 0.012	0.040 ± 0.003
Vanillic acid	1.321 ± 0.064	1.356 ± 0.032	0.576 ± 0.010	0.273 ± 0.011
Vanillin	0.018 ± 0.001	0.001 ± 0.001	0.003 ± 0.001	N/A
Verbascoside	1.624 ± 0.035	0.958 ± 0.018	18.182 ± 0.437	0.909 ± 0.035
Luteolin 7 glucoside	0.010 ± 0.001	0.013 ± 0.001	1.115 ± 0.016	0.096 ± 0.004
Pinoresinol	0.391 ± 0.005	0.333 ± 0.009	0.194 ± 0.006	0.194 ± 0.008
Oleuropein	0.253 ± 0.001	N/A	2.541 ± 0.061	0.221 ± 0.006
Oleacein	314.628 ± 19.535	5.896 ± 0.058	227.273 ± 3.974	23.440 ± 1.257
Apigenin 7 glucoside	N/A	N/A	0.025 ± 0.001	0.015 ± 0.001
1-acetoxypinoresinol	1.673 ± 0.180	0.162 ± 0.027	0.960 ± 0.109	1.627 ± 0.045
Rutin	N/A	N/A	0.022 ± 0.004	0.014 ± 0.003

The extracts of the first collection (October 2022), CL1 and CF1, were particularly plentiful of oleacein (314.628 ± 19.535 and 227.273 ± 3.974 μg/mg, respectively), tyrosol, and hydroxytyrosol (17.602 ± 0.792, 18.808 ± 0.447 μg/mg, and 55.875 ± 1.511, 71.919 ± 3.943 μg/mg, respectively), with almost comparable quantities in both cultivars. Otherwise, CL1 showed a higher amount of oleocanthalic acid than CF1 (10.628 ± 0.484 and 2.149 ± 0.091 μg/mg, respectively) but lower quantities of verbascoside (1.624 ± 0.035 and 18.182 μg/mg, respectively). Caffeic acid was found in both samples, with comparable concentrations (2.791 ± 0.101 and 2.234 ± 0.112 μg/mg, respectively), whereas the concentrations of oleocanthal (10.628 ± 0.484 (CL1) and 2.149 ± 0.091 μg/mg (CF1)) and 1-acetoxypinoresinol (1.673 ± 0.180 and 0.960 ± 0.109 μg/mg, respectively CL1 and CF1) in CL1 were significantly higher than in CF1. Moreover, CF1 contains 10-fold more oleuropein than CL1 (2.541 ± 0.061 and 0.253 ± 0.001 μg/mg, respectively). Other phenolic compounds were present in both samples in amounts < 1 μg/mg.

Analyzing the data of the second collection (November 2022) CL2 and CF2 samples, a considerable reduction in all phenolic compounds in comparison to CL1 and CF1 was evident. In fact, almost all polyphenols decreased in concentration (or even disappeared, as for oleuropein in CL and vanillin in CF) in the extract belonging to the second collection (CL2 and CF2) as a result of a decreased amount of polyphenols during the maturation process of olive fruits over the time. In particular, the most evident reduction was detectable for oleacein, the concentration of which had a 100-fold drop in CL OMWW samples (314.628 ± 19.535 vs. 5.896 ± 0.058 μg/mg, mean diff = 308.7, $p < 0.0001$), and a 10-fold decline in CF OMWW samples (227.273 ± 3.974 vs. 23.440 ± 1.257 μg/mg, mean diff = 203.8, $p < 0.0001$).

3.2. Antioxidant/Antiradical Properties

It is well known that EVOO polyphenols display antioxidant and antiradical properties which affect the oxidative stress damage by inhibiting the production of radicals or decreasing the oxidation processes [27]. The high content of nutraceutical polyphenols presented in OMWW prompts the investigation of the nutraceutical properties and, in particular, the antioxidant and antiradical properties. The antiradical properties of CL1, CL2, CF1, and CF2 were tested in vitro by DPPH and ABTS assays, whereas the antioxidant properties were evaluated by the ferric reducing antioxidant power (FRAP) assay. As shown in Table 3, the results revealed that all samples had a good scavenging effect on the DPPH· and ABTS radicals, and the relationship was dose-dependent. The IC_{50} values were lower than 1 mg/mL for all samples, while CF2 presented the best IC_{50} values, ranging the 0.1 mg/mL. Regarding the FRAP assay, the samples of cultivar Frantoio (CF1 and CF2) showed the most interesting activity, in particular CF1 (103.1 ± 5.1 mmol Trolox/g).

Table 3. Antioxidant and antiradical capacities of CL1, CL2, CT1, and CT2 measured by DPPH, ABTS, and FRAP assays.

Sample	DPPH IC_{50} mg/mL	ABTS IC_{50} mg/mL	FRAP mmol Trolox/g
CL1	0.93 ± 0.03 ^{a}	0.31 ± 0.015 ^{b}	37.75 ± 2.42 ^{c}
CL2	0.73 ± 0.01 ^{b}	0.43 ± 0.013 ^{a}	34.21 ± 0.94 ^{c}
CF1	0.50 ± 0.02 ^{c}	0.21 ± 0.006 ^{c}	103.1 ± 5.1 ^{a}
CF2	0.11 ± 0.01 ^{d}	0.16 ± 0.002 ^{d}	76.36 ± 5.51 ^{b}

Values are the average of three determinations ± standard error. Different letters (a, b, c, d) in the same column indicate significant differences between samples ($p < 0.05$).

3.3. Anti-Inflammatory Properties

Inflammation is a fundamental defense mechanism of the organism, sustained by the immune system, that recognizes and eliminates harmful agents and infected cells, promoting tissue repair and restoring body homeostasis with different mechanisms [28–31]. Polyphenols promote healthy effects, even expressing anti-inflammatory effects with various pathways [32]. EVOO polyphenols are recognized to modulate some inflammatory pathways through a reduction in the expression levels of MMP-9, prostaglandin, and thromboxane (TX) by inhibiting COX-2 and COX-1 enzymes [33,34]. In particular, oleocanthal, oleacein, and hydroxytyrosol induced the inhibition of COX-1 and COX-2 [35,36]; moreover, tyrosol and hydroxytyrosol displayed an inhibitory effect on the arachidonate cascade and the eicosanoid synthesis (PGE2 and LTB4) in cultured macrophages [37].

Considering the data reported above, the high content of some of the most representative polyphenols studied for their anti-inflammatory properties, such as oleacein, directed our interest to the investigation of the anti-inflammatory effects of OMWW extracts. CT1, CT2, CF1, and CF2 were evaluated for their ability to inhibit COX enzymes, deeply involved in inflammation cascade.

All the samples were tested at 1 mg/mL, revealing more than 40% inhibition, excluding CL2, which showed about 25% of inhibition. Noteworthy, CL1 and CF1 displayed high inhibition values, especially in COX-2 percentage inhibition, which reached more than 80% (Table 4). Then, COX inhibition percentages for the samples CL1 and CF1 were very promising, especially with respect to COX2 inhibition. The IC_{50} of CF1 and CL1 on the selected enzymes was performed (Table 5). The results revealed a COX-1 IC_{50} ranging 0.5–0.4 mg/mL and a surprising COX-2 IC_{50} value of 0.08 mg/mL for both the extracts.

Table 4. Inhibitory effects of CL1, CL2, CF1, and CF2 on COX-1 and COX-2.

Sample	COX-1 % Inhibition	COX-2 % Inhibition
CL1	39.46 ± 0.002	83.6 ± 0.001
CL2	25.30 ± 0.004	24.40 ± 0.002
CF1	48.06 ± 0.002	88.14 ± 0.001
CF2	40.04 ± 0.003	55.80 ± 0.001

Values are the average of three determinations ± standard error.

Table 5. IC_{50} values of CL1 and CF1 on COX-1 and COX-2.

Sample	COX-1 IC_{50} (mg/mL)	COX-2 IC_{50} (mg/mL)
CL1	0.563 ± 0.165	0.088 ± 0.008
CF1	0.418 ± 0.145	0.082 ± 0.010

Values are the average of three determinations ± standard error.

4. Discussion

In world globalization, the management of different types of waste deriving from food production affects environmental sustainability. For these reasons, the re-valorization of food by-products to produce value-added products is becoming a challenge in the nutraceutical field.

The olive of *Olea europaea* L. is widely spread in the Mediterranean area, where it accounts for almost 96% of global olive production (FAOSTAT Food and Agriculture Data), and where the majority of EVOO is produced. EVOO is a functional food, rich in exclusive phenolic compounds with diverse health benefits. Unfortunately, EVOO production generates a significant quantity of residues, such as OMWW, olive leaves, and pomace. Nonetheless, a large amount of OMWW remains without application, since only small quantities are used as a source of biomass fuel, and the residual part must be managed with high-cost procedures, accounting environmental concerns. On the other hand, these residues are a precious low-cost starting material from which to produce extracts of dietary polyphenolic molecules that could be used in various nutraceutical fields [38]. However, similarly to EVOO [39], the quality of nutraceutical polyphenols in OMWW differs according to the olive cultivar variety, the technological process of olive oil production, collection time, and climatic conditions [40].

For these reasons, in this work, we studied OMWWs produced in a three-phase EVOO extraction system obtained by processing olives of two different Tuscan cultivars, collected in different periods of the EVOO production season, with the aim of evaluating the different polyphenolic content in relation to the cultivars and the olive harvest period. After collection, OMWWs were treated via liquid–liquid extraction in order to recovery dietary polyphenols resulting in dry residue extracts. The resultant OMWW extracts have been analyzed in their polyphenolic profile by the HPLC/MS technique, identifying 18 polyphenols. As expected, a variation in polyphenol composition is clearly noticeable in the analysis of the two different CL and CF OMWW extracts collected at different timepoints. The quantitative analysis of OMWW extracts (Table 2) indicated that all the samples presented an elevated polyphenols content. The samples CF1 and CL1 showed the higher polyphenolic content, probably due to the green maturation level of olive in October. In fact, two states of maturation are evidenced in olive fruits, namely, green and black maturation, that usually correlate with higher and lower polyphenolic content, respectively [41]. During the first collection, the olive fruit were in the green maturation stage, presenting an increased quantity of polyphenols compared to the second collection time (November 2022) in which the olives were black. The polyphenol analysis revealed a high amount of tyrosol, hydroxytyrosol, and phenolic acids for all the samples. Interestingly, an elevated amount of oleacein was quantified in the CL1 and CF1 OMWW extracts (314.628 ± 19.535 and 227.273 ± 3.974 μg/mg, respectively), a very promising polyphenol rarely found in OMWW and generally present in fresh EVOO.

Considering the rich phenolic compound content of these OMWW extracts, their nutraceutical properties were further investigated in order to better comprehend their nutraceutical potential. The DPPH, ABTS, and FRAP assays were used to estimate the anti-radical/antioxidant capacity of OMWW extracts (Table 3). The results of the DPPH radical scavenging activity showed that the CF2 extract exhibited the highest antiradical activity, as indicated by the lower IC_{50} value (0.11 mg/mL). Similarly, the data analysis of the ABTS assay showed that the CF2 extract gave the best activity, with an IC_{50} = 0.155 mg/mL. From the results of FRAP, the best antioxidant extracts were CF1 and CF2, presenting a FRAP value ranging from 5 to 6 mmol of Trolox/kg. According to De Marco et al. [23], the phenolic compounds commonly presented in OMWW conferred a strong antioxidant potential. Therefore, the present data clearly showed a positive correlation between the total polyphenols content and the antioxidant/antiradical capacity evaluated by the DPPH, ABTS, and FRAP assays. In fact, our results describe CT1, CT2, CF1, and CF2 as excellent antioxidant extracts, reporting values of inhibition of antioxidant/antiradical significantly lower than the OMWW extracts reported by Belaqziz et al. [42] (IC_{50} DPPH = 16–261 mg/mL) and comparable to the antioxidant properties of other promising OMWW extracts [12]. It is important to demonstrate that the antioxidant activity of these OMWW extracts is significantly improved compared to the antioxidant activity of EVOO extracts [24,43,44], highlighting the nutraceutical potential of these byproducts.

The high content of oleacein and hydroxytyrosol, polyphenols deeply studied for their anti-inflammatory properties [36], suggested the evaluation of OMWW extracts in their ability to inhibit cyclooxygenase enzymes. The inhibitory activity of the two isoforms of cyclooxygenase, COX-1 and COX-2, were evaluated. The COX-1 isoform is a constitutive isoform involved in tissue homeostasis, whereas the COX-2 isoform is an inducible isoform induced by pro-inflammatory stimuli. Surprisingly, all the extracts revealed more than 40% inhibition for both enzymes, except for CL2, which is probably due to the lower polyphenols content and the near absence of oleacein compared to other OMWW extracts. However, CL1 and CF1 reported more than 80% inhibition of COX-2, and these data encouraged further investigations. Therefore, the IC_{50} of CL1 and CF1 were evaluated (Table 5), resulting in significantly low values of COX-2 inhibition for both extracts (0.088 mg/mL for CL1 and 0.082 mg/mL for CF1) and almost 5-fold higher IC_{50} values for COX-1 (0.563 μg/mL for CL1 and 0.418 μg/mL for CF1). The interesting anti-inflammatory values of these extracts are probably due to the high concentration of oleacein and hydroxytyrosol already tested in vitro as potent COX inhibitors (oleacein IC_{50} COX-1 = 0.47 μg/mL and IC_{50} COX-2 = 0.41 μg/mL; hydroxytyrosol IC_{50} COX-1 = 0.02 μg/mL and IC_{50} COX-2 = 0.37 μg/mL) [36]. The high inhibitory potential of these two polyphenols may confer a promising anti-inflammatory ability to CL1 and CF1, rich in oleacein and hydroxytyrosol, probably due to the synergic effects of different phenolic compounds.

5. Conclusions

Nowadays, research has increasingly focused on the valorization and reuse of EVOO by-products because they constitute an environmental concern due to their high production and the related costs of disposal procedures. OMWWs represent one of the most produced byproducts in EVOO production, but at the same time, they are regarded as inexpensive and abundant raw materials rich in bioactive compounds with health-related properties. In this study, we investigated the nutraceutical potential of Tuscan OMWWs produced by two different cultivars of olives (cultivar Leccino (CL) and Frantoio (CF)) and collected in different periods of the olive mill season. The recovery of dietary polyphenolic compounds from OMWWs via liquid–liquid extraction generated dry residue polyphenolic extracts, which were further analyzed via HPLC/MS analysis in order to ascertain their polyphenolic profiles. The OMWW extracts showed high polyphenols content, providing valuable insights into the analysis and variability of OMWW phenolic composition depending on the cultivar and collection time. Overall, oleacein, rarely identified in OMWW, and hydroxytyrosol are the most abundant polyphenols. Furthermore, the investigation of nutraceutical properties revealed potent antioxidant, antiradical, and anti-inflammatory activities.

The findings underscore the importance of the proper management and valorization of this waste product, which can contribute to the development of sustainable and eco-friendly strategies in the food, pharmaceutical, and cosmetics industries. In fact, the polyphenolic extracts of OMWW, similarly to EVOO polyphenolic extracts, are rich in bioactive compounds and could be used as functional ingredients. The revalorization and reuse of OMWW, recovering its dietary polyphenols, focus on the chance to identify valuable molecules with different biological properties (e.g., antioxidant and anti-inflammatory activities) and are devoted to the further application of these products in food technologies such as in the nutraceutical, cosmeceutical, and pharmaceutical fields. In particular, OMWW polyphenolic extracts, when added in specific formulations as functional ingredients, might increase the nutritional profile of food products, or else they could be innovative natural additives used to improve food properties or active ingredients or additives in cosmetic and pharmaceutical formulations, exploiting their nutraceutical properties.

Further research and optimization of the extraction methods, such via green extraction, are warranted in order to harness the full potential of OMWW and maximize its value in the nutraceutical industry.

Supplementary Materials: The following supporting information can be downloaded at https://www.mdpi.com/article/10.3390/nu15173746/s1; Table S1: Calibration curves concentrations for each analyte in ESI-MS analysis.

Author Contributions: Conceptualization M.D., D.C. and M.M., methodology, M.D., D.C., A.B. and A.S., validation, D.C. and A.B., investigation, M.D., D.C., A.B., S.B., A.S., C.R., M.G.C. and S.D., writing—original draft preparation, M.D., D.C. and A.B., writing—review and editing, M.D., S.B., A.S. and M.M.; supervision, M.D., S.B., A.S., S.D. and M.M.; funding acquisition, M.M. All authors have read and agreed to the published version of the manuscript.

Funding: This research was funded by the Ministry of University and Research (MUR) as part of the PON 2014–2020 "Research and Innovation" resources—Green/Innovation Action—DM MUR 1062/2021—Title of research "Sviluppo di una piattaforma tecnologica per lo studio delle proprietà nutraceutiche di biomolecole e biomateriali presenti negli scarti derivanti dalla filiera dei prodotti alimentari".

Institutional Review Board Statement: Not applicable.

Informed Consent Statement: Not applicable.

Data Availability Statement: The datasets generated and analysed during the current study are available from the authors on reasonable request.

Acknowledgments: The authors acknowledge Consorzio Olio Toscano IGP, Coldiretti Toscana, and "Fratelli Caprai" and "Terra di Luce" olive mills for their collaboration and for providing the OMWW samples for this study. Moreover, the Centre for Instrumentation Sharing of the University of Pisa (CISUP) is kindly acknowledged for providing the Sciex QTrap 6500+ mass spectrometer used for the spectrometric assays.

Conflicts of Interest: The authors declare no conflict of interest.

References

1. Gorzynik-Debicka, M.; Przychodzen, P.; Cappello, F.; Kuban-Jankowska, A.; Marino Gammazza, A.; Knap, N.; Wozniak, M.; Gorska-Ponikowska, M. Potential Health Benefits of Olive Oil and Plant Polyphenols. *Int. J. Mol. Sci.* **2018**, *19*, 686. [CrossRef]
2. Dominguez, L.J.; Di Bella, G.; Veronese, N.; Barbagallo, M. Impact of Mediterranean Diet on Chronic Non-Communicable Diseases and Longevity. *Nutrients* **2021**, *13*, 2028. [CrossRef]
3. Foti, P.; Romeo, F.V.; Russo, N.; Pino, A.; Vaccalluzzo, A.; Caggia, C.; Randazzo, C.L. Olive Mill Wastewater as Renewable Raw Materials to Generate High Added-Value Ingredients for Agro-Food Industries. *Appl. Sci.* **2021**, *11*, 7511. [CrossRef]
4. Zahi, M.R.; Zam, W.; El Hattab, M. State of Knowledge on Chemical, Biological and Nutritional Properties of Olive Mill Wastewater. *Food Chem.* **2022**, *381*, 132238. [CrossRef]
5. Shabir, S.; Ilyas, N.; Saeed, M.; Bibi, F.; Sayyed, R.Z.; Almalki, W.H. Treatment Technologies for Olive Mill Wastewater with Impacts on Plants. *Environ. Res.* **2023**, *216*, 114399. [CrossRef]
6. Barbera, A.C.; Maucieri, C.; Cavallaro, V.; Ioppolo, A.; Spagna, G. Effects of Spreading Olive Mill Wastewater on Soil Properties and Crops, a Review. *Agric. Water Manag.* **2013**, *119*, 43–53. [CrossRef]
7. Sierra, J.; Martí, E.; Garau, M.A.; Cruañas, R. Effects of the Agronomic Use of Olive Oil Mill Wastewater: Field Experiment. *Sci. Total Environ.* **2007**, *378*, 90–94. [CrossRef]
8. Tajini, F.A.T.M.A.; Ouerghi, A.B.I.D.; Hosni, K.A.R.I.M. Effect of Irrigation with Olive-Mill Waste-Water on Physiological and Biochemical Parameters as Well as Heavy-Metal Accumulation in Common Bean (*Phaseolus vulgaris* L.). *J. New Sci.* **2019**, *66*, 4193–4203.
9. Caporaso, N.; Formisano, D.; Genovese, A. Use of Phenolic Compounds from Olive Mill Wastewater as Valuable Ingredients for Functional Foods. *Crit. Rev. Food Sci. Nutr.* **2018**, *58*, 2829–2841. [CrossRef]
10. Abbattista, R.; Ventura, G.; Calvano, C.D.; Cataldi, T.R.I.; Losito, I. Bioactive Compounds in Waste By-Products from Olive Oil Production: Applications and Structural Characterization by Mass Spectrometry Techniques. *Foods* **2021**, *10*, 1236. [CrossRef] [PubMed]
11. Pinho, I.A.; Lopes, D.V.; Martins, R.C.; Quina, M.J. Phytotoxicity Assessment of Olive Mill Solid Wastes and the Influence of Phenolic Compounds. *Chemosphere* **2017**, *185*, 258–267. [CrossRef]
12. Leouifoudi, I.; Zyad, A.; Ali, A.; Moulay, A.; Oukerrou, M.A.; Ait Mouse, H.; Mbarki, M. Identification and Characterisation of Phenolic Compounds Extracted from Moroccan Olive Mill Wastewater. *Food Sci. Technol.* **2014**, *34*, 249–257. [CrossRef]
13. Di Mauro, M.; Fava, G.; Spampinato, M.; Aleo, D.; Melilli, B.; Saita, M.; Centonze, G.; Maggiore, R.; D'Antona, N. Polyphenolic Fraction from Olive Mill Wastewater: Scale-Up and in Vitro Studies for Ophthalmic Nutraceutical Applications. *Antioxidants* **2019**, *8*, 462. [CrossRef] [PubMed]
14. Khwaldia, K.; Attour, N.; Matthes, J.; Beck, L.; Schmid, M. Olive Byproducts and Their Bioactive Compounds as a Valuable Source for Food Packaging Applications. *Compr. Rev. Food Sci. Food Saf.* **2022**, *21*, 1218–1253. [CrossRef]
15. Silvan, J.M.; Pinto-Bustillos, M.A.; Vásquez-Ponce, P.; Prodanov, M.; Martinez-Rodriguez, A.J. Olive Mill Wastewater as a Potential Source of Antibacterial and Anti-Inflammatory Compounds against the Food-Borne Pathogen Campylobacter. *Innov. Food Sci. Emerg. Technol.* **2019**, *51*, 177–185. [CrossRef]
16. Borzì, A.M.; Biondi, A.; Basile, F.; Luca, S.; Vicari, E.S.D.; Vacante, M. Olive Oil Effects on Colorectal Cancer. *Nutrients* **2018**, *11*, 32. [CrossRef]
17. Barbara Bassani, T.R. Effect of a Purified Extract of Olive Mill Waste Water on Endothelial Cell Proliferation, Apoptosis, Migration and Capillary-Like Structure in Vitro and in Vivo. *J. Bioanal. Biomed.* **2015**, *12*, 1–8. [CrossRef]
18. Rodrigues, F.; Pimentel, F.B.; Oliveira, M.B.P.P. Olive By-Products: Challenge Application in Cosmetic Industry. *Ind. Crops Prod.* **2015**, *70*, 116–124. [CrossRef]
19. Nunes, M.A.; Pimentel, F.B.; Costa, A.S.G.; Alves, R.C.; Oliveira, M.B.P.P. Olive By-Products for Functional and Food Applications: Challenging Opportunities to Face Environmental Constraints. *Innov. Food Sci. Emerg. Technol.* **2016**, *35*, 139–148. [CrossRef]
20. Carpi, S.; Polini, B.; Manera, C.; Digiacomo, M.; Salsano, J.E.; Macchia, M.; Scoditti, E.; Nieri, P. MiRNA Modulation and Antitumor Activity by the Extra-Virgin Olive Oil Polyphenol Oleacein in Human Melanoma Cells. *Front. Pharmacol.* **2020**, *11*, 574317. [CrossRef]
21. Gabbia, D.; Carpi, S.; Sarcognato, S.; Cannella, L.; Colognesi, M.; Scaffidi, M.; Polini, B.; Digiacomo, M.; Esposito Salsano, J.; Manera, C.; et al. The Extra Virgin Olive Oil Polyphenol Oleocanthal Exerts Antifibrotic Effects in the Liver. *Front. Nutr.* **2021**, *8*, 715183. [CrossRef]
22. Tasioula-Margari, M.; Tsabolatidou, E. Extraction, Separation, and Identification of Phenolic Compounds in Virgin Olive Oil by HPLC-DAD and HPLC-MS. *Antioxidants* **2015**, *4*, 548–562. [CrossRef] [PubMed]
23. De Marco, E.; Savarese, M.; Paduano, A.; Sacchi, R. Characterization and Fractionation of Phenolic Compounds Extracted from Olive Oil Mill Wastewaters. *Food Chem.* **2007**, *104*, 858–867. [CrossRef]
24. Cuffaro, D.; Bertini, S.; Macchia, M.; Digiacomo, M. Enhanced Nutraceutical Properties of Extra Virgin Olive Oil Extract by Olive Leaf Enrichment. *Nutrients* **2023**, *15*, 1073. [CrossRef]
25. Borges, T.H.; Cabrera-Vique, C.; Seiquer, I. Antioxidant Properties of Chemical Extracts and Bioaccessible Fractions Obtained from Six Spanish Monovarietal Extra Virgin Olive Oils: Assays in Caco-2 Cells. *Food Funct.* **2015**, *6*, 2375–2383. [CrossRef]

26. Pellegrini, N.; Del Rio, D.; Colombi, B.; Bianchi, M.; Brighenti, F. Application of the 2,2′-Azinobis(3-Ethylbenzothiazoline-6-Sulfonic Acid) Radical Cation Assay to a Flow Injection System for the Evaluation of Antioxidant Activity of Some Pure Compounds and Beverages. *J. Agric. Food Chem.* **2003**, *51*, 260–264. [CrossRef]
27. Cuffaro, D.; Pinto, D.; Silva, A.M.; Bertolini, A.; Bertini, S.; Saba, A.; Macchia, M.; Rodrigues, F.; Digiacomo, M. Insights into the Antioxidant/Antiradical Effects and In Vitro Intestinal Permeation of Oleocanthal and Its Metabolites Tyrosol and Oleocanthalic Acid. *Molecules* **2023**, *28*, 5150. [CrossRef] [PubMed]
28. Chen, L.; Deng, H.; Cui, H.; Fang, J.; Zuo, Z.; Deng, J.; Li, Y.; Wang, X.; Zhao, L. Inflammatory Responses and Inflammation-Associated Diseases in Organs. *Oncotarget* **2017**, *9*, 7204–7218. [CrossRef] [PubMed]
29. Nuti, E.; Rossello, A.; Cuffaro, D.; Camodeca, C.; Van Bael, J.; van der Maat, D.; Martens, E.; Fiten, P.; Pereira, R.V.S.; Ugarte-Berzal, E.; et al. Bivalent Inhibitor with Selectivity for Trimeric MMP-9 Amplifies Neutrophil Chemotaxis and Enables Functional Studies on MMP-9 Proteoforms. *Cells* **2020**, *9*, 1634. [CrossRef]
30. Tangney, C.; Rasmussen, H.E. Polyphenols, Inflammation, and Cardiovascular Disease. *Curr. Atheroscler. Rep.* **2013**, *15*, 324. [CrossRef]
31. Lucas, L.; Russell, A.; Keast, R. Molecular Mechanisms of Inflammation. Anti-Inflammatory Benefits of Virgin Olive Oil and the Phenolic Compound Oleocanthal. *Curr. Pharm. Des.* **2011**, *17*, 754–768. [CrossRef] [PubMed]
32. Bucciantini, M.; Leri, M.; Nardiello, P.; Casamenti, F.; Stefani, M. Olive Polyphenols: Antioxidant and Anti-Inflammatory Properties. *Antioxidants* **2021**, *10*, 1044. [CrossRef]
33. Rosignoli, P.; Fuccelli, R.; Fabiani, R.; Servili, M.; Morozzi, G. Effect of Olive Oil Phenols on the Production of Inflammatory Mediators in Freshly Isolated Human Monocytes. *J. Nutr. Biochem.* **2013**, *24*, 1513–1519. [CrossRef]
34. Carluccio, M.A.; Siculella, L.; Ancora, M.A.; Massaro, M.; Scoditti, E.; Storelli, C.; Visioli, F.; Distante, A.; De Caterina, R. Olive Oil and Red Wine Antioxidant Polyphenols Inhibit Endothelial Activation: Antiatherogenic Properties of Mediterranean Diet Phytochemicals. *Arterioscler. Thromb. Vasc. Biol.* **2003**, *23*, 622–629. [CrossRef]
35. Beauchamp, G.K.; Keast, R.S.J.; Morel, D.; Lin, J.; Pika, J.; Han, Q.; Lee, C.-H.; Smith, A.B.; Breslin, P.A.S. Phytochemistry: Ibuprofen-like Activity in Extra-Virgin Olive Oil. *Nature* **2005**, *437*, 45–46. [CrossRef]
36. Costa, V.; Costa, M.; Videira, R.A.; Andrade, P.B.; Paiva-Martins, F. Anti-Inflammatory Activity of Olive Oil Polyphenols-The Role of Oleacein and Its Metabolites. *Biomedicines* **2022**, *10*, 2990. [CrossRef]
37. Moreno, J.J. Effect of Olive Oil Minor Components on Oxidative Stress and Arachidonic Acid Mobilization and Metabolism by Macrophages RAW 264.7. *Free Radic. Biol. Med.* **2003**, *35*, 1073–1081. [CrossRef] [PubMed]
38. Araújo, M.; Pimentel, F.B.; Alves, R.C.; Oliveira, M.B.P.P. Phenolic Compounds from Olive Mill Wastes: Health Effects, Analytical Approach and Application as Food Antioxidants. *Trends Food Sci. Technol.* **2015**, *45*, 200–211. [CrossRef]
39. Esposito Salsano, J.; Digiacomo, M.; Cuffaro, D.; Bertini, S.; Macchia, M. Content Variations in Oleocanthalic Acid and Other Phenolic Compounds in Extra-Virgin Olive Oil during Storage. *Foods* **2022**, *11*, 1354. [CrossRef]
40. Tapia-Quirós, P.; Montenegro-Landívar, M.F.; Reig, M.; Vecino, X.; Alvarino, T.; Cortina, J.L.; Saurina, J.; Granados, M. Olive Mill and Winery Wastes as Viable Sources of Bioactive Compounds: A Study on Polyphenols Recovery. *Antioxidants* **2020**, *9*, 1074. [CrossRef] [PubMed]
41. Ryan, D.; Robards, K.; Lavee, S. Changes in Phenolic Content of Olive during Maturation. *Int. J. Food Sci. Technol.* **1999**, *34*, 265–274. [CrossRef]
42. Belaqziz, M.; Tan, S.P.; El-Abbassi, A.; Kiai, H.; Hafidi, A.; O'Donovan, O.; McLoughlin, P. Assessment of the Antioxidant and Antibacterial Activities of Different Olive Processing Wastewaters. *PLoS ONE* **2017**, *12*, e0182622. [CrossRef] [PubMed]
43. Negro, C.; Aprile, A.; Luvisi, A.; Nicolì, F.; Nutricati, E.; Vergine, M.; Miceli, A.; Blando, F.; Sabella, E.; De Bellis, L. Phenolic Profile and Antioxidant Activity of Italian Monovarietal Extra Virgin Olive Oils. *Antioxidants* **2019**, *8*, 161. [CrossRef] [PubMed]
44. Ballus, C.A.; Meinhart, A.D.; de Souza Campos, F.A., Jr.; Godoy, H.T. Total Phenolics of Virgin Olive Oils Highly Correlate with the Hydrogen Atom Transfer Mechanism of Antioxidant Capacity. *J. Am. Oil Chem. Soc.* **2015**, *92*, 843–851. [CrossRef]

nutrients

MDPI

Article

Ellagitannins from *Castanea sativa* Mill. Leaf Extracts Impair *H. pylori* Viability and Infection-Induced Inflammation in Human Gastric Epithelial Cells

Stefano Piazza [1,†], Giulia Martinelli [1,†], Marco Fumagalli [1], Carola Pozzoli [1], Nicole Maranta [1], Flavio Giavarini [1], Luca Colombo [2], Giovanna Nicotra [3], Silvia Francesca Vicentini [3], Francesca Genova [1], Emma De Fabiani [1,*], Enrico Sangiovanni [1,*] and Mario Dell'Agli [1]

[1] Department of Pharmacological and Biomolecular Sciences "Rodolfo Paoletti", University of Milan, 20133 Milan, Italy; stefano.piazza@unimi.it (S.P.); giulia.martinelli@unimi.it (G.M.); marco.fumagalli3@unimi.it (M.F.); carola.pozzoli@unimi.it (C.P.); nicole.maranta@unimi.it (N.M.); flavio.giavarini@unimi.it (F.G.); genova.francesca@hsr.it (F.G.); mario.dellagli@unimi.it (M.D.)

[2] Consorzio Castanicoltori di Brinzio, Orino e Castello Cabiaglio, Società Cooperativa Agricola-Varese, 21100 Varese, Italy; info@consorziocastanicoltori.it

[3] Estratti Piante Officinali (E.P.O.) S.r.l, 20141 Milan, Italy; gnicotra@eposrl.com (G.N.); svicentini@eposrl.com (S.F.V.)

* Correspondence: emma.defabiani@unimi.it (E.D.F.); enrico.sangiovanni@unimi.it (E.S.)

† The authors contributed equally to this work.

Citation: Piazza, S.; Martinelli, G.; Fumagalli, M.; Pozzoli, C.; Maranta, N.; Giavarini, F.; Colombo, L.; Nicotra, G.; Vicentini, S.F.; Genova, F.; et al. Ellagitannins from *Castanea sativa* Mill. Leaf Extracts Impair *H. pylori* Viability and Infection-Induced Inflammation in Human Gastric Epithelial Cells. *Nutrients* **2023**, *15*, 1504. https://doi.org/10.3390/nu15061504

Academic Editors: Maria Digiacomo and Doretta Cuffaro

Received: 8 March 2023
Revised: 15 March 2023
Accepted: 18 March 2023
Published: 21 March 2023

Abstract: *Helicobacter pylori* (*H. pylori*) is an etiologic factor of peptic ulcer disease and gastric cancer. Virulent strains of *H. pylori* are correlated with the severity of gastritis, due to NF-κB activation and IL-8 expression at the epithelial level. Ellagitannins have been documented for antibacterial and anti-inflammatory activities, thus suggesting their potential use in gastritis. Recently, several authors, including our group, demonstrated that tannin-rich extracts from chestnut byproducts, at present considered agricultural waste, display promising biological activities. In this work, we detected high levels of polyphenols in hydroalcoholic extracts from chestnut leaves (*Castanea sativa* L.). Among polyphenols, the ellagitannin isomers castalagin and vescalagin (about 1% *w/w* of dry extract) were identified as potential bioactive compounds. In GES-1 cells infected by *H. pylori*, leaf extract and pure ellagitannins inhibited IL-8 release ($IC_{50} \approx 28$ μg/mL and 11 μM, respectively). Mechanistically, the anti-inflammatory activity was partly due to attenuation of NF-κB signaling. Moreover, the extract and pure ellagitannins reduced bacterial growth and cell adhesion. A simulation of the gastric digestion suggested that the bioactivity might be maintained after oral administration. At the transcriptional level, castalagin downregulated genes involved in inflammatory pathways (NF-κB and AP-1) and cell migration (Rho GTPase). To the best of our knowledge, this is the first investigation in which ellagitannins from plant extracts have demonstrated a potential role in the interaction among *H. pylori* and human gastric epithelium.

Keywords: *Castanea sativa* Mill.; gastritis; chestnut; tannins; ellagitannins; castalagin; *Helicobacter pylori*

1. Introduction

The discovery of *Helicobacter pylori* (*H. pylori*) infection by Warren and Marshall in 1983 is relatively recent, especially considering the subsequent efforts to convince the scientific community regarding its causal association with peptic ulcers, which led to the Nobel prize awarded in 2005 [1,2]. Gastric inflammation and oxidative stress are the clinical traits of acute and chronic gastritis, in which neutrophils, macrophages, and lymphocytes infiltrate the mucosa following chemokine attraction (for a comprehensive review, see Naito et al. [3]). The major inflammatory signal is the release of IL-8, whose levels correlate with *H. pylori* infection and gastric disease severity [4–7], but other mediators such as

IL-6 and TNFα also play pivotal roles during disease progression [8]. Analysis of clinical biopsies revealed that activated NF-κB colocalizes with IL-8 in the *H. pylori*-infected mucosa of patients with gastritis [9], making the chemokine and NF-κB signaling key targets in the search for new therapies to treat *H. pylori*-induced gastritis and ulcers in humans. In the search for new therapies, the therapeutic approach to gastritis changed from mild diet interventions and prescription of antiacids to antibiotic therapy. Meanwhile, alternative approaches have been proposed to prevent or adjuvate the use of antibiotics, such as vaccines, probiotics, and food supplements [10,11].

Accordingly, whereas tannin-rich plants have been traditionally used against gastric diseases for long time, their direct anti-*H. pylori* effect has only been investigated during the last 20 years. To date, the hypothesis of potential benefits from tannin supplementation is mostly derived from rodent models of gastric ulcer (ethanol or FANS injury). The presence of ellagitannins in dietary products, such as fruits from *Punica granatum* L. and *Rubus* spp., contributes to their preclinical efficacy against gastric ulcer [12–15]. The antioxidant and anti-inflammatory properties of ellagitannins, observed through different experimental models [16,17], are supposed to contribute to the gastroprotective effect. Moreover, Funatogawa and colleagues suggested that ellagitannins also inhibit the growth of clinical strains of *H. pylori* at μM concentrations in microbiological tests [18]. Thus, the role of ellagitannins with reference to *H. pylori*-related gastritis still requires pharmacological validation.

Castanea sativa Mill. is a traditional source of polyphenols, mostly tannins. The chemical nature of chestnut tannins has recently been reported by several authors and our group; hydrolyzable tannins (ellagitannins and gallotannins) have been documented in bark [19,20], burs [21], flours [22], and leaves [23,24], while chestnut peel only contains condensed tannins [25]. The isomers castalagin and vescalagin were also identified as the main ellagitannins present in the bark [20]. Of note, other authors suggested that castalagin may counteract the inflammatory activity of human neutrophils and reduce the ethanol-induced ulcer formation in vivo [{Khennouf, 2003 #195}{Piwowarski, 2015 #234}], but its antibacterial and anti-inflammatory effect in *H. pylori* infection was poorly investigated.

Our previous work suggested the potential nutraceutical use in gastritis of chestnut peel from two botanical varieties (*Castanea sativa* Mill. var. venegon, and var. verdesa), collected from a specific area of Northern Italy (Campo dei Fiori, Varese) [25]. Similarly to chestnut peel, leaves may represent a sustainable plant material for pharmaceutical purposes in comparison to the bark. Of note, leaves have been traditionally used to treat gastrointestinal ailments, but the pharmacological properties and contribution of each component to the biological activity exerted by the extract still require a deep scientific assessment.

The research on anti-gastritis compounds is affected by the challenge of creating reliable models of *H. pylori* infection, which naturally occurs in humans only [26–28]. For this reason, an in vitro model mimicking the human gastric epithelium–*H. pylori* interaction represents a useful tool for the evaluation of novel anti-gastritis candidates.

In this work, we addressed the presence of polyphenols in *Castanea sativa* Mill. leaf extracts (var. venegon, and var. verdesa), with particular focus on ellagitannins, as a valuable polyphenol class for potential gastroprotective properties. Castalagin and vescalagin, commercially available as analytical standards, were identified and quantified in hydroalcoholic leaf extracts by LC–MS. Thus, we deeply investigated the anti-inflammatory and antibacterial activity of leaf extracts, in comparison with pure castalagin, an ellagitannin typically present in the leaves of *Castanea sativa* Mill., in a model of human gastric epithelium (GES-1) infected by *H. pylori*.

2. Materials and Methods

2.1. Materials

RPMI 1640 medium, penicillin, streptomycin, L glutamine, and trypsin-EDTA were purchased from Gibco (Life Technologies Italia, Monza, Italy). Fetal bovine serum (FBS), and disposable materials for cell culture were purchased from Euroclone (Euroclone S.p.A., Pero-Milan, Italy). SV-40 immortalized GES-1 cells from human gastric epithelium were kindly donated by Dr. Dawit Kidane-Mulat, The University of Texas at Austin. Mueller–Hinton Broth, Brucella Broth, and glycerol from BD (BD, Franklin Lakes, NJ, USA), agar from Merck Life Science (Milan, Italy), and defibrinated sheep blood from Thermo Fischer Scientific (Oxoid™ Hampshire, Basingstoke, UK) were used to cultivate and store *Helicobacter pylori*, cag+ strain 26695 from ATCC (ATCC 700392™, Manassas, VA, USA). The reagents 3-(4,5-dimethylthiazol-2-yl)-2,5-diphenyltetrazolium bromide (MTT) and fluorescein-5-isothiocyanate (FITC) were purchased from Merck Life Science (Milan, Italy). Lipofectamine® 3000 and carboxyfluorescein succinimidyl ester (CFSE) 5 mM (CellTrace™, Cell Proliferation kits) and ActinRed™ 555 ReadyProbes™ reagent were from Invitrogen (Thermo Fisher Scientific, Waltham, MA, USA). Britelite™ Plus reagent was from Perkin Elmer (Milano, Italy). The ellagitannins castalagin and vescalagin (certified purity, >95%) were from Phytolab GmbH & Co. KG (Vestenbergsgreuth, Germany). All reagents used for the biological assays were HPLC-grade. Human TNFα and human IL-8 ELISA Development Kits were from Peprotech Inc. (London, UK). All chromatographic solvents were HPLC-grade or LC–MS-grade for MS experiments. Acetonitrile, methanol, ethanol, formic acid, hydrochloric acid, vanillin, and iron sulfate were from Merck Life Science (Milan, Italy).

2.2. Extract Preparation and Phytochemical Characterization

Leaves from *Castanea sativa* Mill. var. venegon (VN) and *Castanea sativa* Mill. var. verdesa (VR) were collected at the end of the blossoming period, from the regional natural park "Campo dei Fiori" (from Castello Cabiaglio and Brinzio area, respectively for VN and VR, Varese, Italy). The plant material was naturally dried for 48 h at room temperature, and then extracted following previously reported methods [25]. In brief, 2.5 g of milled leaves were extracted twice with 50 mL of ethanol/water 50:50 (hydroalcoholic extract) for 4 and 16 h, respectively, at room temperature under dark conditions. Then, plant debris was removed using Supervelox filter paper; the extracts obtained were frozen at −80 °C overnight, lyophilized, and maintained at −20 °C. Before the biological activity evaluation, the extracts were dissolved in sterilized distilled water and DMSO (H_2O:DMSO 50:50, 25 mg/mL), before being stored in aliquots at −20 °C.

2.3. LC–MS/MS Analysis

Ellagitannins castalagin and vescalagin were detected in the methanol solution of leaf extracts, and further quantified by UPLC–MS analysis using Exion LCTM AC System (AB Sciex, Foster City, CA, USA) with a Synergi 4 µm Hydro-RP 80 A LC Column 150 × 4.6 mm (Phenomenex, Torrance, CA, USA), coupled to a Triple Quad™ 3500 system (AB Sciex, Foster City, CA, USA) with an ESI (−) source. The mobile-phase solvents consisted of water 0.1% formic acid (A) and methanol (B) set as 95% A/5% B (0–5 min), 100% B (5–8 min), and 95% A/5% B (8–15 min), with a flow rate of 0.8 mL/min.

2.4. Total Phenol Content Assay

Total polyphenol content was determined according to Folin–Ciocâlteu's method, as reported by Singleton and Rossi [29]. Freeze-dried leaf extracts (1 mg) were solubilized in 1 mL of water. Aliquots of 300 µL from different samples were mixed in test tubes with 1.5 mL of Folin–Ciocâlteu's reagent diluted 10 times and 1.2 mL of 7.5% (*w*/*v*) sodium carbonate. The absorbance was measured at 765 nm using a UV–Vis spectrophotometer (Victor™ X3, Perkin Elmer, Walthman, MA, USA). Gallic acid was used as the reference standard for the calibration curve (0–30 µg/mL). Results were expressed as the weight of gallic acid equivalents (GA eq.) per weight of dry extract (% *w*/*w*).

2.5. Cell Culture

Human gastric epithelial (GES-1) cells were cultivated in RPMI 1640 medium, supplemented with penicillin 100 units/mL, streptomycin 100 mg/mL, L-glutamine 2 mM, and 10% heat-inactivated FBS. Cells were incubated at 37 °C, 5% CO_2, in a humidified atmosphere. The subculture in new flasks and fresh medium (1×10^6 cells) was repeated every 48–72 h upon reaching confluency, by detaching cells with trypsin ethylenediaminetetraacetic acid (EDTA) 0.25% solution.

2.6. Bacterial Culture

H. pylori cag+ strain 26695 was cultured in Petri dishes with Mueller–Hinton Broth medium, 5% agar, and 25% blood, for 72 h, under a microaerophilic atmosphere (5% O_2, 10% CO_2, and 85% N_2) at 37 °C, 100% humidity. For the biological experiments, the bacterium was collected from the Petri dishes and counted according to its optical density at 600 nm (an OD value of 5 corresponds to 2×10^8 bacteria).

2.7. Cell Treatment

GES-1 cells were seeded in a 24-well plate (3×10^4 cells/well) or 12-well plate (6×10^4 cells/well), according to the specific investigation. Cells were treated with the proinflammatory stimulus TNFα (10 ng/mL) or *H. pylori* (bacterium-to-cell ratio of 50:1) for 1 h or 6 h, along with the leaf extracts (*Castanea sativa* Mill. var. verdesa and var. venegon) at different concentrations. The day before the bacterial infection, serum starvation was performed using 0.5% serum medium, supplemented with 1% L-glutamine and 1% penicillin/streptomycin. All the infection treatments were conducted with serum- and antibiotic-free medium, while TNFα treatments were conducted only with serum-free medium. During the treatment, cells were maintained in an incubator at 37 °C and 5% CO_2. EGCG (50 μM) and procyanidin A2 (500 μM) were used as reference inhibitors of inflammatory markers and bacterial adhesion, respectively [11,30–32].

2.8. Cytotoxicity Assay

The integrity of the cell morphology before and after treatment was assessed by light microscope inspection. Cell viability was measured, after 6 h treatment (24-well plate, 3×10^4 cells/well), using the 3,4,5-dimethylthiazol-2-yl-2-5-diphenyltetrazolium bromide (MTT) method [33]. This method evaluates the activity of a mitochondrial enzyme, which is an index of cell viability. Briefly, the medium was discarded; then, 200 μL of MTT solution (0.1 mg/mL, phosphate-buffered saline (PBS) 1×) was added to each well (45 min, 37 °C) and kept in darkness. Then, MTT solution was discarded, the remaining purple salt was dissolved in isopropanol/dimethyl sulfoxide (DMSO) (90:10 *v*/*v*), and the absorbance was read at 595 nm (Victor™ X3, Perkin Elmer, Walthman, MA, USA). The absorbance of the solvent (blank) was subtracted from each value.

Ellagitannins and leaf extracts had no significant impact on cell viability up to the concentrations of 50 μM and 100 μg/mL, respectively (Figure S1). The impact on cell viability up to the concentrations used for 1 h treatments, i.e., 200 μM and 250 μg/mL, respectively, was also excluded.

2.9. Measurement of IL-8 Release

Cells were seeded at the density of 3×10^4 cells/well (24-well plate) for 48 h. The chemokine IL-8 was quantified in cell medium in at least three independent experiments, after 6 h treatments with TNFα or *H. pylori* and the extracts, using a sandwich enzyme-linked immunosorbent assay (human IL-8 ABTS ELISA Development Kit, Peprotech). The assay was conducted according to manufacturer's instructions, as previously reported [34]. The absorbance of the sample at 405 nm was compared to the absorbance of human recombinant standard IL-8 (0–1000 pg/mL). Data (mean ± SEM of at least three experiments) were expressed as the percentage relative to stimulated control, to which the value of 100% was arbitrarily assigned.

2.10. NF-κB Activation

The NF-κB pathway was evaluated by immunofluorescence and plasmid transfection, to measure the translocation of p65 subunit into cell nuclei or κB element-driven transcription, respectively, in at least three independent experiments.

2.10.1. Immunofluorescence

The immunofluorescence technique was applied for the evaluation of NF-κB (p65 subunit) translocation into GES-1 nuclei; cells were challenged with TNFα or *H. pylori* and treated with the extracts, as previously described [34]. Briefly, cells were cultivated (3×10^4/well) on coverslips placed in 24-well plates for 24 h. *H. pylori* was stained with CFSE 5 mM (2 μL of CFSE: 5×10^8 bacteria) and incubated for 20 min at 37 °C; then, the bacterial suspension was supplemented with FBS, washed three times (PBS 1×), and centrifuged at 3150× *g* for 5 min to remove the excess of CFSE not bound to the bacterium. After 1 h treatment, the coverslips were washed (PBS 1×) and fixed with 4% formaldehyde solution for 15 min at r.t. A blocking solution (5% BSA) was added to the well and incubated at room temperature for 1 h. Cells were incubated with the primary antibody (NF-κB p65 (D14E12) XP® Rabbit mAb #8242, Cell Signaling Technology, Danvers, MA, USA) diluted 1:400 *v/v* overnight at 4 °C, washed three times (PBS 1×), and then incubated with the secondary antibody (Alexa Fluor 647 conjugated with anti-rabbit immunoglobulin G (IgG) (heavy + light (H + L)), F(ab')2 Fragment #4414, Cell Signaling Technology, Danvers, MA, USA) diluted 1:1000 *v/v*. A drop of ActinRed reagent, previously diluted 1:5 in PBS 1×, was added 30 min before the end of incubation. After 2 h, coverslips were washed with PBS and mounted on slides with a drop of ProLong Gold Antifade Reagent with 4′,6-diamidino-2-phenylindole (DAPI) (#8961, Cell Signaling Technology, Danvers, MA, USA), and then imaged using a confocal laser scanning microscope (LSM 900, Zeiss, Oberkochen, Germany).

2.10.2. Measurement of the NF-κB-Driven Transcription

GES-1 cells were seeded in 24-well plates for 48 h (3×10^4 cells), and then transiently transfected with a reporter plasmid, in which the luciferase gene was under the control of the E-selectin promoter containing three κB elements responsive to NF-κB (50 ng per well). Lipofectamine® 3000 reagent was used for the transfection assays, according to the manufacturer's instructions. The plasmid was a gift from Dr. N. Marx (Department of Internal Medicine-Cardiology, University of Ulm; Ulm, Germany). The day after, cells were treated with TNFα in addition to the extracts for 6 h. At the end of the treatment, the luciferase activity in cells was measured using Britelite™ Plus reagent, according to the manufacturer's instructions, as previously described [25]. Results (mean ± SEM of at least three experiments) were expressed as the percentage relative to the stimulated control, to which a value of 100% was arbitrarily assigned.

2.11. Bacterial Adhesion to Cells

The bacterial adhesion was measured using a cytofluorimetric method adapted from Messing and colleagues [30]. GES-1 cells were seeded in 12-well plates for 48 h (6×10^4 cells) before the infection with FITC-labeled *H. pylori*. FITC (2 μL) solution (1% in DMSO) was added to 10^8 bacteria suspended in PBS 1× and incubated for 45 min at 37 °C; then, the bacterial suspension was centrifuged (3150× *g*, 5 min) and washed twice (PBS 1×) to remove the excess probe, before resuspending in PBS 1× for cell infection. Cell treatment (at least three independent experiments) was carried out for 1 h with extracts and a reference inhibitor (procyanidin A2, 500 μM), before (pretreatment) or after (cotreatment) infection. After 1 h (37 °C), cells were washed twice with PBS 1×, collected through a scraper in PBS/EDTA 2 mM, and centrifuged (3150× *g* for 5 min); then, they were fixed by formaldehyde (4% in PBS) and incubated in an ice bath for 10 min. Finally, cells were centrifuged, washed, and resuspended in 0.5% BSA (PBS/EDTA 2 mM) for analysis using the NovoCyte flow cytometer (ACEA Biosciences, San Diego, CA, USA) and NovoExpress software (ACEA Biosciences).

2.12. Minimum Inhibitory Concentration (MIC)

The microbroth dilution method was performed according to the recommendations of the Clinical and Laboratory Standards Institute (CLSI) [35] and was used to evaluate the MIC value, in at least three independent experiments. Extracts at different concentrations and the positive control (tetracycline 0.125 μg/mL) were prepared in 5% FBS Brucella broth; next, 100 μL of each sample were placed in a 96-well U-bottom plate. Then, 100 μL of *H. pylori* suspension (OD = 0.1) prepared in the same medium was added to each well. After mixing well, the 96-well plate was incubated at 37 °C in a 5% CO_2 incubator under microaerophilic conditions. After 72 h, the rate of bacterial growth was measured using a microplate reader at 600 nm (Victor™ X3, PerkinElmer, Waltham, MA, USA).

2.13. In Vitro Simulated Gastric Digestion

The gastric digestion was mimicked by an in vitro simulation, as previously described [34]. In brief, the extracts (100 mg) were incubated for 5 min at 37 °C with reconstituted saliva (6 mL); then, gastric juice (12 mL) was added, and the sample was incubated for 2 h at 37 °C under shaking. Finally, the suspension was centrifuged (5 min, 3000× *g*), and the supernatant was freeze-dried.

2.14. RNA Sequencing

GES-1 cells were seeded in a 12-well plate at a density of 6×10^4 cells/well for 48 h and mRNA isolated using miRNeasy Mini Kit (Qiagen, Hilden, Germany). Cells were treated for 6 h with *H. pylori*, with or without castalagin (10 μM). Three replicates were used for each condition (ctrl, *H. pylori* treatment, *H. pylori*, and castalagin treatment).

The quality and the concentration of the mRNA were assessed with RNA Screen Tape (Agilent Technologies, Santa Clara, CA, USA). The mRNA (500 ng) was fragmented and converted into complementary DNA (cDNA) through Illumina® Stranded mRNA Prep (Illumina, San Diego, CA, USA) according to the manufacturer's instructions. The quality and concentration of the final dual-indexed libraries were checked with D1000 Screen Tape (Agilent Technologies, Santa Clara, CA, USA). The NextSeq 500/550 High-Output Kit v2.5 was used to perform the sequencing through NextSeq 550 instrument (Illumina, San Diego, CA, USA).

2.15. RNA Sequencing Data Analysis

For RNA sequencing data analysis, Illumina blc2fastq software was used to generate the Fastq files. The quality of each Fastq was inspected individually using the FASTQC tool vs. 0.11.9 [36], and Multiqc vs. 1.10.1 was then used to assess the overall good sequencing quality [37]. After the quality check, reads were aligned to the human reference genome GRCh38.p13 using STAR 2.7.9a [38], while Feature Counts 2.0.1 was used to obtain the counts for each sample [39] (see Table S1). The rlog function was applied to transform the count data, reducing differences among samples, and normalizing for the library size. Transformed data were used to perform principal component analysis (PCA) to evaluate the sample distribution and their clustering within groups representing the same condition. Normalization and differential gene expression analyses were carried out using the Bioconductor package DeSeq2 [40]. Following normalization, differentially expressed genes (DEGs) were detected. The Wald test in DESeq2 is the default test used for hypothesis testing. Genes were considered differentially expressed when they had AdjPval < 0.05 and $\log_2$FoldChange $\geq$ 0.58, indicating a fold change > 1.5 in either direction. The DEGs were then used to perform the enrichment analysis with the R package "EnrichR" [41]. The enriched pathways were determined using the GO_Biological_Process_2021 database.

2.16. Statistical Analysis

All biological results were expressed as the mean ± SEM of at least three independent experiments; the confidence interval related to the half-maximal inhibitory concentrations (IC_{50}) calculation is reported in Section 3. Data were elaborated through an unpaired ANOVA test and Bonferroni post-hoc analysis. Statistical assessment and IC_{50} calculation were conducted using GraphPad Prism 8.0 software (GraphPad Software Inc., San Diego, CA, USA). Values of $p < 0.05$ were considered statistically significant.

3. Results

3.1. Phytochemical Characterization

Castanea sativa Mill. leaf is a well-known source of polyphenols, which include tannins and flavonoids [23,24]. According to our previous work on chestnut-derived extracts [25], we selected leaves from two varieties (*Castanea sativa* Mill. var. venegon and *Castanea sativa* Mill. var. verdesa) to perform a polar extraction with hydroalcoholic solvent (50:50 water/ethanol). The extraction yields were 16.24% and 21.88% (wt. of dry extract/wt. of dry plant material), respectively.

According to the literature, tannins represent 6–8% *w*/*w* of polar extracts from leaves [23]. The chemical nature of tannins from leaves has been poorly investigated in comparison to the bark, which should contain high levels of hydrolyzable tannins (up to 10%), including ellagitannins, such as the isomers castalagin and vescalagin, and their hydrolysis products, castalin and vescalin [20]. For this reason, we firstly characterized the extracts from leaves using LC–MS analysis, thus confirming the presence of castalagin and vescalagin.

The analytes, castalagin and vescalagin, were identified by comparing the retention time and the m/z with the respective standard compounds, using the following mass transitions: 933/631 (castalagin) and 933/613 (vescalagin) (Figure S2). The quantitative analysis was validated in terms of LOD and LOQ; thus, the ellagitannins were measured using the linear regression curve of the respective analytical standards (Table 1).

Table 1. Results of linear regression analysis of the LC-MS/MS.

Compound	Equation	R^2	LOD (ng/μL)	LOQ (ng/μL)
Castalagin	y = 74284x − 727.79	0.9999	0.012	0.015
Vescalagin	y = 7929.8x − 46.85	0.9998	0.026	0.053

LOD, limit of detection; LOQ, limit of quantification.

In parallel, we measured the total phenolic content in the extracts from venegon and verdesa varieties, which was 26.59% and 28.97% of weight GA eq./weight of dried extract, respectively (Table 2). The analysis was repeated after an in vitro simulated gastric digestion (see Section 2), and the process slightly diminished the total phenol level in both varieties, with statistically significant difference just for verdesa variety.

Table 2. Total phenolic content (wt.% GA eq./wt.% d.e.) of extracts from *Castanea sativa* Mill. leaves.

	Before Gastric Digestion	After Gastric Digestion
Total Phenol Content	GA eq. (% *w*/*w*) ± SEM	GA eq. (% *w*/*w*) ± SEM
VR	28.97 ± 0.47	19.1 ± 0.81 ***
VN	26.59 ± 1.05	23.26 ± 1.28

VN, *Castanea sativa* Mill. var. venegon; VR, *Castanea sativa* Mill. var. verdesa; GA eq., gallic acid equivalents; d.e., dry extract. *** $p < 0.001$; $n = 4$.

Quantitative analysis using LC–MS/MS of castalagin and vescalagin showed that the two isomers overall represented 1.07% (10.69 μg/mg) and 1.24% (12.36 μg/mg) of the extracts from verdesa and venegon varieties, respectively. To gain more insight into the gastric stability of the two ellagitannins, their amount was also measured following the simulated gastric digestion as well. Reduction was statistically significant for both verdesa and venegon varieties: a fraction of castalagin (around 40% and 70%, respectively) was preserved, while vescalagin was not detectable after gastric digestion. The results are summarized in Table 3.

Table 3. Characterization of ellagitannins from *Castanea sativa* Mill. leaf extracts.

	Before Gastric Digestion	After Gastric Digestion
Ellagitannins (Castalagin)	Mean ± SEM (μg/mg)	Mean ± SEM (μg/mg)
VR	5.79 ± 0.2	2.26 ± 0.04 ***
VN	5.88 ± 0.1	4.07 ± 0.08 ***
Ellagitannins (Vescalagin)	Mean ± SEM (μg/mg)	Mean ± SEM (μg/mg)
VR	4.90 ± 0.27	N.D. ***
VN	6.48 ± 0.27	N.D. ***

VN, *Castanea sativa* Mill. var. venegon; VR, *Castanea sativa* Mill. var. verdesa; GA eq., gallic acid equivalents; d.e., dry extract. *** $p < 0.001$, $n = 3$.

3.2. *Castanea sativa Mill. Extracts Impair Inflammatory Markers Typically Increased during Gastritis*

H. pylori infection and antibacterial effectors, such as TNFα, induce the release of IL-8 from gastric epithelial cells through the modulation of pathways converging in the NF-κB activation [6,42,43]. Thus, we carried out experiments to investigate the bioactivity of leaf extracts in both TNFα- and *H. pylori*-induced GES-1 cells (Figure 1). Firstly, we evaluated the effect of the leaf extracts on IL-8 release in cells challenged with TNFα (10 ng/mL): both the extracts inhibited IL-8 release in a concentration-dependent fashion (Figure 1A), with IC_{50} lower than 10 μg/mL. Then, the effect was also demonstrated on *H. pylori*-infected cells at slightly higher concentrations (IC_{50} < 30 μg/mL) (Figure 1B).

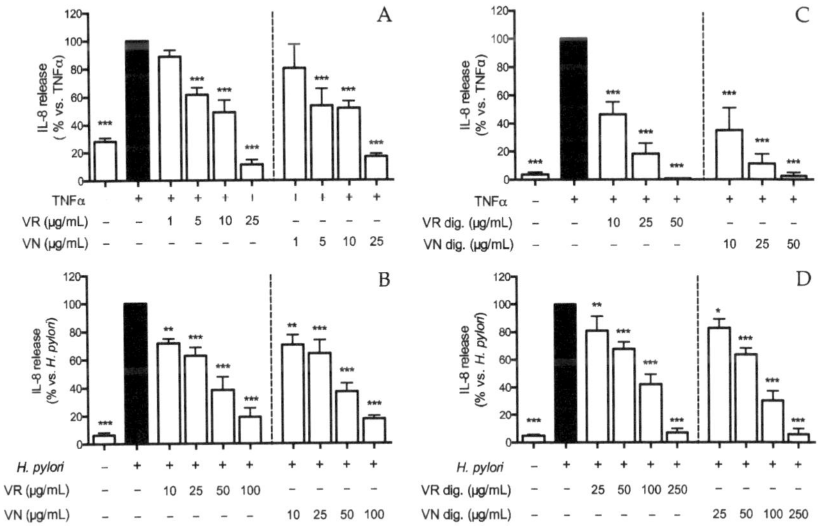

Figure 1. Effect of *Castanea sativa* Mill. leaf extracts on IL-8 release. GES-1 cells were treated for 6 h with TNFα (10 ng/mL) or *H. pylori* (ratio 50:1, bacteria/cell), in addition to leaf extracts before (**A**,**B**) and after in vitro simulated digestion (**C**,**D**). IL-8 was measured by ELISA assay. Results are expressed as the mean ± SEM ($n = 3$) of the relative percentage in comparison to stimulus (black bar), to which the value of 100% was arbitrarily assigned. * $p < 0.05$, ** $p < 0.01$, and *** $p < 0.001$ vs. stimulus. VR, *Castanea sativa* Mill. var. verdesa leaf extract; VN, *Castanea sativa* Mill. var. venegon leaf extract; dig., in vitro simulated digestion.

The in vitro simulated gastric digestion showed a moderate impact on the inhibitory activity (Figure 1C,D), as evident by the comparison of the IC_{50} values reported in Table 4.

Table 4. IC_{50} (μg/mL) of the *Castanea sativa* Mill. extracts on IL-8 release.

	Before Gastric Digestion		After Gastric Digestion	
Extracts vs. TNFα	IC_{50} (μg/mL)	CI (95%)	IC_{50} (μg/mL)	CI (95%)
VR	6.81	4.64 to 9.99	9.31	6.51 to 13.33
VN	4.30	2.11 to 8.79	7.09	2.96 to 16.74
Extracts vs. *H. pylori*	IC_{50} (μg/mL)	CI (95%)	IC_{50} (μg/mL)	CI (95%)
VR	28.22	19.77 to 40.28	70.76	52.14 to 96.03
VN	27.76	18.14 to 42.48	60.57	49.94 to 73.47

IC_{50}, 50% inhibitory concentration; CI (%), confidence interval; VN, *Castanea sativa* Mill. var. venegon; VR, *Castanea sativa* Mill. var. verdesa.

Since castalagin and vescalagin were found in leaf extracts, and castalagin was partially stable after the simulated digestion, we addressed their potential effect on IL-8 secretion triggered by inflammatory stimuli or *H. pylori* infection. Ellagitannins strongly inhibited TNFα-induced IL-8 release with comparable IC_{50} (0.27 and 0.22 μM, respectively) (Figure 2A). Notably, they also inhibited *H. pylori*-induced IL-8 release, albeit at higher concentrations (Figure 2B). The IC_{50} values of castalagin and vescalagin with these parameters are summarized in Table 5.

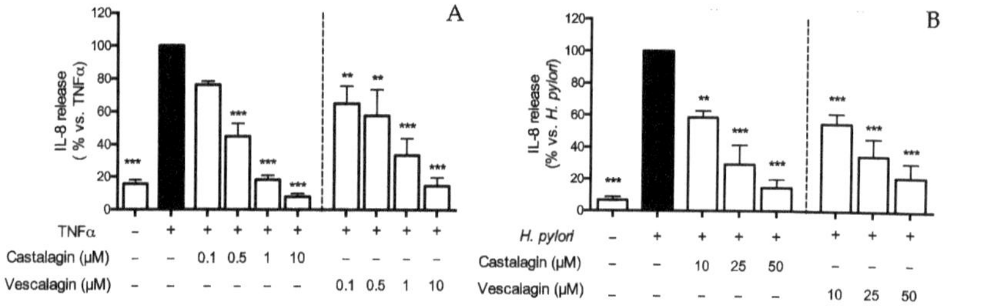

Figure 2. Effect of ellagitannins from *Castanea sativa* Mill. on IL-8 release. GES-1 cells were treated for 6 h with TNFα (10 ng/mL) (**A**) or *H. pylori* (ratio 50:1, bacteria:cell) (**B**), in the presence of castalagin or vescalagin. IL-8 was measured by ELISA assay. Results are expressed as the mean ± SEM (n = 3) of the relative percentage in comparison to stimulus (black bar), to which the value of 100% was arbitrarily assigned. ** $p < 0.01$, and *** $p < 0.001$ vs. stimulus.

Table 5. IC_{50} (μM) of IL-8 release: ellagitannins from *Castanea sativa* Mill.

Ellagitannins vs. TNFα	**IC_{50} (μM)**	**CI (95%)**
Castalagin	0.27	0.20 to 0.38
Vescalagin	0.22	0.07 to 0.66
Ellagitannins vs. *H. pylori*	**IC_{50} (μM)**	**CI (95%)**
Castalagin	11.70	7.65 to 17.89
Vescalagin	10.86	5.28 to 22.35

IC_{50}: 50% inhibitory concentration; CI: confidence interval.

Our group previously reported that ellagitannins can interfere with the NF-κB pathway in gastric epithelial models [25,44]. Consequently, we wondered whether the impairment of NF-κB pathway contributes to the IL-8 inhibition observed in the presence of the leaf's extracts and pure ellagitannins. Firstly, we tested the underlying mechanism using the plasmid transfection assay, observing that extracts (25–100 μg/mL) and pure ellagitannins (0.5–10 μM) counteracted the NF-κB-driven transcription induced by TNFα (Figure S3).

The bioactivity of the extracts was maintained after the simulated digestion, albeit at higher IC_{50} (>100 µg/mL), thus remarking that gastric environment may cause a partial reduction in chemical stability, in line with data regarding IL-8 release (Table 5). However, castalagin and vescalagin exhibited IC_{50} lower than 1 µM toward NF-κB-driven transcription, thus reflecting the inhibitory effect observed for leaf extracts (IC_{50} values are summarized in Table S2).

The impact on the NF-κB pathway was further evaluated by confocal microscopy experiments. At the selected concentration (200 µg/mL), the extracts interfered with NF-κB (subunit p65) translocation into the nucleus, during both TNFα (Figure 3) and *H. pylori* challenges (Figure 4). At the same concentration, the bioactivity of the extracts subjected to simulated gastric digestion was moderately affected, as previously observed.

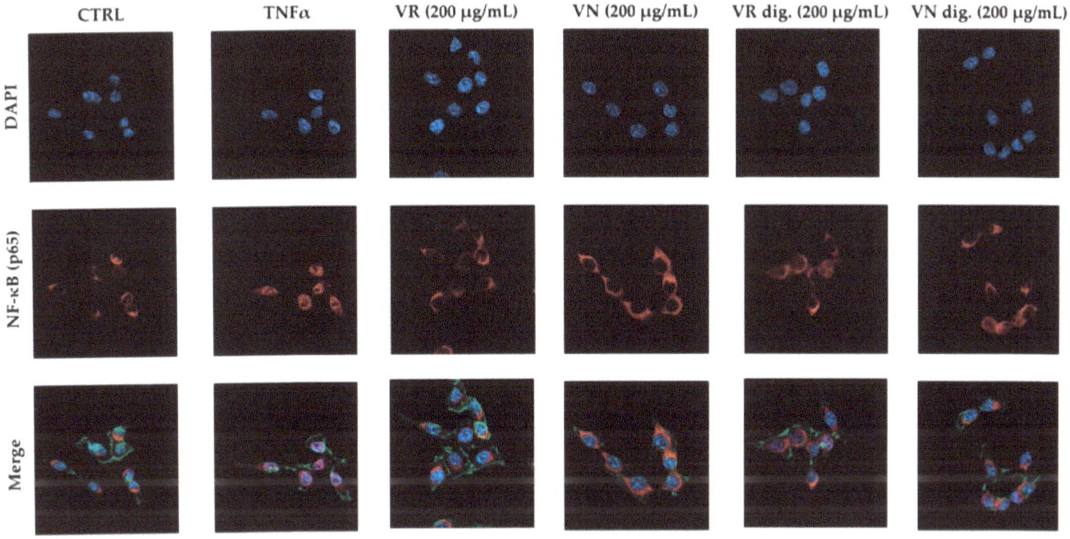

Figure 3. Effect of *Castanea sativa* Mill. leaf extracts on NF-κB (p65) nuclear translocation induced by TNFα. GES-1 cells were treated for 1 h with TNFα (10 ng/mL) in addition to leaf extracts (200 µg/mL) before and after in vitro simulated digestion. NF-κB (red) nuclear translocation from the cell cytoplasm into the nuclei (DAPI, blue) was detected by confocal immunofluorescence microscopy (63× objective). Merged images include β-actin staining (green). VR, *Castanea sativa* Mill. var. verdesa leaf extract; VN, *Castanea sativa* Mill. var. venegon leaf extract; dig., in vitro simulated digestion.

In this case, we also confirmed the parallelism between the biological effects of the extracts and pure ellagitannins during *H. pylori* infection. As expected, castalagin and vescalagin impaired the nuclear translocation of p65 at the highest concentration responsible for the total inhibition of IL-8 release (50 µM) (Figure 5). The overall data sustain that the extracts may counteract IL-8 release by interfering with the NF-κB pathway, and that pure ellagitannins present in leaves may be responsible for this inhibitory mechanism.

For this reason, we proceeded by comparing leaf extracts and ellagitannins for potential antibacterial effect against *H. pylori*, such as bacterial proliferation and adhesion to gastric cells.

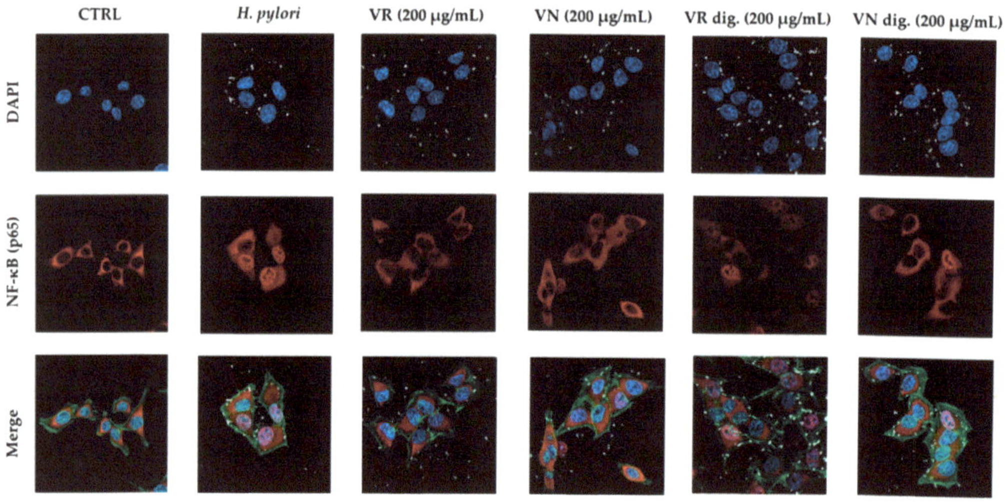

Figure 4. Effect of *Castanea sativa* Mill. leaf extracts on NF-κB (p65) nuclear translocation induced by *H. pylori*. GES-1 cells were treated for 1 h with *H. pylori* (ratio 50:1, bacteria/cell) in addition to leaf extracts (200 µg/mL) before and after in vitro simulated digestion. NF-κB (red) nuclear translocation from the cell cytoplasm into the nuclei (blue) was detected by confocal immunofluorescence microscopy (63× objective). Merged images include β-actin (green) and bacterial DNA (CFSE, white) staining. VR, *Castanea sativa* Mill. var. verdesa leaf extract; VN, *Castanea sativa* Mill. var. venegon leaf extract; dig., in vitro simulated digestion.

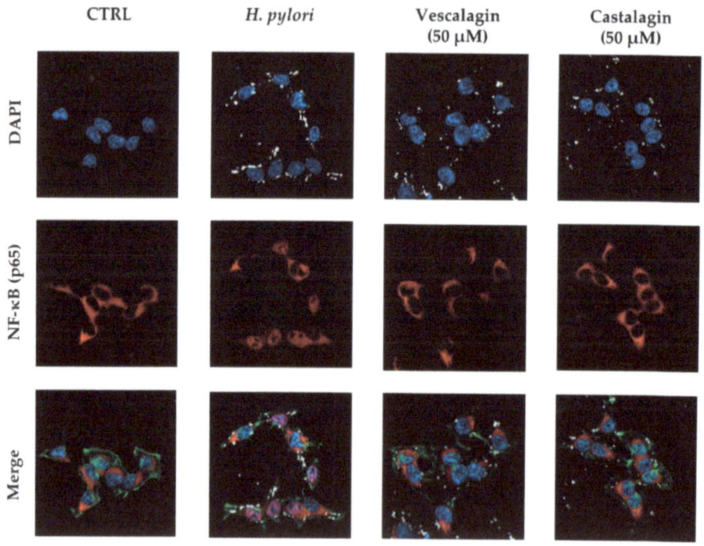

Figure 5. Effect of ellagitannins from *Castanea sativa* Mill. on NF-κB (p65) nuclear translocation induced by *H. pylori*. GES-1 cells were treated for 1 h with *H. pylori* (ratio 50:1, bacteria/cell) in addition to the ellagitannin isomers castalagin and vescalagin (50 µM). NF-κB (red) nuclear translocation from the cell cytoplasm into the nuclei (blue) was detected by confocal immunofluorescence microscopy (63× objective). Merged images include β-actin (green) and bacterial DNA (CFSE, white) staining. Cast., castalagin; Vesc., vescalagin.

3.3. Antibacterial Effect

A previous paper demonstrated the direct antibacterial effect of common monomeric ellagitannins on *H. pylori*, with MIC values below 25 µM [18]. However, to the best of our knowledge, castalagin and vescalagin have never been investigated for their antibacterial properties against *H. pylori*. Moreover, the antibacterial effect in coculture models of *H. pylori* infection has never been previously reported. For this reason, we decided to evaluate the impact of leaf extracts and their ellagitannins on bacterial growth and adhesion to gastric epithelial cells.

Both the extracts impaired bacterial growth with MICs of 100 µg/mL, although the simulated digestive process caused a significant decrease in the bioactivity, increasing the MICs to 200 µg/mL (Figure 6A,B). Of note, castalagin and vescalagin showed MIC values of 25 µM (Figure 6C), similarly to other monomeric ellagitannins [18].

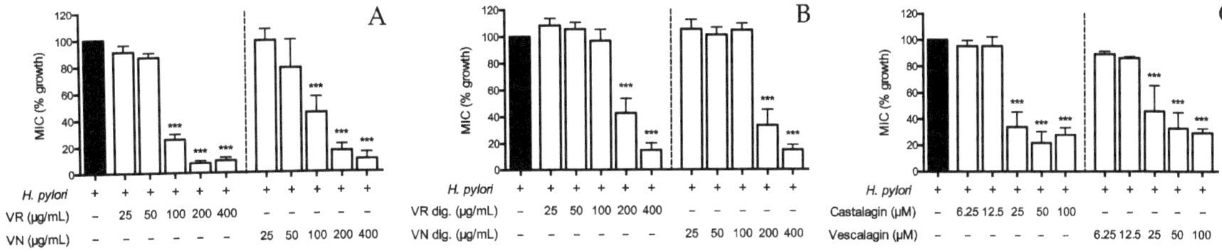

Figure 6. Effect of *Castanea sativa* Mill. leaf extracts and ellagitannins on *H. pylori* growth. *H. pylori* (OD = 0.1) was treated for 72 h with leaf extracts before (**A**) and after (**B**) in vitro simulated digestion, or ellagitannins (**C**). The rate of bacterial growth was measured as optical density (600 nm) using a photometer. Results are expressed as the mean ± SEM ($n = 3$) of the relative percentage in comparison to *H. pylori* growth (black bar), to which the value of 100% was arbitrarily assigned. *** $p < 0.001$ vs. *H. pylori*. MIC, minimum inhibitory concentration; VR, *Castanea sativa* Mill. var. verdesa leaf extract; VN, *Castanea sativa* Mill. var. venegon leaf extract; dig., in vitro simulated digestion.

H. pylori adhesion to gastric epithelial cells is a crucial event for the pathogenesis of gastritis. Accordingly, we wondered whether leaf extracts and pure ellagitannins may counteract the bacterial adhesion through a direct effect on *H. pylori* or by acting on the adhesion machinery of gastric epithelial cells. Thus, we treated GES-1 cells for 1 h after (cotreatment) or before (pretreatment) *H. pylori* infection; the first experimental setting involved the interaction of the natural compounds with either the bacterium or human cell, whereas the second setting excluded the direct interaction with the bacteria, thus testing the involvement of variables related to the host cells.

Following the first experimental setting, the treatment with leaf extracts from verdesa and venegon varieties impaired the bacterial adhesion in a concentration-dependent fashion; the highest inhibitory effect was observed at 200 µg/mL (−42.5% and −30.1%, respectively), with comparable values after the simulation of gastric digestion (−44.5% and 52.8%, respectively) (Figure 7A,B). Castalagin and vescalagin showed an inhibitory activity within the concentration range of 50–200 µM (−46.5% and −41.4%, respectively, at the concentration of 200 µM) (Figure 7C). Of note, the inhibitory activity was observed at concentrations comparable to the MIC values, thus suggesting that antiadhesive properties may be due, at least in part, to the antibacterial activity.

When experiments were carried out following the pretreatment setting, the extracts showed significant antiadhesive properties at 200 µg/mL (−53.8% and −27.5%, respectively) (Figure 8A), thus suggesting not only a direct effect on *H. pylori* during infection but also a potential impact on host cell variables. However, the simulated digestion abolished this effect (Figure 8B). Accordingly, castalagin and vescalagin were ineffective until the highest concentration tested (200 µM) (Figure 8C). This observation led us to hypothesize that direct antibacterial activity might represent the main explanation for the antiadhesive properties. However, it is not possible to exclude potential interference with the plethora of virulence factors involved during bacterial adhesion to host cells, as further mechanisms. This intriguing aspect is still poorly investigated referring to ellagitannins and related plant sources, thus demanding further host–pathogen interaction studies.

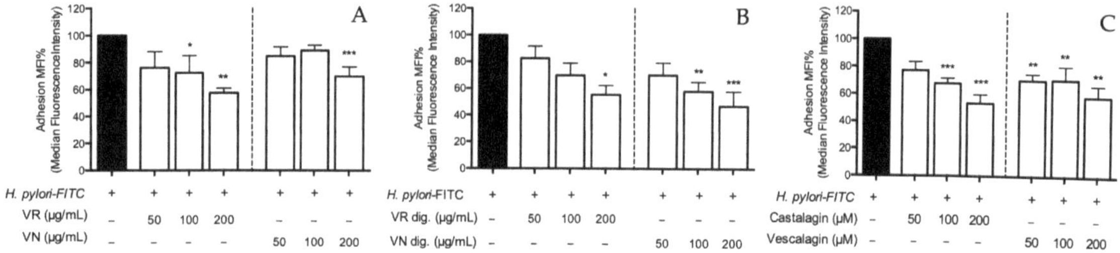

Figure 7. Effect of *Castanea sativa* Mill. leaf extracts and pure ellagitannins on *H. pylori* adhesion. GES-1 cells were treated for 1 h with *H. pylori*-FITC (ratio 50:1, bacteria/cell), in addition to leaf extracts before (**A**) and after (**B**) in vitro simulated digestion, or ellagitannins (**C**). The bacterial adhesion to GES-1 cells was measured as fluorescence intensity using a cytofluorimeter. Results re expressed as the mean ± SEM (n = 3) of the relative percentage in comparison to *H. pylori* infection (black bar), to which the value of 100% was arbitrarily assigned. * $p < 0.05$, ** $p < 0.01$, and *** $p < 0.001$ vs. *H. pylori*. MFI%, median fluorescence intensity; FITC, fluorescein isothiocyanate; VR, *Castanea sativa* Mill. var. verdesa leaf extract; VN, *Castanea sativa* Mill. var. venegon leaf extract; dig., in vitro simulated digestion.

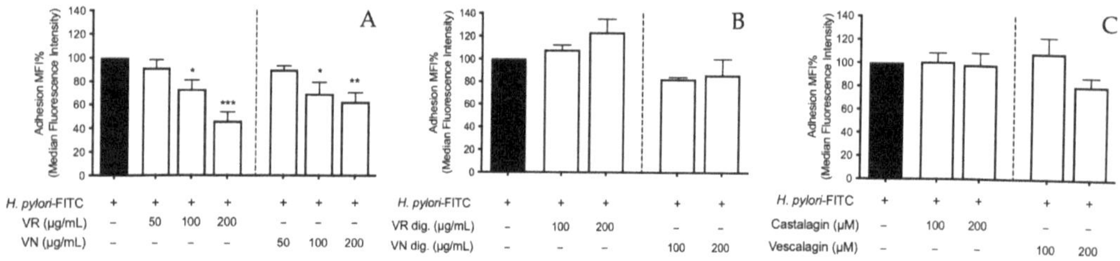

Figure 8. Preventive effect of *Castanea sativa* Mill. leaf extracts and ellagitannins on *H. pylori* adhesion. GES-1 cells were treated for 1 h with leaf extracts before (**A**) and after (**B**), or ellagitannins (**C**) in vitro simulated digestion, prior to *H. pylori*-FITC infection (ratio 50:1, bacteria/cell). The bacterial adhesion to GES-1 cells was measured as fluorescence intensity by cytofluorimeter. Results are expressed as the mean ± SEM (n = 3) of the relative percentage in comparison to *H. pylori* growth (black bar), to which the value of 100% was arbitrarily assigned. * $p < 0.05$, ** $p < 0.01$, and *** $p < 0.001$ vs. *H. pylori*. MFI%, median fluorescence intensity; FITC, fluorescein isothiocyanate; VR, *Castanea sativa* Mill. var. verdesa leaves extract; VN, *Castanea sativa* Mill. var. venegon leaves extract; dig., in vitro simulated digestion.

3.4. RNA-Seq Analysis

We then wondered whether the gastroprotective properties of ellagitannins may be partly due to or be mediated by changes in the epithelial cell transcriptome. Therefore, we carried out RNA-seq analysis of GES-1 cells infected with *H. pylori* without or upon treatment with castalagin (10 µM).

The principal component analysis (Figure S4) showed that replicates clustered very well according to their condition (*H. pylori* vs. castalagin); in fact, the analysis showed a clear separation between replicates of infected cells and replicates of infected cells subjected to castalagin treatment. The normalized counts obtained through DESeq2 analysis are reported in the Table S1.

A total of 36 differentially expressed genes were identified with three upregulated and 33 downregulated genes in RNA samples from *H. pylori*-infected cells treated with castalagin compared to RNA samples from infected cells with no other treatment (Figure 9A,B).

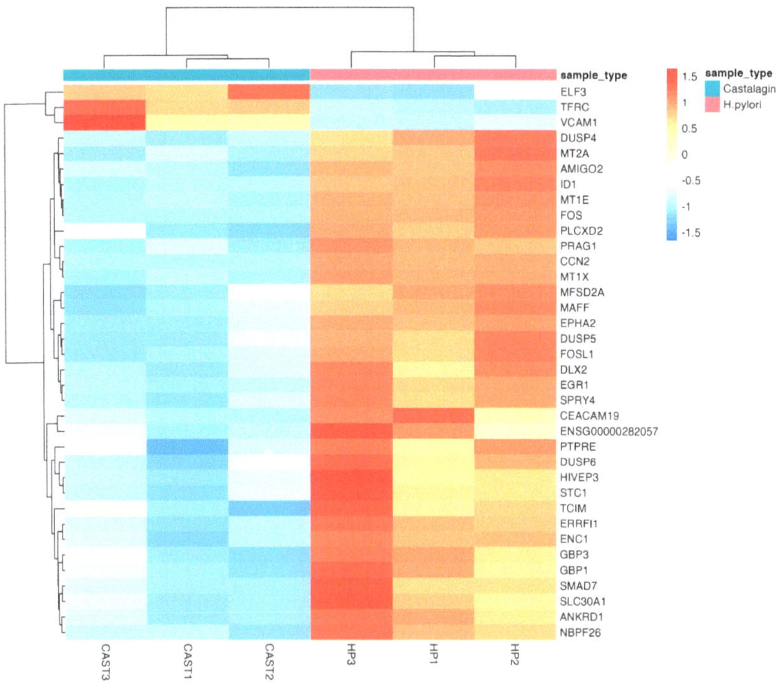

(A)

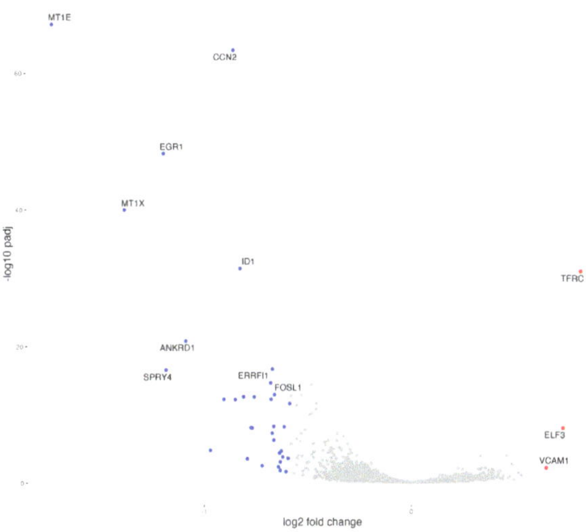

(B)

Figure 9. Heatmap showing the trend of the 36 differentially expressed genes between samples treated with *H. pylori* and samples treated with both *H. pylori* and castalagin (**A**). Volcano plot representing the downregulated and upregulated genes with blue and red dots, respectively (**B**).

Of note, considering the same subgroup of genes, most of them (29 out of 36) were regulated by *H. pylori* versus control uninfected cells, thus indicating that castalagin was able to specifically revert some of the transcriptional changes induced by *H. pylori* infection (Table 6).

Table 6. Table showing the 36 DEGs from RNA-seq analysis. Genes are reported with both the Ensembl ID and Gene Symbol.

		H. pylori vs. Castalagin		*H. pylori* vs. CTRL	
Gene ID	**Gene Symbol**	**Log_2FoldChange**	**Padj**	**Log_2FoldChange**	**Padj**
ENSG00000169715	MT1E	−1.725	0.000	1.142	0.000
ENSG00000118523	CCN2	−0.853	0.000	1.069	0.000
ENSG00000120738	EGR1	−1.190	0.000	3.966	0.000
ENSG00000187193	MT1X	−1.377	0.000	0.777	0.000
ENSG00000125968	ID1	−0.823	0.000	No change	-
ENSG00000072274	TFRC	0.814	0.000	−0.682	0.000
ENSG00000148677	ANKRD1	−1.087	0.000	0.956	0.000
ENSG00000116285	ERRFI1	−0.669	0.000	2.540	0.000
ENSG00000187678	SPRY4	−1.181	0.000	2.585	0.000
ENSG00000175592	FOSL1	−0.677	0.000	3.425	0.000
ENSG00000142627	EPHA2	−0.659	0.000	1.805	0.000
ENSG00000120875	DUSP4	−0.808	0.000	1.559	0.000
ENSG00000115844	DLX2	−0.757	0.000	1.947	0.000
ENSG00000171617	ENC1	−0.675	0.000	0.699	0.000
ENSG00000117228	GBP1	−0.902	0.000	1.952	0.000
ENSG00000275342	PRAG1	−0.848	0.000	No change	-
ENSG00000185022	MAFF	−0.586	0.000	4.396	0.000
ENSG00000138166	DUSP5	−0.663	0.000	2.125	0.000
ENSG00000139211	AMIGO2	−0.612	0.000	0.892	0.000
ENSG00000125148	MT2A	−0.773	0.000	0.856	0.000
ENSG00000170345	FOS	−0.767	0.000	0.954	0.000
ENSG00000163435	ELF3	0.731	0.000	No change	-
ENSG00000101665	SMAD7	−0.670	0.000	No change	-
ENSG00000168389	MFSD2A	−0.663	0.000	2.679	0.000
ENSG00000170385	SLC30A1	−0.969	0.000	2.064	0.000
ENSG00000127124	HIVEP3	−0.626	0.000	1.717	0.000
ENSG00000176907	TCIM	−0.636	0.000	1.969	0.000
ENSG00000132334	PTPRE	−0.620	0.000	1.424	0.000
ENSG00000117226	GBP3	−0.593	0.000	1.082	0.000
ENSG00000139318	DUSP6	−0.791	0.000	No change	-
ENSG00000240891	PLCXD2	−0.632	0.001	0.796	0.000
ENSG00000159167	STC1	−0.720	0.004	1.767	0.000

Table 6. *Cont.*

		H. pylori vs. Castalagin		*H. pylori* vs. CTRL	
Gene ID	**Gene Symbol**	**Log_2FoldChange**	**Padj**	**Log_2FoldChange**	**Padj**
ENSG00000159167	STC1	−0.720	0.004	1.767	0.000
ENSG00000282057	ENSG00000282057	−0.641	0.005	No change -	-
ENSG00000162692	VCAM1	0.650	0.007	2.183	0.000
ENSG00000186567	CEACAM19	−0.632	0.018	2.463	0.000
ENSG00000273136	NBPF26	−0.604	0.027	No change	-

Lastly, the 33 downregulated genes were used to perform an enrichment analysis to evaluate which pathways were over-represented; the identified significant pathways with the corresponding number of genes are reported in Table 7. The enriched pathway analysis suggested that the main transcriptional targets of castalagin are as follows: signals involved in response to cytokines, including transcription factors (AP-1 and NF-κB) and MAPKs (ERK1/2), transport and signals related to divalent ions, and DNA-regulatory proteins, such as targets of the polymerase II (Figure S5).

Table 7. Table showing the identified enriched pathways using the 33 downregulated DEGs.

Term	Overlap	*p* Value	Padj	Genes
Cellular divalent inorganic cation homeostasis (GO:0072503)	5	1.01×10^{-8}	8.93×10^{-9}	MT2A; STC1; MT1X; SLC30A1; MT1E
Cellular response to cytokine stimulus (GO:0071345)	7	1.12×10^{-9}	0.0005	EGR1; MT2A; ANKRD1; MT1X; FOS; GBP1; GBP3
Regulation of ERK1 and ERK2 cascade (GO:0070372)	5	4.14×10^{-9}	0.001	DUSP4; CCN2; GBP1; DUSP6; EPHA2
Regulation of transcription, DNA-templated (GO:0006355)	10	0.002	0.025	FOSL1; EGR1; DLX2; TCIM; ID1; MAFF; ANKRD1; HIVEP3; FOS; SMAD7
Regulation of transcription by RNA polymerase II (GO:0006357)	9	0.008	0.055	FOSL1; EGR1; DLX2; ID1; MAFF; ANKRD1; HIVEP3; FOS; SMAD7
Positive regulation of transcription, DNA-templated (GO:0045893)	6	0.012	0.062	FOSL1; EGR1; MAFF; HIVEP3; FOS; SMAD7
Positive regulation of transcription by RNA polymerase II (GO:0045944)	5	0.012	0.066	FOSL1; EGR1; MAFF; FOS; SMAD7
Cellular protein modification process (GO:0006464)	5	0.025	0.081	DUSP4; DUSP5; PTPRE; PRAG1; DUSP6

4. Discussion

Castanea sativa Mill. is a well-known source of polyphenols, including gallotannins and ellagitannins, with potential gastroprotective properties. It is remarkable that byproducts from chestnut harvesting, such as chestnut peel and leaves, represent a source of bioactive compounds with potential nutraceutical use. We previously reported that the outer tissues from chestnuts, belonging to var. venegon and var. verdesa (*Castanea sativa* Mill.), are rich in proanthocyanidins, showing anti-inflammatory effects on gastric epithelial cells [25]; however, chestnuts are devoid of ellagitannins, contrarily to what reported for the bark [19,20].

Ellagitannins have been reported to possess anti-ulcer and antibacterial properties [12–14,18], but their potential role in *H. pylori*-related gastritis is still poorly investigated. In this work, we demonstrated the presence of ellagitannin isomers, castalagin and vescalagin, in hydroalcoholic extracts from leaves belonging to both varieties from *Castanea sativa* Mill. We validated their pharmacological activity in a model of the nontumoral gastric epithelium (GES-1), challenged with TNFα or *H. pylori* infection.

Extracts from leaf and pure ellagitannins that are typically found in *Castanea sativa* Mill. inhibited the release of IL-8 from GES-1 cells, with more pronounced activity against TNFα challenge (Tables 5 and 6). The mechanism of action may be ascribed, at least in part, to the impairment of the NF-κB pathway (Figures 3–5 and S3), a known crossroad of TNFα- and *H. pylori*-induced molecular signaling [43]. However, it is plausible that other inflammatory targets were involved since, in our experiments, IL-8 release was always inhibited at lower concentrations than those impairing the NF-κB cascade.

The same parallelism among leaf extracts and pure ellagitannins was observed in respect to the direct antibacterial and the antiadhesive properties against *H. pylori*. Our results are in line with those obtained by other authors, reporting that casuarinin, structurally similar to castalagin and vescalagin, could inhibit the growth of *H. pylori* at 13.35–26.70 μM [18]. Of note, the same authors suggested a higher specificity for *H. pylori* than *E. coli* (MIC > 100 μg/mL). Further investigations should clarify the antibacterial mechanism of action and the potential selectivity.

Regarding the possibility to attribute the bioactivity of leaf extracts to castalagin and vescalagin as main compounds, it is worth mentioning that the anti-inflammatory properties of the extracts showed a clear parallelism with the activities exhibited by ellagitannin isomers. The latter were selected as available compounds, representative for the class of ellagitannins, widely reported in *Castanea sativa* Mill. Indeed, especially during *H. pylori* challenge, pure ellagitannins exhibited their inhibitory effects at concentrations higher than those present in the extracts, thus suggesting the contribution of other hydrolyzable tannins, including ellagitannins, to the biological effect of the extract.

To obtain a more comprehensive profile of the biological activities of ellegitannins, castalagin (10 μM) was selected as a candidate compound for RNA-seq experiments. This high-throughput approach allows capturing transcriptional changes and identifying target genes and pathways. A small set of genes (n = 36) was reverted by castalagin treatment with respect to *H. pylori* infection; most of them were downregulated (33), thus suggesting the prevalence of an inhibitory effect on transcription. Of note, a subset of downregulated genes (*tfrc*, *spry4*, *gbp1*, *gbp3*, *prag1*, *amigo2*, and *ptpre*) were related to the Rho GTPase pathway; among them, *amigo2* has been suggested as a biomarker of poor prognosis in gastric adenocarcinoma [45–47]. Despite the restricted number of modulated genes, PCA showed that the group of data related to castalagin treatment clustered from *H. pylori* infection, thus reflecting a significant difference among groups.

The impact of ellagitannins on the Rho GTPase pathway is still poorly described, although other sources of ellagitannins have been investigated for their anti-inflammatory and antitumoral effect at the gut level [48]. Regarding inflammation, we could not exclude, on the basis of the literature concerning ellagitannins [16], that post-transcriptional and enzymatic mechanisms may have contributed to IL-8 inhibition in our setting.

Altogether, our data suggest that the use of the whole extract, accurately standardized for bioactive compounds such as castalagin and vescalagin, may exert more beneficial activity with respect to single compounds in gastritis. Other polyphenols, whose presence has been documented in *Castanea sativa* Mill. leaves, such as gallotannins or flavonol-derivatives [23,24], have previously been reported for anti-inflammatory and antibacterial properties at the gastric level [14,49]. In summary, our work attributed a relevant role to ellagitannins in the impairment of *H. pylori* viability and infection-induced inflammation in human gastric epithelial cells.

5. Conclusions

Our work sustains, for the first time, the characterization of ellagitannins as bioactive principles in *Castanea sativa* Mill. leaf extracts. We suggest that leaves should be taken into consideration as suitable plant material to produce sustainable and bioactive extracts. Moreover, to the best of our knowledge, this is the first study in which the bioactivity of ellagitannins and related plant extracts has been validated in a model mimicking the gastric epithelium–*H. pylori* interaction.

Supplementary Materials: The following supporting information can be downloaded at https://www.mdpi.com/article/10.3390/nu15061504/s1: Figure S1. Effect of *Castanea sativa* Mill. leaf extracts and ellagitannins on GES-1 viability (MTT test); Table S1. Total number of raw and normalized counts of the RNA-seq analysis for each sample; Figure S2. LC-MS chromatograms identifying the presence of castalagin and vescalagin in leaf extracts; Figure S3. Effect of *Castanea sativa* Mill. leaf extracts and ellagitannins on NF-κB driven transcription; Table S2. IC_{50} (μg/mL) of NF-κB driven transcription: extracts and ellagitannins from *Castanea sativa* Mill. leaves; Figure S4. Transcriptional effect of castalagin on *H. pylori*-induced gene expression in GES-1 cells. Enriched pathway analysis (A), volcano plot (B), and PCA (C); Figure S5. Summary of the relevant pathways according to P adjusted values (Padj).

Author Contributions: Conceptualization, M.D., S.P., E.S., E.D.F., G.N., S.F.V. and L.C.; investigation, G.M., S.P., N.M., F.G. (Flavio Giavarini), F.G. (Francesca Genova), M.F. and C.P.; supervision, E.S., M.D. and E.D.F.; data analysis, G.M., S.P., F.G. (Francesca Genova), M.F. and C.P., writing and editing, S.P., M.D., E.S. and E.D.F. All authors have read and agreed to the published version of the manuscript.

Funding: The present work was partially funded by EPO s.r.l. (MI) Italy. This research was supported by grants from MIUR "Progetto Eccellenza" 2018–2022.

Institutional Review Board Statement: Not applicable.

Informed Consent Statement: Not applicable.

Data Availability Statement: Data are available on request from the corresponding authors, E.D. and E.S. (emma.defabiani@unimi.it; enrico.sangiovanni@unimi.it).

Acknowledgments: The authors thank Dawit Kidane-Mulat (University of Texas, Austin, TX, USA) for providing GES-1 cells.

Conflicts of Interest: The authors declare no conflict of interest. Giovanna Nicotra (G.N.) and Silvia Francesca Vicentini (S.F.V.) are, respectively, the scientific director and the R&D officer responsible from EPO s.r.l. However, this paper does not necessarily reflect the company's views of its future policy on this area. G.N. and S.F.V. were involved in the idea of the study, but had no role in the design, collection, analysis, or interpretation of data.

References

1. Marshall, B. *Helicobacter pylori*—A Nobel pursuit? *Can. J. Gastroenterol.* **2008**, *22*, 895–896. [CrossRef]
2. Marshall, B.J.; Windsor, H.M. The relation of *Helicobacter pylori* to gastric adenocarcinoma and lymphoma: Pathophysiology, epidemiology, screening, clinical presentation, treatment, and prevention. *Med. Clin. N. Am.* **2005**, *89*, 313–344, viii. [CrossRef]
3. Naito, Y.; Yoshikawa, T. Molecular and cellular mechanisms involved in *Helicobacter pylori*-induced inflammation and oxidative stress. *Free Radic. Biol. Med.* **2002**, *33*, 323–336. [CrossRef]
4. Ando, T.; Kusugami, K.; Ohsuga, M.; Shinoda, M.; Sakakibara, M.; Saito, H.; Fukatsu, A.; Ichiyama, S.; Ohta, M. Interleukin-8 activity correlates with histological severity in *Helicobacter pylori*-associated antral gastritis. *Am. J. Gastroenterol.* **1996**, *91*, 1150–1156.
5. Crabtree, J.E.; Peichl, P.; Wyatt, J.I.; Stachl, U.; Lindley, I.J. Gastric interleukin-8 and IgA IL-8 autoantibodies in *Helicobacter pylori* infection. *Scand. J. Immunol.* **1993**, *37*, 65–70. [CrossRef]
6. Crabtree, J.E.; Farmery, S.M.; Lindley, I.J.; Figura, N.; Peichl, P.; Tompkins, D.S. CagA/cytotoxic strains of *Helicobacter pylori* and interleukin-8 in gastric epithelial cell lines. *J. Clin. Pathol.* **1994**, *47*, 945–950. [CrossRef]
7. Uemura, N.; Oomoto, Y.; Mukai, T.; Okamoto, S.; Yamaguchi, S.; Mashiba, H.; Taniyama, K.; Sasaki, N.; Sumii, K.; Haruma, K.; et al. Gastric corpus IL-8 concentration and neutrophil infiltration in duodenal ulcer patients. *Aliment. Pharmacol. Ther.* **1997**, *11*, 793–800. [CrossRef]
8. Basso, D.; Plebani, M.; Kusters, J.G. Pathogenesis of *Helicobacter pylori* infection. *Helicobacter* **2010**, *15* (Suppl. 1), 14–20. [CrossRef]
9. Isomoto, H.; Mizuta, Y.; Miyazaki, M.; Takeshima, F.; Omagari, K.; Murase, K.; Nishiyama, T.; Inoue, K.; Murata, I.; Kohno, S. Implication of NF-kappaB in *Helicobacter pylori*-associated gastritis. *Am. J. Gastroenterol.* **2000**, *95*, 2768–2776.
10. Malfertheiner, P.; Selgrad, M.; Wex, T.; Romi, B.; Borgogni, E.; Spensieri, F.; Zedda, L.; Ruggiero, P.; Pancotto, L.; Censini, S.; et al. Efficacy, immunogenicity, and safety of a parenteral vaccine against *Helicobacter pylori* in healthy volunteers challenged with a Cag-positive strain: A randomised, placebo-controlled phase 1/2 study. *Lancet Gastroenterol. Hepatol.* **2018**, *3*, 698–707. [CrossRef]

11. Gotteland, M.; Andrews, M.; Toledo, M.; Munoz, L.; Caceres, P.; Anziani, A.; Wittig, E.; Speisky, H.; Salazar, G. Modulation of *Helicobacter pylori* colonization with cranberry juice and Lactobacillus johnsonii La1 in children. *Nutrition* **2008**, *24*, 421–426. [CrossRef]
12. Lai, S.; Zhou, Q.; Zhang, Y.; Shang, J.; Yu, T. Effects of pomegranate tannins on experimental gastric damages. *Zhongguo Zhong Yao Za Zhi* **2009**, *34*, 1290–1294.
13. Samardzic, S.; Arsenijevic, J.; Bozic, D.; Milenkovic, M.; Tesevic, V.; Maksimovic, Z. Antioxidant, anti-inflammatory and gastroprotective activity of *Filipendula ulmaria* (L.) Maxim. and *Filipendula vulgaris* Moench. *J. Ethnopharmacol.* **2018**, *213*, 132–137. [CrossRef]
14. Khennouf, S.; Benabdallah, H.; Gharzouli, K.; Amira, S.; Ito, H.; Kim, T.H.; Yoshida, T.; Gharzouli, A. Effect of tannins from *Quercus suber* and *Quercus coccifera* leaves on ethanol-induced gastric lesions in mice. *J. Agric. Food Chem.* **2003**, *51*, 1469–1473. [CrossRef]
15. Sangiovanni, E.; Vrhovsek, U.; Rossoni, G.; Colombo, E.; Brunelli, C.; Brembati, L.; Trivulzio, S.; Gasperotti, M.; Mattivi, F.; Bosisio, E.; et al. Ellagitannins from Rubus berries for the control of gastric inflammation: In vitro and in vivo studies. *PLoS ONE* **2013**, *8*, e71762. [CrossRef]
16. Piazza, S.; Fumagalli, M.; Martinelli, G.; Pozzoli, C.; Maranta, N.; Angarano, M.; Sangiovanni, E.; Dell'Agli, M. Hydrolyzable Tannins in the Management of Th1, Th2 and Th17 Inflammatory-Related Diseases. *Molecules* **2022**, *27*, 7593. [CrossRef]
17. Colombo, E.; Sangiovanni, E.; Dell'agli, M. A review on the anti-inflammatory activity of pomegranate in the gastrointestinal tract. *Evid. Based Complement. Alternat Med.* **2013**, *2013*, 247145. [CrossRef]
18. Funatogawa, K.; Hayashi, S.; Shimomura, H.; Yoshida, T.; Hatano, T.; Ito, H.; Hirai, Y. Antibacterial activity of hydrolyzable tannins derived from medicinal plants against *Helicobacter pylori*. *Microbiol. Immunol.* **2004**, *48*, 251–261. [CrossRef]
19. Ricci, A.; Lagel, M.C.; Parpinello, G.P.; Pizzi, A.; Kilmartin, P.A.; Versari, A. Spectroscopy analysis of phenolic and sugar patterns in a food grade chestnut tannin. *Food Chem.* **2016**, *203*, 425–429. [CrossRef]
20. Comandini, P.; Lerma-Garcia, M.J.; Simo-Alfonso, E.F.; Toschi, T.G. Tannin analysis of chestnut bark samples (*Castanea sativa* Mill.) by HPLC-DAD-MS. *Food Chem.* **2014**, *157*, 290–295. [CrossRef]
21. Esposito, T.; Celano, R.; Pane, C.; Piccinelli, A.L.; Sansone, F.; Picerno, P.; Zaccardelli, M.; Aquino, R.P.; Mencherini, T. Chestnut (*Castanea sativa* Miller.) Burs Extracts and Functional Compounds: UHPLC-UV-HRMS Profiling, Antioxidant Activity, and Inhibitory Effects on Phytopathogenic Fungi. *Molecules* **2019**, *24*, 302. [CrossRef] [PubMed]
22. Caleja, C.; Barros, L.; Prieto, M.A.; Bento, A.; Oliveira, M.; Ferreira, I. Development of a natural preservative obtained from male chestnut flowers: Optimization of a heat-assisted extraction technique. *Food Funct.* **2019**, *10*, 1352–1363. [CrossRef] [PubMed]
23. Wagner, H.B.S. Plant Drug Analysis. In *A Thin Layer Chromatography Atlas*; Springer: Berlin/Heidelberg, Germany 1984.
24. Cerulli, A.; Masullo, M.; Mari, A.; Balato, A.; Filosa, R.; Lembo, S.; Napolitano, A.; Piacente, S. Phenolics from *Castanea sativa* leaves and their effects on UVB-induced damage. *Nat. Prod. Res.* **2018**, *32*, 1170–1175. [CrossRef]
25. Sangiovanni, E.; Piazza, S.; Vrhovsek, U.; Fumagalli, M.; Khalilpour, S.; Masuero, D.; Di Lorenzo, C.; Colombo, L.; Mattivi, F.; De Fabiani, E.; et al. A bio-guided approach for the development of a chestnut-based proanthocyanidin-enriched nutraceutical with potential anti-gastritis properties. *Pharmacol. Res.* **2018**, *134*, 145–155. [CrossRef] [PubMed]
26. Marchetti, M.; Arico, B.; Burroni, D.; Figura, N.; Rappuoli, R.; Ghiara, P. Development of a mouse model of *Helicobacter pylori* infection that mimics human disease. *Science* **1995**, *267*, 1655–1658. [CrossRef]
27. Hirayama, F.; Takagi, S.; Iwao, E.; Yokoyama, Y.; Haga, K.; Hanada, S. Development of poorly differentiated adenocarcinoma and carcinoid due to long-term *Helicobacter pylori* colonization in Mongolian gerbils. *J. Gastroenterol.* **1999**, *34*, 450–454. [CrossRef] [PubMed]
28. Ruggiero, P.; Rossi, G.; Tombola, F.; Pancotto, L.; Lauretti, L.; Del Giudice, G.; Zoratti, M. Red wine and green tea reduce *H pylori*- or VacA-induced gastritis in a mouse model. *World J. Gastroenterol.* **2007**, *13*, 349–354. [CrossRef]
29. Singleton, V.L.; Rossi, J.A. Colorimetry of total phenolics with phosphomolybdic-phosphotungstic acid reagents. *Am. J. Enol. Vitic.* **1965**, *16*, 144–158.
30. Messing, J.; Thole, C.; Niehues, M.; Shevtsova, A.; Glocker, E.; Boren, T.; Hensel, A. Antiadhesive properties of *Abelmoschus esculentus* (Okra) immature fruit extract against *Helicobacter pylori* adhesion. *PLoS ONE* **2014**, *9*, e84836. [CrossRef]
31. Gutierrez-Orozco, F.; Stephens, B.R.; Neilson, A.P.; Green, R.; Ferruzzi, M.G.; Bomser, J.A. Green and black tea inhibit cytokine-induced IL-8 production and secretion in AGS gastric cancer cells via inhibition of NF-kappaB activity. *Planta Med.* **2010**, *76*, 1659–1665. [CrossRef]
32. Di Lorenzo, C.; Dell'Agli, M.; Sangiovanni, E.; Dos Santos, A.; Uberti, F.; Moro, E.; Bosisio, E.; Restani, P. Correlation between catechin content and NF-kappaB inhibition by infusions of green and black tea. *Plant Foods Hum. Nutr.* **2013**, *68*, 149–154. [CrossRef]
33. Denizot, F.; Lang, R. Rapid colorimetric assay for cell growth and survival. Modifications to the tetrazolium dye procedure giving improved sensitivity and reliability. *J. Immunol. Methods* **1986**, *89*, 271–277. [CrossRef]
34. Martinelli, G.; Angarano, M.; Piazza, S.; Fumagalli, M.; Magnavacca, A.; Pozzoli, C.; Khalilpour, S.; Dell'Agli, M.; Sangiovanni, E. The Nutraceutical Properties of Sumac (*Rhus coriaria* L.) against Gastritis: Antibacterial and Anti-Inflammatory Activities in Gastric Epithelial Cells Infected with *H. pylori*. *Nutrients* **2022**, *14*, 1757. [CrossRef]

35. Weinstein, M.D.P.; Patel, J.B.; Burnham, C.A.; Campeau, S.; Conville, P.S.; Doern, C.; Eliopolous, G.M.; Galas, M.F.; Humphries, R.M.; Jenkins, S.G. *Methods for Dilution Antimicrobial Susceptibility Tests for Bacteria That Grow Aerobically*; Clinical and Laboratory Standards Intitute: Wayne, PA, USA, 2018.
36. Andrews, S. FASTQC. *A Quality Control Tool for High throughput Sequence Dat*. 2010. Available online: https://www.bioinformatics.babraham.ac.uk/projects/fastqc/ (accessed on 12 November 2022).
37. Ewels, P.; Magnusson, M.; Lundin, S.; Kaller, M. MultiQC: Summarize analysis results for multiple tools and samples in a single report. *Bioinformatics* **2016**, *32*, 3047–3048. [CrossRef]
38. Dobin, A.; Davis, C.A.; Schlesinger, F.; Drenkow, J.; Zaleski, C.; Jha, S.; Batut, P.; Chaisson, M.; Gingeras, T.R. STAR: Ultrafast universal RNA-seq aligner. *Bioinformatics* **2013**, *29*, 15–21. [CrossRef]
39. Liao, Y.; Smyth, G.K.; Shi, W. featureCounts: An efficient general purpose program for assigning sequence reads to genomic features. *Bioinformatics* **2014**, *30*, 923–930. [CrossRef]
40. Love, M.I.; Huber, W.; Anders, S. Moderated estimation of fold change and dispersion for RNA-seq data with DESeq2. *Genome Biol.* **2014**, *15*, 550. [CrossRef]
41. Kuleshov, M.V.; Jones, M.R.; Rouillard, A.D.; Fernandez, N.F.; Duan, Q.; Wang, Z.; Koplev, S.; Jenkins, S.L.; Jagodnik, K.M.; Lachmann, A.; et al. Enrichr: A comprehensive gene set enrichment analysis web server 2016 update. *Nucleic Acids Res.* **2016**, *44*, W90–W97. [CrossRef]
42. Sharma, S.A.; Tummuru, M.K.; Blaser, M.J.; Kerr, L.D. Activation of IL-8 gene expression by *Helicobacter pylori* is regulated by transcription factor nuclear factor-kappa B in gastric epithelial cells. *J. Immunol.* **1998**, *160*, 2401–2407. [CrossRef]
43. Foryst-Ludwig, A.; Naumann, M. p21-activated kinase 1 activates the nuclear factor kappa B (NF-kappa B)-inducing kinase-Ikappa B kinases NF-kappa B pathway and proinflammatory cytokines in *Helicobacter pylori* infection. *J. Biol. Chem.* **2000**, *275*, 39779–39785. [CrossRef]
44. Fumagalli, M.; Sangiovanni, E.; Vrhovsek, U.; Piazza, S.; Colombo, E.; Gasperotti, M.; Mattivi, F.; De Fabiani, E.; Dell'Agli, M. Strawberry tannins inhibit IL-8 secretion in a cell model of gastric inflammation. *Pharmacol. Res.* **2016**, *111*, 703–712. [CrossRef] [PubMed]
45. Zhang, S.; Tang, Q.; Xu, F.; Xue, Y.; Zhen, Z.; Deng, Y.; Liu, M.; Chen, J.; Liu, S.; Qiu, M.; et al. RhoA regulates G1-S progression of gastric cancer cells by modulation of multiple INK4 family tumor suppressors. *Mol. Cancer Res.* **2009**, *7*, 570–580. [CrossRef] [PubMed]
46. Namikawa, K.; Tanaka, N.; Ota, Y.; Takamatsu, M.; Kosugi, M.; Tokai, Y.; Yoshimizu, S.; Horiuchi, Y.; Ishiyama, A.; Yoshio, T.; et al. Genomic features of *Helicobacter pylori*-naive diffuse-type gastric cancer. *J. Pathol.* **2022**, *258*, 300–311. [CrossRef] [PubMed]
47. Goto, K.; Morimoto, M.; Osaki, M.; Tanio, A.; Izutsu, R.; Fujiwara, Y.; Okada, F. The impact of AMIGO2 on prognosis and hepatic metastasis in gastric cancer patients. *BMC Cancer* **2022**, *22*, 280. [CrossRef]
48. Nunez-Sanchez, M.A.; Gonzalez-Sarrias, A.; Garcia-Villalba, R.; Monedero-Saiz, T.; Garcia-Talavera, N.V.; Gomez-Sanchez, M.B.; Sanchez-Alvarez, C.; Garcia-Albert, A.M.; Rodriguez-Gil, F.J.; Ruiz-Marin, M.; et al. Gene expression changes in colon tissues from colorectal cancer patients following the intake of an ellagitannin-containing pomegranate extract: A randomized clinical trial. *J. Nutr. Biochem.* **2017**, *42*, 126–133. [CrossRef]
49. Ivyna de Araujo Rego, R.; Guedes Silvestre, G.F.; Ferreira de Melo, D.; Albino, S.L.; Pimentel, M.M.; Silva Costa Cruz, S.B.; Silva Wurzba, S.D.; Rodrigues, W.F.; Goulart de Lima Damasceno, B.P.; Cancado Castellano, L.R. Flavonoids-Rich Plant Extracts against *Helicobacter pylori* Infection as Prevention to Gastric Cancer. *Front. Pharmacol.* **2022**, *13*, 951125. [CrossRef]

nutrients

Article

Enhanced Nutraceutical Properties of Extra Virgin Olive Oil Extract by Olive Leaf Enrichment

Doretta Cuffaro [1], Simone Bertini [1], Marco Macchia [1,2] and Maria Digiacomo [1,2,*]

[1] Department of Pharmacy, University of Pisa, via Bonanno 6, 56126 Pisa, Italy
[2] Interdepartmental Research Center "Nutraceuticals and Food for Health", University of Pisa, 56100 Pisa, Italy
* Correspondence: maria.digiacomo@unipi.it

Abstract: (1) Background: Nowadays, the health-promoting properties of extra virgin olive oil (EVOO), including the antioxidant and anti-inflammatory actions, are well recognized and mainly attributed to the different polyphenols, such as oleocanthal and oleacein. In EVOO production, olive leaves represent a high value by-product, showing a wide spectrum of beneficial effects due to the presence of polyphenols, especially oleuropein. Here we report the study of olive leaf extract (OLE)-enriched EVOO extracts, obtained by adding different percentages of OLE to EVOO in order to ameliorate their nutraceutical activities. (2) Methods: The polyphenolic content of the EVOO/OLE extracts was analyzed by HPLC and the Folin-Ciocalteau assay. For further biological testing, an 8% OLE-enriched EVOO extract was chosen. Therefore, antioxidant effects were evaluated by three different methods (DPPH, ABTS, and FRAP), and the anti-inflammatory properties were assessed in terms of cyclooxygenase activity inhibition. (3) Results: The antioxidant and anti-inflammatory profiles of the new EVOO/OLE extract are significantly improved compared to those of EVOO extract; (4) Conclusions: The combination of OLE and EVOO extract can lead to an extract enriched in terms of bioactive polyphenols and endowed with better biological properties than the singular EVOO extract. Therefore, it may represent a new complement in the nutraceutical field.

Keywords: extra virgin olive oil; olive leaves; polyphenols; antioxidant activity; anti-inflammatory activity; EVOO enrichment

Citation: Cuffaro, D.; Bertini, S.; Macchia, M.; Digiacomo, M. Enhanced Nutraceutical Properties of Extra Virgin Olive Oil Extract by Olive Leaf Enrichment. *Nutrients* **2023**, *15*, 1073. https://doi.org/10.3390/nu15051073

Academic Editor: Qun Shen

Received: 19 January 2023
Revised: 14 February 2023
Accepted: 17 February 2023
Published: 21 February 2023

1. Introduction

The Mediterranean diet plays a salient role in the primary prevention of chronic disorders, as documented by several studies [1,2]. The healthy properties of the Mediterranean diet are correlated with high consumption of extra virgin olive oil (EVOO), which is rich in polyphenols. In EVOO, the high content of phenolic compounds exerts, among other things, antioxidant, anti-inflammatory, anti-cancer, and antimicrobial effects [3–5]. Nevertheless, polyphenols are contained in EVOO in small quantities, less than 2%, as the main constituents of EVOO are triglycerides. Among EVOO polyphenols, the simple phenols such as tyrosol and hydroxytyrosol and the secoiridoids can be distinguished. Secoiridoids constitute the largest family of polyphenols in EVOO, and the main representative compounds are oleacein and oleocanthal, endowed with important nutraceutical properties [6,7]. Beauchamp et al. reported the anti-inflammatory activity of oleocanthal as comparable to that of ibuprofen, a known non-steroidal anti-inflammatory drug (NSAID) [8]. Moreover, oleocanthal showed other important biological activities, such as anticancer activity [9,10], an anti-Alzheimer effect [11,12], and a protective role in arthropathy [13] and cardiovascular diseases [14]. Oleacein, similarly to oleocanthal, decreases cyclooxygenase activity, thereby reducing inflammation [15,16], and is responsible for the anti-sclerotic effect attributed to EVOO [17]. Furthermore, oleacein revealed in vitro activity against some types of cancer [18] and an anti-estrogenic effect [19].

In the last decade, the EVOO industry's production has grown considerably, causing a significant environmental impact in terms of waste generation. Olive leaves are the primary source of waste generation, accounting for 10% of harvested olives by weight. Olive leaves are an abundant but unavoidable waste that accumulates in high amounts during the tree-pruning process of olive trees [20]. It is estimated that for every liter of produced EVOO, 6.23 kg of pruning residues (consisting of leaves and branches) are generated [21], and for example, in Spain every year an average of 750,000 tons of olive leaves have been gathered [22]. Considering the huge quantities of leaves, many efforts to use them and obtain energy or nutraceutical molecules from them have been made. Likewise, olive leaves represent the main site of plant metabolism, where photosynthesis takes place, generating a valuable source of primary and secondary plant-derived products. Therefore, bioactive compounds, including polyphenols such as oleuropein, hydroxytyrosol, and verbascoside, and flavonoids such as luteolin and apigenin, are amply present in olive leaves. Olive leaves have been shown to have health benefits such as antioxidant, antimicrobial, and anti-atherosclerosis effects, owing to the presence of oleuropein [7,23,24]. Indeed, in vitro and in vivo studies underlined a plethora of properties for oleuropein, such as the antioxidant, antimicrobial, antifungal, anti-tumoral, hypolipidemic, and especially hypotensive, anticancer, and cardioprotective actions [25–29]. Moreover, the anti-inflammatory effect of oleuropein in terms of pro-inflammatory agent inhibition, and especially cyclooxygenase (COX) inhibition, is consistently reported by several studies [30,31]. Oleuropein content is significantly high in olive leaves, in sharp contrast to EVOO, where it is present only marginally. Therefore, enriching the polyphenol content of EVOO extract by adding olive leaf extract (OLE) as a source of oleuropein may promote synergistic interactions between nutraceutical polyphenols, increasing EVOO's health promoting properties. In fact, it is known that the combination of extracts can lead to better chemotherapeutic and chemopreventive effects than either extract used alone, thanks to synergistic interactions between the components [32].

Starting from these evidences, in this work we propose the study of new EVOO/OLE extracts obtained by adding an oleouropein-rich OLE to EVOO extract in different ratios, with the aim of increasing the polyphenol content and ameliorating the biological activities of EVOO. In the literature, some studies investigating the variation in polyphenolic composition after the addition of olive leaves to EVOO have already been reported, focusing mainly on the extraction method, composition, and quality of EVOO [33,34]. In this work, our attention is focused on the evaluation of how the nutraceutical properties of EVOO could be affected by the addition of OLE. In particular, we investigate the antioxidant and anti-inflammatory properties of EVOO/OLE 8% extract, which presents a high content of some of the most important nutraceutical polyphenols: oleocanthal, oleacein, and oleuropein (Figure 1). This new approach might be a proper strategy to valorize olive leaves as a sustainable alternative use, converting them into higher-value by-products. This study well aligns with the purposes of the Recovery and Resilience Plan regarding the re-evaluation of biowastes derived from the food supply chain, which significantly affect the environment.

Figure 1. Chemical structures of the principal polyphenols of EVOO and OLE extracts.

2. Materials and Methods

2.1. Chemicals and Standards

Solvents used for extraction procedures and HPLC analyses were purchased from Merck (Merck srl, Milan, Italy). The pure standards oleocanthal and oleacein were obtained through EVOO extraction and purification using the method described in our previous work [35]. The following commercial compounds are used as analytical standards: oleuropein, *p*-hydroxyphenylacetic acid, and Trolox, purchased from Merck (Merck srl, Milan, Italy); tyrosol and hydroxytyrosol purchased from TCI (Zwijndrecht, Belgium); and luteolin-7-*O*-glucoside and apigenin-7-*O*-glucoside purchased from Extrasynthese (Lyon, France). Folin–Ciocalteu reagent (FCR), 2,2-diphenyl-1-picrylhydrazyl (DPPH) 2,2′-azino-bis(3-ethylbenzothiazoline-6-sulfonic acid (ABTS), and 2,4,6-Tris(2-pyridyl)-s-triazine (TPTZ) were purchased from Merk (Darmstadt, Germany).

2.2. Preparation of Samples

2.2.1. Extra Virgin Olive Oil Extract

An EVOO sample (Moraiolo, Frantoio, and Leccino varieties) produced in the 2020/2021 crop season was used as an EVOO sample. The EVOO extract was prepared as previously described [10]. Briefly, to the EVOO sample (3 g) n-hexane (12 mL) and acetonitrile (15 mL) were added. After homogenization in a vortex mixer for 30 s and a rotary shaker for 30 min, the mixture was centrifugated at 4000 rpm for 5 min. Then the acetonitrile phase was separated and evaporated to afford EVOO extract.

2.2.2. Olive Leaf Extract

Olive leaves were derived from Olivastra seggianese groves located at CNR-IVALSA, Follonica (GR), Italy. The collection was performed manually in September 2019 and stored at 25 °C. After harvesting, 20 g of leaves were put in liquid nitrogen and crushed manually. Afterwards, water was added to the powdered leaves and the solution was sonicated and mixed by vortex. Then the solution was centrifugated at 4000 rpm for 5 min at 25 °C, and the water phase was filtered and freeze-dried, affording OLE.

2.2.3. Extra Virgin Olive Oil/Olive Leaf Extract

A solution of EVOO extract in MeOH (20 mg/mL) and of OLE in water (2 mg/mL) have been prepared. 8%, 4%, 2%, and 1% of OLE water solution have been added to a methanolic EVOO solution, obtaining the corresponding EVOO/OLE extracts.

2.3. Determination of Total Phenolic Content

The total phenolic content (TPC) of the extracts was evaluated using FCR as previously described [36]. Briefly, 2.5 g of the EVOO sample were dissolved in 5 mL of n-hexane and subsequently extracted with 5 mL of MeOH (80% *v*/*v*). The methanolic phase was collected, obtaining 10 mL of methanolic extract. EVOO-OLE 8% and OLE sample extracts have been prepared as reported in Sections 2.2.2 and 2.2.3, respectively. After evaporation, the resulting dry extract was dissolved in 1 mL of a solution of methanol (80% *v*/*v*).

A total of 0.25 mL of FCR and 1.5 mL of Na_2CO_3 (20% *w*/*v*) were added to 1 mL of the methanolic solution in a volumetric flask, and then distilled water was added up to 10 mL. The solution was incubated for 45 min at 25 °C, and then the absorbance was read at $\lambda = 725$ nm. TPC was calculated using a gallic acid calibration curve (2.5–40.0 μg/mL) and expressed as mg of gallic acid (GA) equivalent/kg of sample (ppm). mg gallic acid equivalent). Analyses were performed in triplicate, and the mean value was calculated for each sample.

2.4. Analysis of Phenolic Compounds

HPLC analysis of samples, prepared as illustrated in Section 2.3, was carried out using a slightly revised method reported in our previous studies [35,37]. The extracts were injected as a mixture of $MeOH/H_2O$ (1:1 *v*/*v*) in a Shimadzu HPLC Nexera series (model CBM-40D), which consisted of a binary pump (LC-40D XR), a degassing unit (DGU-405), and a diode array detector (SPD-M40) (Shimadzu, OR, USA). The data processing was performed on the Shimadzu LabSolutions software LC-GC. HPLC analysis was performed using a Phenomenex Gemini reverse-phase C18 column (250 × 4.6 mm, 5 μm particle size; Phenomenex, Castel Maggiore, Italy). A *p*-hydroxyphenylacetic acid was chosen as an internal standard. In the mobile phase, a mixture of H_2O/AcOH (97.5:2.5 *v*/*v*) (A) and ACN/MeOH (1:1 *v*/*v*) (B) was used. The linear gradient progressed from 5% (B) to 30% (B) in 45 min; it changed to 70% (B) during 20 min (65 min total time); in 5 min it changed to 80% (B) (70 min total time); it remained at 80% (B) for 15 min (85 min total time); it changed to 100% (B) in 5 min (90 min total time); after re-equilibration for 5 min (95 min total time) to initial composition, it remained at 5% (B) for 10 min (105 min total time). The flow rate was 1 mL/min, and the injected volume was 20.0 μL.

2.5. *DPPH• Radical Scavenging Activity*

The antioxidant activity of OLE, EVOO, and EVOO/OLE extracts was assessed through the DPPH• free radical scavenging assay, which was modified slightly from Brand–Willians [38]. A methanol solution of DPPH• (40.0 μg/mL) was added to the samples solubilized in MeOH at different concentrations (1 mg/mL–6 mg/mL for EVOO extract, 0.3 mg/mL–1 mg/mL for EVOO/OLE extract, and 0.3–0.025 mg/mL for OLE). After 45 min of incubation at room temperature and in the dark, the absorbance was read at 517 nm in a SPECTROstarNano (200–1000 nm) UV/Vis spectrophotometer (BMG Labtec, Germany). MeOH was used as a blank, and Trolox® was used as a positive control (0.015–0.0005 mg/mL) and treated under the same conditions as the samples. The percent of antioxidant activity (%AA) was calculated according to the following formula:

$$\%AA = (Abs_{DPPH}) - (Abs_{sample})/Abs_{DPPH} \times 100$$

Abs_{DPPH} = absorbance of DPPH solution, subtracted from the absorbance of MeOH

Abs_{sample} = absorbance of DPPH solution including the test compound subtracted from the absorbance of test compound solution without DPPH.

The results were expressed as the efficient inhibitory concentration of antioxidants necessary to decrease the initial DPPH concentration by 50% (EC_{50}). EC_{50} has been calculated by linear regression, as already reported [39]. All experiments were performed in triplicate.

2.6. *ABTS•+ Radical Scavenging Activity*

The free radical scavenging activity of OLE and EVO, EVO-OLE 8% extracts was assessed by using ABTS radical cation decolouration assay, which was modified from the protocol reported by Pellegrini et al. [40]. Briefly, an ABTS solution was prepared by mixing an aqueous solution of ABTS (7 mM) with potassium persulfate (2.45 mM) in a 1:1 ratio. The solution was incubated for 12 h in the dark at room temperature, then it was diluted with water to an absorbance of 0.7 at 750 nm. 190 μL of ABTS solution were mixed with 10 μL of sample dissolved in EtOH. The solution was incubated for 5 min at room temperature, and the final absorbance was read at 734 nm in a SPECTROstarNano (200–1000 nm) UV/Vis spectrophotometer (BMG Labtec, Germany). Calculations were performed to evaluate the percentage of inhibition of the ABTS radical cation as follows:

$$\%\ \text{scavenging ability} = (Abs_{ABTS} - Abs_{sample})/Abs_{ABTS} \times 100$$

Abs_{ABTS} = the absorbance of the ABTS solution

Abs_{sample} = the absorbance of the ABTS solution containing the test compound. (1)

The percentage of scavenging ability was calculated against the sample concentration to give the efficient concentration to decrease the initial ABTS concentration by 50% (EC_{50}). All experiments were performed in triplicate.

2.7. Ferric Reducing Antioxidant Power Assay

The method described by Borges et al. [41], was used to assess the antioxidant activity. Ferric Reducing Antioxidant Power (FRAP) assay measures the ferric-reducing ability of a sample in an acid medium (pH 3.6) through the formation of a specific blue color as the ferric tripyridyltriazine (Fe^{3+}–TPTZ) complex, caused by reduction to the ferrous (Fe^{2+}) form. The FRAP reagent was obtained by mixing acetate sodium buffer (0.3 M) at pH 3.6, ferric chloride (20 mM), and TPTZ (10 mM) in HCl (40 mM) in a ratio of 10:1:1. 20 μL of extracts (OLE, EVO, or EVO/OLE 8%) were mixed with FRAP solution (280 μL), and the mixture was incubated at 37 °C for 30 min. The absorbance of the reaction mixture was read at 595 nm in a SPECTROstarNano (200–1000 nm) UV/Vis spectrophotometer (BMG Labtec, Ortenberg, Germany). The calibration curve was built using different concentrations of Trolox® (0.01–0.2 mg mL^{-1}), and the results are expressed as mmol of Trolox equivalents per kg of the sample. All experiments were performed in triplicate.

2.8. Cyclooxygenase Enzyme Inhibitory Assay

The ability of OLE, EVOO, and EVOO/OLE extracts to inhibit COX-1 and COX-2 was evaluated using a COX-1 (ovine) and COX-2 (human)-inhibitor screening assay (kit No. 701050 from Cayman Chemical Co. Michigan, USA), following the manufacturer's protocols. An initial COX-1/COX-2 inhibitory evaluation test was performed at 225 μg/mL for all the extracts (OLE, EVO, and EVO/OLE 8%). Then, increasing concentrations of each sample (45–225 μg/mL) have been tested. Arachidonic acid at 1.1 mM was the substrate, and ibuprofene was used as a control. The peroxidase activity was examined colorimetrically at 590 nm after an incubation of 120 min at room temperature using a SPECTROstarNano (200–1000 nm) UV/Vis spectrophotometer (BMG Labtec, Germany). All tests were performed three times. The percent (%) inhibition of COX-1 and COX-2 is derived from the following formula:

$$\% \text{ inhibition} = (\text{EAA} - \text{AIA})/\text{EAA} \times 100$$

$$\text{EAA} = \text{Enzyme test activity absorbance}$$

$$\text{AIA} = \text{Activity inhibition test absorbance.} \quad (2)$$

Results were expressed as % of inhibition or inhibitory concentration at 50% (IC_{50}) calculated by least-squares regression analysis of inhibition versus concentration, as already reported [8,42,43]. All experiments were performed in triplicate.

2.9. Statistical Analysis

Data were presented as the mean ± standard deviation of three independent experiments. Graphpad 9.0 has been used to investigate the statistical differences among results. A one-way analysis of variance (ANOVA) was applied to determine the differences between samples, and Turkey's multiple-comparison test was used as a post hoc comparison of the means. A denoting significance was accepted for $p < 0.05$.

3. Results

3.1. Phenolic Compound Content of Extracts

The phenolic contents of EVOO, OLE, and EVOO/OLE have been evaluated by qualitative and quantitative HPLC analysis. The results are reported in Table 1. From HPLC analysis of EVOO, it emerged that it contains high amounts of oleacein (11.05 mg/g EVOO extract) and oleocanthal (20.97 mg/g EVOO extract) and a small quantity of hydroxytyrosol and tyrosol (0.72 and 1.02 mg/g EVOO extract, respectively) as expected for a fresh EVOO [37]. Regarding OLE, the most representative phenolic compound is oleuropein (35.58 mg/g OLE), followed by luteolin-7-*O*-glucoside and apigenin-7-*O*-glucoside (6.70 and 1.80 mg/g OLE, respectively), as confirmed by the literature [44]. The enriched EVOO/OLE extracts maintained high quantities of oleocanthal and oleacein with increasing amounts of oleuropein and other polyphenols usually not present in EVOO, such as luteolin-7-O-glucoside and apigenin-7-O-glucoside. EVOO/OLE 8% presented the best profile in terms of phenolic compound content, with a balanced amount of oleacein (10.84 mg/g), oleocanthal (19.03 mg/g), and oleuropein (8.46 mg/g). Moreover, luteolin-7-O-glucoside, apigenin-7-O-glucoside, tyrosol, and hydroxytyrosol are present.

Table 1. Phenolic compound contents (mg/g of extract) of EVOO/OLE, EVOO and OLE.

Phenolic Compounds	EVOO/OLE				EVOO	OLE
	EVOO/OLE 8%	EVOO/OLE 4%	EVOO/OLE 2%	EVOO/OLE 1%		
tyrosol	0.99 ± 0.058	0.89 ± 0.097	0.92 ± 0.016	0.95 ± 0.006	1.02 ± 0.04	
hydroxytyrosol	0.55 ± 0.029	0.51 ± 0.015	0.49 ± 0.01	0.56± 0.033	0.72 ± 0.025	
oleacein	10.84 ± 0.72	10.77 ± 0.097	10.02 ± 0.46	10.45 ± 0.097	11.05 ± 0.04	
oleocanthal	19.03 ± 0.42	19.98 ± 1.16	20.29 ± 0.16	19.65 ± 0.66	20.97 ± 0.89	
oleuropein	8.46 ± 0.16	2.75 ± 0.003	1.55 ± 0.064	0.80 ± 0.026		35.58 ± 2.21
luteolin-7-O-glucoside	1.80 ± 0.15	0.48 ± 0.028	0.23 ± 0.016	0.13 ± 0.011		6.70 ± 0.61
apigenin 7-O-glucoside	0.65 ± 0.019	0.25 ± 0.021	0.20 ± 0.001	0.15 ± 0.007		1.80 ± 0.035

As reported in Figure 2, TPC concentration of EVOO is 330 mg GA eq/Kg oil. The addition of 8% OLE to the EVOO extract provided an increase in TPC, presenting a new value of 360 mg GA eq/kg oil. Notably, the total phenolic content of OLE is 58.47 mg GA equivalents/g of OLE. It is worth noting that the addition of OLE caused a 10% increase in TPC, confirmed by the HPLC phenolic characterization.

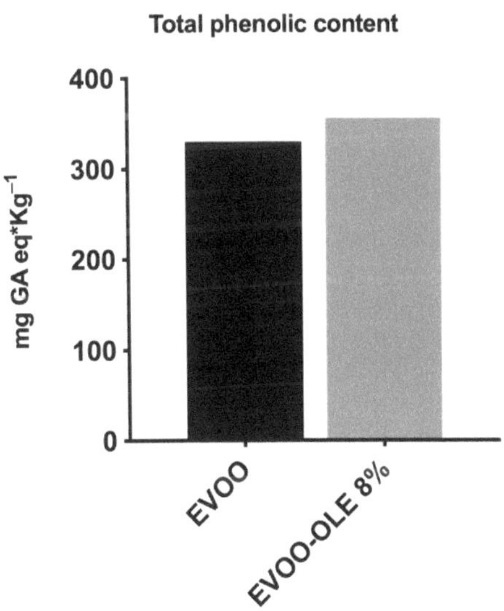

Figure 2. Total phenolic content (TPC) of EVOO and EVOO/OLE 8% extracts.

Considering the role of these polyphenols in the influence of antioxidant and anti-inflammatory properties, EVOO/OLE 8% have been further investigated.

3.2. Antioxidant Activity

Several assays have recently been developed to evaluate the antioxidant capacity of foods. These methods differ both on the basis of antioxidant measurements, which include the formation of different radicals and/or the detection of reduced metal cations, and on how end points are measured. Moreover, considering that the antioxidant compounds may act in vivo through different mechanisms and that quite often the antioxidant effect is due to a combination of the actions of diverse antioxidant polyphenols, a single method can be inadequate to completely evaluate the antioxidant capacity of food [45]. Three in vitro established antioxidant systems (DPPH, ABTS, and FRAP assays) were used to evaluate the antioxidant capacities of OLE, EVOO, and the enriched EVOO/OLE 8% extracts.

The DPPH is normally considered a radical scavenging assay and is one of the most frequently employed single electron transfer-based antioxidant procedures because of its ease of performance, rapidness, automation potential, reproducibility, and usability at ambient temperature [46]. EVOO, OLE, and EVOO/OLE 8% samples were tested in the proper range of concentration, reporting a dose response relationship. The EC_{50} values for DPPH scavenging of all three extracts are reported in Table 2. As expected, OLE showed a good ability to reduce DPPH radical scavenging with an EC_{50} of 0.147 mg/mL (Table 2). The EVOO/OLE 8% extract demonstrated an improvement in radical scavenging activity compared to the EVOO extract. In fact, the EC_{50} of EVOO/OLE 8% is 11 times greater than that of single EVOO. Following the addition of only 8% of OLE, the antioxidant capacity was significantly increased.

Table 2. Antioxidant capacity of EVOO and EVOO/OLE 8% measured by DPPH, ABTS, and FRAP assays.

Sample	DPPH [a] EC_{50} mg/mL	ABTS [b] EC_{50} mg/mL	FRAP [c] mmol Trolox/kg
EVOO	5.55 ± 0.29	1.21 ± 0.13	2.55 ± 0.006
EVOO/OLE 8%	0.50 ± 0.001	0.419 ± 0.04	2.90 ± 0.08
OLE	0.147 ± 0.001	0.055 ± 0.003	1.90 ± 0.04

[a] DPPH: radical scavenging activity assay; [b] ABTS: radical scavenging activity assay. [c] FRAP: ferric reducing ability power assay. [b] Values are the average of three determinations ± standard deviation. Different letters in the same column indicate a significant difference (Turkey test $p > 0.05$).

The ABTS assay measures the efficiency in scavenging the radical cation $ABTS^{\bullet +}$, which was reduced to ABTS, and the data are reported as the efficient concentration to decrease the initial ABTS concentration by 50% (EC_{50}). Similar to DPPH results, OLE reported the strongest antioxidant effects (EC_{50} = 0.055 mg/mL), while the EC_{50} of EVOO/OLE 8% showed an increase of 2-fold with respect to EVOO extract.

The FRAP assay shows a different mechanism of action since it does not implicate a reaction regarding free radicals but rather evaluates the ability to reduce ferric ions (Fe^{3+}) to ferrous ions (Fe^{2+}) [47]. The FRAP value of samples ranged from 1.90 to 2.90 mmol Trolox/kg using this assay. Even in this assay, the addition of OLE conferred improved antioxidant activity to EVOO extract, reporting an enhancement of ferric reducing ability compared to the antioxidant activity of EVOO (Table 2). This improvement is quite interesting because the redox ability of OLE (1.90 mmol Trolox/kg) is similar to that of EVOO (2.55 mmol Trolox/kg).

These results highlight the importance of using different methods to assess the antioxidant activity of an extract. In fact, as already discussed [48], the evaluation of different processes such as radical scavenging activity (DPPH and ABTS) and redox reducing ability (FRAP) offers a complete panel of fundamental antioxidant processes.

3.3. Anti-Inflammatory Profile

In the present study, OLE, EVOO, and EVOO/OLE 8% extracts have been evaluated for their ability to inhibit the catalytic activities of COX-1 and COX-2. The assay allowed the measurement of the enzymatic activity of the peroxidase component of COXs, through the use of a commercial colorimetric kit (Cayman Chemicals). The method was based on the detection of the oxidation of the chromogenic substrate N,N,N′,N′-tetramethyl-*p*-phenylenediamine (TMPD) at $\lambda = 590$ nm. The catalytic activities of the two enzymes were quantified by measuring TMPD formation in the presence of various concentrations of the test compounds, revealing a dose-response inhibition.

OLE, EVOO, and EVOO/OLE 8% extracts were initially tested at 225 μg/mL, evaluating an improved COX-1 and COX-2 anti-inflammatory effect for the enriched EVOO/OLE 8% compared to the EVOO extract and OLE, as reported in Figure 3.

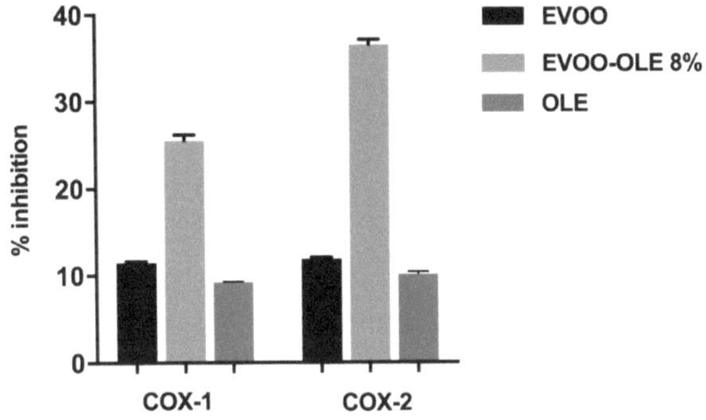

Figure 3. Anti-inflammatory effect at 225 μg/mL of EVO and EVOO/OLE 8% extracts.

The percentage of COX-1 inhibitory activity at 225 μg/mL resulted of 11.5% for EVOO extract, 9.25% for OLE, and 25.5% for EVOO/OLE 8% (Figure 3). Likewise, the percentage of COX-2 inhibitory activity at 225 μg/mL was 11.9% for EVOO extract, 10.1% for OLE, and 36.5% for EVOO/OLE 8%.

For both enzymes, although OLE and EVOO extract demonstrated about 10% inhibition, the EVOO/OLE 8% extract reported more than 25% activity with an enhancement of about 3-fold.

Regarding IC_{50} data, the EVOO/OLE 8% IC_{50} value (IC_{50} COX-1 = 0.475 mg/mL; IC_{50} COX-2 = 0.383 mg/mL, Table 3) is improved by 4-fold in COX-1 inhibition and 2-fold in COX-2 inhibition with respect to the IC_{50} of the EVOO extract (IC_{50} COX-1 = 1.90 mg/mL; IC_{50} COX-2 = 0.90 mg/mL, Table 3) as reported in Table 3.

Table 3. Inhibitory effects of EVOO and EVOO/OLE 8% on COX-1 and COX-2.

Sample	COX 1 IC_{50} mg/mL	COX-2 IC_{50} mg/mL
EVOO	1.90 ± 0.17	0.9 ± 0.01
EVOO/OLE 8%	0.475 ± 0.06	0.383 ± 0.009

Values are the average of three determinations ± standard deviation.

4. Discussion

EVOO represents a functional food with health properties mainly attributed to the presence of phenolic compounds. The most abundant phenolic compounds in EVOO, endowed with nutraceutical properties, are the secoiridoids: oleocanthal and oleacein, and the simple alcohols: tyrosol and hydroxytyrosol. Oleuropein is a secoiridoid with several beneficial health properties and is the most representative polyphenol in olive leaves, a waste by-product derived from the harvesting of olive trees, but unfortunately it is poorly present in EVOO [23]. Therefore, the addition of oleuropein-rich OLE to EVOO extract could represent a high value for EVOO, using oleuropein enrichment in order to improve its nutraceutical properties. The advantage in terms of nutraceutical properties for the EVOO/OLE extract thus obtained could be further enhanced due to beneficial synergistic interactions between the different polyphenols [32]. In this work, we evaluated some extracts obtained by adding different percentages of oleuropein-rich OLE to an EVOO extract. From the qualitative and quantitative study of these extracts, performed by HPLC analysis, we were able to select the EVOO/OLE 8% extract, which has a balanced quantity of the most representative polyphenols (oleocanthal, oleacein, and oleuropein) of almost 10 mg/g of extract, as a starting point to investigate a hypothetical synergistic effect as already reported in other studies [49,50]. The high content of oleacein, oleocanthal, and oleuropein in the 8% extract of EVOO/OLE 8% allowed us to evaluate how the biological activity might be influenced by the interaction of nutraceutical polyphenols.

The EVOO/OLE 8% extract was subjected to further investigations to evaluate its antioxidant and anti-inflammatory properties.

The antioxidant effect is fundamental as a nutraceutical property of a food extract, as it delays oxidative processes and can contribute to the prevention of numerous chronic diseases [51]. EVOO has antioxidant properties that are linked to polyphenols like oleacein and oleocanthal [52]. However, oleuropein, contained in olive leaves, is characterized by strong antioxidant activity, particularly as a free radical scavenger [53]. In this study, we investigated the antioxidant activity in terms of the evaluation of the radical scavenging effect on the DPPH• and ABTS•$^+$ radicals of OLE, EVOO, and EVO/OLE 8% extracts. In both in vitro assays (DPPH and ABTS) used to evaluate the antioxidant capacity, the EVOO/OLE 8% extract displayed an increased antioxidant effect compared to EVOO extract, demonstrating the additional antioxidant action of oleuropein. As expected, OLE reported a high EC_{50} value, as demonstrated by the literature, but it should be noted that with only the 8% of OLE addition, the antioxidant effect of EVOO/OLE 8% is significantly improved over EVOO extract.

It is well known that the antioxidant activity could be due to a combination of different mechanisms; for this reason, we tested OLE, EVOO, and EVOO/OLE 8% extracts also for their redox ability to reduce ferric ions by the FRAP assay. In this assay, all the extracts demonstrated a similar activity, but the improvement of the EVOO/OLE 8% antioxidant effect compared to EVOO extract and OLE was confirmed.

In addition to oxidative stress, several studies assert the role of inflammation in the onset of a wide variety of age-related disorders such as diabetes, cardiovascular disease, cancer, central nervous system-related and autoimmune diseases [54,55]. As previously mentioned, oleocanthal possesses anti-inflammatory properties as it is able to inhibit inflammatory mediators COX-1 and COX-2 similarly to ibuprofen, a known anti-inflammatory drug [8]. Moreover, recently, the anti-inflammatory activity of oleacein in terms of COX inhibition was reported [16]. Consequently, in this study, we investigated the anti-inflammatory properties of oleocanthal-rich EVOO, OLE, and EVOO/OLE 8% extracts, evaluating their capability to inhibit COX-1 and COX-2 enzymes. Preliminarily, all the extracts have been evaluated at 225 μg/mL. The percentage of inhibition reported by OLE and EVOO for both enzymes was quite similar, close to 10% of inhibition (Figure 3). Surprisingly, the EVOO/OLE 8% extract reported a 25.5% COX-1 inhibition and a 36.5% COX-2 inhibition (Figure 3). The interaction among the polyphenols presented in EVOO extract and OLE significantly affects the inhibition of these proinflammatory enzymes.

The high presence of the three most important polyphenols (oleocanthal, oleacein, and oleuropein) in EVOO/OLE 8% extract gave a positive contribution to the antioxidant and anti-inflammatory properties compared to EVOO extract.

It is very important to underline that a small addition of OLE to EVOO extract, which does not much affect the total amount of polyphenols, significantly ameliorates the nutraceutical properties of EVOO.

5. Conclusions

In this study, we investigated EVOO/OLE extracts obtained by adding different percentages of oleuropein-rich OLE to the EVOO extract. The extracts thus obtained are enriched with oleuropein, maintaining a high level of oleocanthal and oleacein typical of EVOO. This addition was found to be significant in terms of nutraceutical properties, as the newly selected EVOO/OLE 8% extract demonstrated improved antioxidant and anti-inflammatory properties compared to the singular EVOO extract.

This new approach represents a strategy for the further enhancement of olive leaves as by-products of great economic and environmental value.

Author Contributions: Conceptualization, M.D. and M.M.; methodology, D.C., M.D. and S.B.; validation, D.C., M.D. and S.B.; formal analysis, D.C., M.D. and S.B.; investigation, D.C., M.D. and S.B.; writing—original draft preparation, D.C.; writing—review and editing, M.D. and S.B.; supervision, M.D. and M.M.; funding acquisition, M.M. All authors have read and agreed to the published version of the manuscript.

Funding: This work is supported by the University of Pisa under the "PRA Progetti di Ricerca di Ateneo" (Institutional Research Grants)—Project No. PRA_2020–2021 "Agenti innovativi e nanosistemi per target molecolari nell'ambito dell'oncologia", and by the Ministry of University and Research (MUR) as a part of the PON 2014–2020 "Research and Innovation" resources—Green/Innovation Action—DM MUR 1062/2021—Title of research "Sviluppo di una piattaforma tecnologica per lo studio delle proprietà nutraceutiche di biomolecole e biomateriali presenti negli scarti derivanti dalla filiera dei prodotti alimentari".

Institutional Review Board Statement: Not applicable.

Informed Consent Statement: Not applicable.

Data Availability Statement: Not applicable.

Conflicts of Interest: The authors declare no conflict of interest.

References

1. Schwingshackl, L.; Morze, J.; Hoffmann, G. Mediterranean Diet and Health Status: Active Ingredients and Pharmacological Mechanisms. *Br. J. Pharmacol.* **2020**, *177*, 1241–1257. [CrossRef]
2. Wahrburg, U.; Kratz, M.; Cullen, P. Mediterranean Diet, Olive Oil and Health. *Eur. J. Lipid Sci. Technol.* **2002**, *104*, 698–705. [CrossRef]
3. Parkinson, L.; Cicerale, S. The Health Benefiting Mechanisms of Virgin Olive Oil Phenolic Compounds. *Molecules* **2016**, *21*, 1734. [CrossRef]
4. Gorzynik-Debicka, M.; Przychodzen, P.; Cappello, F.; Kuban-Jankowska, A.; Marino Gammazza, A.; Knap, N.; Wozniak, M.; Gorska-Ponikowska, M. Potential Health Benefits of Olive Oil and Plant Polyphenols. *Int. J. Mol. Sci.* **2018**, *19*, 686. [CrossRef]
5. Rigacci, S.; Stefani, M. Nutraceutical Properties of Olive Oil Polyphenols. An Itinerary from Cultured Cells through Animal Models to Humans. *Int. J. Mol. Sci.* **2016**, *17*, 843. [CrossRef]
6. Lozano-Castellón, J.; López-Yerena, A.; Rinaldi de Alvarenga, J.F.; Romero del Castillo-Alba, J.; Vallverdú-Queralt, A.; Escribano-Ferrer, E.; Lamuela-Raventós, R.M. Health-Promoting Properties of Oleocanthal and Oleacein: Two Secoiridoids from Extra-Virgin Olive Oil. *Crit. Rev. Food Sci. Nutr.* **2020**, *60*, 2532–2548. [CrossRef]
7. Romani, A.; Ieri, F.; Urciuoli, S.; Noce, A.; Marrone, G.; Nediani, C.; Bernini, R. Health Effects of Phenolic Compounds Found in Extra-Virgin Olive Oil, By-Products, and Leaf of *Olea europaea* L. *Nutrients* **2019**, *11*, 1776. [CrossRef]
8. Beauchamp, G.K.; Keast, R.S.J.; Morel, D.; Lin, J.; Pika, J.; Han, Q.; Lee, C.-H.; Smith, A.B.; Breslin, P.A.S. Phytochemistry: Ibuprofen-like Activity in Extra-Virgin Olive Oil. *Nature* **2005**, *437*, 45–46. [CrossRef]
9. LeGendre, O.; Breslin, P.A.; Foster, D.A. (-)-Oleocanthal Rapidly and Selectively Induces Cancer Cell Death via Lysosomal Membrane Permeabilization. *Mol. Cell. Oncol.* **2015**, *2*, e1006077. [CrossRef]
10. Fogli, S.; Arena, C.; Carpi, S.; Polini, B.; Bertini, S.; Digiacomo, M.; Gado, F.; Saba, A.; Saccomanni, G.; Breschi, M.C.; et al. Cytotoxic Activity of Oleocanthal Isolated from Virgin Olive Oil on Human Melanoma Cells. *Nutr. Cancer* **2016**, *68*, 873–877. [CrossRef]
11. Qosa, H.; Batarseh, Y.S.; Mohyeldin, M.M.; Sayed, K.A.E.; Keller, J.N.; Kaddoumi, A. Oleocanthal Enhances Amyloid-β Clearance from the Brains of TgSwDI Mice and in Vitro across a Human Blood-Brain Barrier Model. *ACS Chem. Neurosci.* **2015**, *6*, 1849. [CrossRef]
12. Monti, M.C.; Margarucci, L.; Riccio, R.; Casapullo, A. Modulation of Tau Protein Fibrillization by Oleocanthal. *J. Nat. Prod.* **2012**, *75*, 1584–1588. [CrossRef]
13. Scotece, M.; Gómez, R.; Conde, J.; Lopez, V.; Gómez-Reino, J.J.; Lago, F.; Smith, A.B.; Gualillo, O. Further Evidence for the Anti-Inflammatory Activity of Oleocanthal: Inhibition of MIP-1α and IL-6 in J774 Macrophages and in ATDC5 Chondrocytes. *Life Sci.* **2012**, *91*, 1229–1235. [CrossRef]
14. Agrawal, K.; Melliou, E.; Li, X.; Pedersen, T.L.; Wang, S.C.; Magiatis, P.; Newman, J.W.; Holt, R.R. Oleocanthal-Rich Extra Virgin Olive Oil Demonstrates Acute Anti-Platelet Effects in Healthy Men in a Randomized Trial. *J. Funct. Foods* **2017**, *36*, 84. [CrossRef]
15. Rosignoli, P.; Fuccelli, R.; Fabiani, R.; Servili, M.; Morozzi, G. Effect of Olive Oil Phenols on the Production of Inflammatory Mediators in Freshly Isolated Human Monocytes. *J. Nutr. Biochem.* **2013**, *24*, 1513–1519. [CrossRef]

16. Costa, V.; Costa, M.; Videira, R.A.; Andrade, P.B.; Paiva-Martins, F. Anti-Inflammatory Activity of Olive Oil Polyphenols—The Role of Oleacein and Its Metabolites. *Biomedicines* **2022**, *10*, 2990. [CrossRef]
17. Naruszewicz, M.; Czerwinska, M.E.; Kiss, A.K. Oleacein. Translation from Mediterranean Diet to Potential Antiatherosclerotic Drug. *Curr. Pharm. Des.* **2015**, *21*, 1205–1212. [CrossRef] [PubMed]
18. Fabiani, R. Anti-Cancer Properties of Olive Oil Secoiridoid Phenols: A Systematic Review of in Vivo Studies. *Food Funct.* **2016**, *7*, 4145–4159. [CrossRef]
19. Keiler, A.M.; Djiogue, S.; Ehrhardt, T.; Zierau, O.; Skaltsounis, L.; Halabalaki, M.; Vollmer, G. Oleocanthal Modulates Estradiol-Induced Gene Expression Involving Estrogen Receptor α. *Planta Med.* **2015**, *81*, 1263–1269. [CrossRef]
20. Khwaldia, K.; Attour, N.; Matthes, J.; Beck, L.; Schmid, M. Olive Byproducts and Their Bioactive Compounds as a Valuable Source for Food Packaging Applications. *Compr. Rev. Food Sci. Food Saf.* **2022**, *21*, 1218–1253. [CrossRef]
21. Avraamides, M.; Fatta, D. Resource Consumption and Emissions from Olive Oil Production: A Life Cycle Inventory Case Study in Cyprus. *J. Clean. Prod.* **2008**, *16*, 809–821. [CrossRef]
22. Manzanares, P.; Ruiz, E.; Ballesteros, M.; Negro, M.J.; Gallego, F.J.; López-Linares, J.C.; Castro, E. Residual Biomass Potential in Olive Tree Cultivation and Olive Oil Industry in Spain: Valorization Proposal in a Biorefinery Context. *Span. J. Agric. Res.* **2017**, *15*, e0206. [CrossRef]
23. Qabaha, K.; AL-Rimawi, F.; Qasem, A.; Naser, S.A. Oleuropein Is Responsible for the Major Anti-Inflammatory Effects of Olive Leaf Extract. *J. Med. Food* **2018**, *21*, 302–305. [CrossRef] [PubMed]
24. Omar, S.H. Oleuropein in Olive and Its Pharmacological Effects. *Sci. Pharm.* **2010**, *78*, 133–154. [CrossRef] [PubMed]
25. Cicerale, S.; Lucas, L.J.; Keast, R.S.J. Antimicrobial, Antioxidant and Anti-Inflammatory Phenolic Activities in Extra Virgin Olive Oil. *Curr. Opin. Biotechnol.* **2012**, *23*, 129–135. [CrossRef]
26. Özcan, M.M.; Matthäus, B. A Review: Benefit and Bioactive Properties of Olive (*Olea europaea* L.) Leaves. *Eur. Food Res. Technol.* **2017**, *243*, 89–99. [CrossRef]
27. Salem, M.B.; Affes, H.; Ksouda, K.; Sahnoun, Z.; Zeghal, K.M.; Hammami, S. Pharmacological Activities of *Olea europaea* Leaves. *J. Food Process. Preserv.* **2015**, *39*, 3128–3136. [CrossRef]
28. De la Ossa, J.G.; Felice, F.; Azimi, B.; Salsano, J.E.; Digiacomo, M.; Macchia, M.; Danti, S.; Di Stefano, R. Waste Autochthonous Tuscan Olive Leaves (*Olea europaea Var. Olivastra seggianese*) as Antioxidant Source for Biomedicine. *Int. J. Mol. Sci.* **2019**, *20*, 5918. [CrossRef]
29. Mao, X.; Xia, B.; Zheng, M.; Zhou, Z. Assessment of the Anti-Inflammatory, Analgesic and Sedative Effects of Oleuropein from *Olea europaea* L. *Cell. Mol. Biol.* **2019**, *65*, 52–55. [CrossRef]
30. Larussa, T.; Oliverio, M.; Suraci, E.; Greco, M.; Placida, R.; Gervasi, S.; Marasco, R.; Imeneo, M.; Paolino, D.; Tucci, L.; et al. Oleuropein Decreases Cyclooxygenase-2 and Interleukin-17 Expression and Attenuates Inflammatory Damage in Colonic Samples from Ulcerative Colitis Patients. *Nutrients* **2017**, *9*, 391. [CrossRef]
31. Procopio, A.; Alcaro, S.; Nardi, M.; Oliverio, M.; Ortuso, F.; Sacchetta, P.; Pieragostino, D.; Sindona, G. Synthesis, Biological Evaluation, and Molecular Modeling of Oleuropein and Its Semisynthetic Derivatives as Cyclooxygenase Inhibitors. *J. Agric. Food Chem.* **2009**, *57*, 11161–11167. [CrossRef]
32. Caesar, L.K.; Cech, N.B. Synergy and Antagonism in Natural Product Extracts: When 1 + 1 Does Not Equal 2. *Nat. Prod. Rep.* **2019**, *36*, 869–888. [CrossRef] [PubMed]
33. Japón-Luján, R.; Luque de Castro, M.D. Liquid–Liquid Extraction for the Enrichment of Edible Oils with Phenols from Olive Leaf Extracts. *J. Agric. Food Chem.* **2008**, *56*, 2505–2511. [CrossRef] [PubMed]
34. Malheiro, R.; Casal, S.; Teixeira, H.; Bento, A.; Pereira, J.A. Effect of Olive Leaves Addition during the Extraction Process of Overmature Fruits on Olive Oil Quality. *Food Bioprocess Technol.* **2013**, *6*, 509–521. [CrossRef]
35. Palla, M.; Digiacomo, M.; Cristani, C.; Bertini, S.; Giovannetti, M.; Macchia, M.; Manera, C.; Agnolucci, M. Composition of Health-Promoting Phenolic Compounds in Two Extra Virgin Olive Oils and Diversity of Associated Yeasts. *J. Food Compos. Anal.* **2018**, *74*, 27–33. [CrossRef]
36. Alessandri, S.; Ieri, F.; Romani, A. Minor Polar Compounds in Extra Virgin Olive Oil: Correlation between HPLC-DAD-MS and the Folin-Ciocalteu Spectrophotometric Method. *J. Agric. Food Chem.* **2014**, *62*, 826–835. [CrossRef]
37. Esposito Salsano, J.; Digiacomo, M.; Cuffaro, D.; Bertini, S.; Macchia, M. Content Variations in Oleocanthalic Acid and Other Phenolic Compounds in Extra-Virgin Olive Oil during Storage. *Foods* **2022**, *11*, 1354. [CrossRef]
38. Brand-Williams, W.; Cuvelier, M.E.; Berset, C. Use of a Free Radical Method to Evaluate Antioxidant Activity. *LWT-Food Sci. Technol.* **1995**, *28*, 25–30. [CrossRef]
39. Marxen, K.; Vanselow, K.H.; Lippemeier, S.; Hintze, R.; Ruser, A.; Hansen, U.-P. Determination of DPPH Radical Oxidation Caused by Methanolic Extracts of Some Microalgal Species by Linear Regression Analysis of Spectrophotometric Measurements. *Sensors* **2007**, *7*, 2080–2095. [CrossRef]
40. Pellegrini, N.; Del Rio, D.; Colombi, B.; Bianchi, M.; Brighenti, F. Application of the 2,2′-Azinobis(3-Ethylbenzothiazoline-6-Sulfonic Acid) Radical Cation Assay to a Flow Injection System for the Evaluation of Antioxidant Activity of Some Pure Compounds and Beverages. *J. Agric. Food Chem.* **2003**, *51*, 260–264. [CrossRef]
41. Borges, T.H.; Cabrera-Vique, C.; Seiquer, I. Antioxidant Properties of Chemical Extracts and Bioaccessible Fractions Obtained from Six Spanish Monovarietal Extra Virgin Olive Oils: Assays in Caco-2 Cells. *Food Funct.* **2015**, *6*, 2375–2383. [CrossRef] [PubMed]

42. Columba-Palomares, M.F.M.C.; Villareal, D.M.L.; Acevedo Quiroz, M.C.M.E.; Marquina Bahena, M.C.S.; Álvarez Berber, D.L.P.; Rodríguez-López, D.V. Anti-Inflammatory and Cytotoxic Activities of Bursera Copallifera. *Pharmacogn. Mag.* **2015**, *11*, S322–S328. [CrossRef] [PubMed]
43. Lopes, A.J.O.; Vasconcelos, C.C.; Garcia, J.B.S.; Pinheiro, M.S.D.; Pereira, F.A.N.; de Sousa Camelo, S.; de Morais, S.V.; Freitas, J.R.B.; da Rocha, C.Q.; de Sousa Ribeiro, M.N.; et al. Anti-Inflammatory and Antioxidant Activity of Pollen Extract Collected by Scaptotrigona Affinis Postica: In Silico, in Vitro, and in Vivo Studies. *Antioxidants* **2020**, *9*, 103. [CrossRef] [PubMed]
44. Guinda, Á.; Castellano, J.M.; Santos-Lozano, J.M.; Delgado-Hervás, T.; Gutiérrez-Adánez, P.; Rada, M. Determination of Major Bioactive Compounds from Olive Leaf. *LWT-Food Sci. Technol.* **2015**, *64*, 431–438. [CrossRef]
45. Em, B.; Lr, N.; Lh, S. Antioxidant Evaluation Protocols: Food Quality or Health Effects. *Eur. Food Res. Technol.* **2004**, *219*, 561–571. [CrossRef]
46. Kedare, S.B.; Singh, R.P. Genesis and Development of DPPH Method of Antioxidant Assay. *J. Food Sci. Technol.* **2011**, *48*, 412–422. [CrossRef]
47. Benzie, I.F.; Strain, J.J. The Ferric Reducing Ability of Plasma (FRAP) as a Measure of "Antioxidant Power": The FRAP Assay. *Anal. Biochem.* **1996**, *239*, 70–76. [CrossRef]
48. Shah, P.; Modi, H. Comparative Study of DPPH, ABTS and FRAP Assays for Determination of Antioxidant Activity. *Int. J. Res. Appl. Sci. Eng. Technol.* **2015**, *3*, 636–641.
49. Hajimehdipoor, H.; Shahrestani, R.; Shekarchi, M. Investigating the Synergistic Antioxidant Effects of Some Flavonoid and Phenolic Compounds. *Res. J. Pharmacogn.* **2014**, *1*, 35–40.
50. Amin, M.U.; Khurram, M.; Khattak, B.; Khan, J. Antibiotic Additive and Synergistic Action of Rutin, Morin and Quercetin against Methicillin Resistant Staphylococcus Aureus. *BMC Complement. Altern. Med.* **2015**, *15*, 59. [CrossRef] [PubMed]
51. Willcox, J.K.; Ash, S.L.; Catignani, G.L. Antioxidants and Prevention of Chronic Disease. *Crit. Rev. Food Sci. Nutr.* **2004**, *44*, 275–295. [CrossRef]
52. Servili, M.; Sordini, B.; Esposto, S.; Urbani, S.; Veneziani, G.; Di Maio, I.; Selvaggini, R.; Taticchi, A. Biological Activities of Phenolic Compounds of Extra Virgin Olive Oil. *Antioxidants* **2013**, *3*, 1–23. [CrossRef] [PubMed]
53. Borjan, D.; Leitgeb, M.; Knez, Ž.; Hrnčič, M.K. Microbiological and Antioxidant Activity of Phenolic Compounds in Olive Leaf Extract. *Molecules* **2020**, *25*, 5946. [CrossRef] [PubMed]
54. Franceschi, C.; Campisi, J. Chronic Inflammation (Inflammaging) and Its Potential Contribution to Age-Associated Diseases. *J. Gerontol. A Biol. Sci. Med. Sci.* **2014**, *69*, S4–S9. [CrossRef] [PubMed]
55. Khansari, N.; Shakiba, Y.; Mahmoudi, M. Chronic Inflammation and Oxidative Stress as a Major Cause of Age-Related Diseases and Cancer. *Recent Pat. Inflamm. Allergy Drug Discov.* **2009**, *3*, 73–80. [CrossRef]

nutrients

Article

Citri Reticulatae Pericarpium (Chenpi) Protects against Endothelial Dysfunction and Vascular Inflammation in Diabetic Rats

Yuehan Wang [1,†], Xutao Zhang [1,†], Chunxiu Zhou [1], Haroon Khan [2], Manqin Fu [3,*] and Wai San Cheang [1,*]

1 Institute of Chinese Medical Sciences, State Key Laboratory of Quality Research in Chinese Medicine, University of Macau, Macao SAR 999078, China
2 Department of Pharmacy, Abdul Wali Khan University Mardan, Mardan 23200, Pakistan
3 Sericultural & Agri-Food Research Institute, Guangdong Academy of Agricultural Sciences/Key Laboratory of Functional Foods, Ministry of Agriculture and Rural Affairs/Guangdong Key Laboratory of Agricultural Products Processing, Guangzhou 510610, China
* Correspondence: fumanqin84@126.com (M.F.); annacheang@um.edu.mo (W.S.C.)
† These authors contributed equally to this work.

Abstract: Dried tangerine peel (*Citri reticulatae Pericarpium*, CRP; Chenpi in Chinese) possesses medicine and food homology with hypolipidemic, anti-inflammatory and antioxidant activities. This study aimed to explore the protective effect of CRP extract on endothelial function and inflammation in type 2 diabetic rats and the related mechanisms. Type 2 diabetes mellitus was induced by high-fat diet (HFD)/streptozotocin (STZ) in male Sprague Dawley rats, and CRP extract was orally administered at 400 mg/kg/day for 4 weeks. Rat and mouse aortas were treated with high glucose and CRP extract ex vivo. The data showed that the ethanolic extract of CRP normalized blood pressure and the plasma lipid profile as well as the plasma levels of liver enzymes in diabetic rats. Impaired endothelium-dependent relaxations in aortas, carotid arteries and renal arteries were improved. CRP extract suppressed vascular inflammatory markers and induced AMPK activation in aortas of diabetic rats. Exposure to high glucose impaired vasodilation in aortas of rats and mice, and this impairment was prevented by co-incubation with CRP extract. In conclusion, our findings suggest that CRP extract protects endothelial function by inhibiting the vascular inflammatory state on activation of AMPK in diabetic rats.

Keywords: tangerine peel; diabetes mellitus; endothelial dysfunction; inflammation; AMPK

Citation: Wang, Y.; Zhang, X.; Zhou, C.; Khan, H.; Fu, M.; Cheang, W.S. *Citri Reticulatae Pericarpium* (Chenpi) Protects against Endothelial Dysfunction and Vascular Inflammation in Diabetic Rats. *Nutrients* **2022**, *14*, 5221. https://doi.org/10.3390/nu14245221

Academic Editors: Maria Digiacomo and Doretta Cuffaro

Received: 10 November 2022
Accepted: 6 December 2022
Published: 7 December 2022

1. Introduction

Cardiovascular disease is the leading cause of mortality and morbidity in patients with diabetes mellitus [1]. In addition, patients with chronic inflammatory diseases have higher chances of developing atherosclerosis and cardiovascular diseases, and diabetes mellitus as an inflammatory disease causes a two-fold increase in the risk of death from coronary heart disease, stroke and other vascular diseases [2]. Endothelial dysfunction and inflammation are commonly considered as interrelated and are crucial factors in the development of cardiovascular complications during diabetes progression [3]. The hyperglycemic, hypertensive, and hyperlipidemic conditions under diabetes contribute to vascular inflammation [4,5]. Inflammatory burden thereby leads to endothelial dysfunction, impairing endothelium-dependent dilatation [6].

High glucose levels in the blood circulation induce the expression of the pro-inflammatory enzymes inducible nitric oxide synthase (iNOS) and cyclooxygenase 2 (COX-2), which leads to endothelial cell apoptosis, disrupting endothelial function [7]. The adhesion molecules intercellular adhesion molecule 1 (ICAM-1) and vascular cell adhesion protein 1 (VCAM-1) are highly expressed in endothelial cells, accompanied by an increased density of inflammatory infiltrating cells during the vascular inflammation associated with the progression of cardiovascular diseases [8]. Adenosine monophosphate-activated protein kinase (AMPK) is a critical energy sensor and bioenergy regulator [9]. Moreover, extensive evidence supports that the activation of AMPK protects vascular function and suppresses vascular inflammation in diabetes through multiple pathways [10–12].

Dried tangerine peel (*Citrus reticulatae Pericarpium*, CRP; *Chenpi* in Chinese), is a traditional Chinese medicine with medicinal and food properties. It has the effects of regulating qi, invigorating the spleen, regulating dryness and dampness, and resolving phlegm [13]. Its main components are volatile oils such as d-limonene and flavonoid components such as hesperidin, naringenin, nobiletin, tangeretin, etc. In addition, polysaccharides, alkaloids and other components are also present [14]. Modern pharmacological studies have demonstrated that tangerine peel has various beneficial effects including hepatoprotective [15], anti-asthma [16], antioxidant [17,18], anti-inflammatory [19], and hypolipidemic activities [20]. In addition, flavonoids from tangerine peel were found to ameliorate metabolic and vascular dysfunction. Flavonoids including naringenin, naringin and nobiletin possess anti-atherogenic effects [21]. Nobiletin also exhibits vasodilatory effects in rat aortas [22]. To date, the possible protective effect of consuming dried tangerine peel in alleviating vascular inflammation and endothelial dysfunction in diabetes remains to be investigated.

Feeding with a high-fat diet (HFD) together with low doses of streptozotocin (STZ) is shown to induce type 2 diabetes in rats [23]. In this study, we aimed to investigate whether CRP extracts have a protective effect against endothelial dysfunction and vascular inflammation associated with diabetes in a rat model.

2. Materials and Methods

2.1. Preparation of CRP Extract

The tangerine peels of *Citrus reticulatae Pericarpium* (CRP) were collected in Xinhui, Guangdong Province, China, in 2016. The fresh peels were dried in a heat pump dryer (GHRH-20, Guangdong Agri-machinery Research Institute, Guangzhou, China) at 45 °C. Then the milled powder of tangerine peels was ultrasonically extracted 3 times with 80% EtOH (1:10 *w*/*v*) for 30 min each time. The filtrates were combined, and the solvent was recovered by rotary evaporation under reduced pressure, followed by lyophilisation with a Virtis Freeze Dryer (Te Virtis Company, New York, NY, USA) to obtain the final freeze-dried powder. Characterization of chemical components in CRP extract was determined by ultra-performance liquid chromatography-mass spectrometry (UPLC-MS) and the contents of hesperidin and hesperetin were quantified as reported in our previous study [19].

2.2. Animal Protocols

A total of 15 male Sprague Dawley (SD) rats at 8 weeks old were randomly divided into 3 groups, 5 rats for each group. All the rats were housed at a constant temperature (22–23 °C) with alternating 12-h light/12-h dark cycle with water ad libitum. The rats in the control group were fed a normal rodent diet; while rats in the diabetic (DM) and treatment (DM + CRP) groups were fed on a high-fat diet containing 45 kcal% fat for 12 weeks plus a single injection of low-dose streptozotocin (STZ, 35 mg/kg body weight, i.p.) at week 10 to induce type 2 diabetes. Then, rats in the treatment group were given CRP aqueous solution (400 mg/kg body weight) once daily for 4 weeks by oral gavage, whereas rats in DM and control groups were given distilled water. The body weights of the rats were measured biweekly. The weights of normal chow diet and high-fat diet consumed by each group of rats were also measured.

2.3. Oral Glucose Tolerance Test (OGTT)

The rats were loaded with glucose aqueous solution (2 g/kg body weight) by oral gavage after 12-h fasting. Afterwards, blood was taken from the tails of the rats to measure the blood glucose levels at 0, 15, 30, 60, 90 and 120 min by commercial glucometer (Jiangsu Yuyue Medical Equipment and Supply Co. Ltd., Zhenjiang, China). OGTT was performed at the end point of the experiment (after 4-week CRP administration).

2.4. Blood Pressure Measurement

Systolic (SBP) and diastolic (DBP) blood pressures were measured by the CODA non-invasive blood pressure system (a tail-cuff method, Kent Scientific Corporation, Torrington, CT, USA) in the three groups of conscious rats at the end of the experiment. The values of SBP and DBP were reported as the average of five successive measurements.

2.5. Determination of Plasma Lipid Profile

After the rats were sacrificed by CO_2 inhalation, blood was drawn from the inferior vena cava and collected in heparin-coated microcentrifuge tubes. Plasma was obtained after centrifugation at 3000 rpm at 4 °C for 10 min and stored at −80 °C until further analysis. Plasma levels of total cholesterol (TC), triglyceride (TG), high-density lipoprotein cholesterol (HDL-C) and low-density lipoprotein cholesterol (LDL-C) were quantified using commercial kits according to the manufacturer's instructions (Nanjing Jiancheng Bioengineering Institute, Nanjing, China, catalog numbers A111-1-1, A110-1-1, A112-1-1 and A113-1-1).

2.6. Determination of Liver Function Biomarkers in Plasma

The plasma levels of alanine aminotransferase (ALT) and aspartate aminotransferase (AST) as biomarkers for liver damage were measured using commercial kits according to the manufacturer's instructions (Nanjing Jiancheng Bioengineering Institute, Nanjing, China, catalog numbers C009-2-1 and C010-2-1).

2.7. Ex vivo Culture of Rat and Mouse Aortas

Rat thoracic aortic segments were isolated and dissected free of adhering connective tissue in sterile phosphate buffered saline (PBS) and subsequently incubated in RPMI 1640 medium with additions of 10% fetal bovine serum (FBS) and 1% penicillin/streptomycin (Gibco, Grand Island, NY, USA), at 37 °C in a 5% CO_2 environment. The aortic segments were supplemented with high glucose (44 mM) for 5 h to mimic hyperglycemic conditions in diabetes, while some were co-treated with 400 μg/mL CRP extract. The control group was cultured at the normal glucose level of the culture medium (11 mM) with addition of mannitol at the same volume as the osmotic control.

Thoracic aortic segments from C57BL/6J mice were also isolated in sterile PBS and subsequently incubated in Dulbecco's Modified Eagle's Medium (DMEM) with 10% FBS and 1% penicillin/streptomycin at 37 °C in a 5% CO_2 environment. Some of the aortas were additionally supplemented with 30 mM glucose to simulate a high-glucose environment, with or without the co-treatment of 400 μg/mL CRP extract. The control group was cultured at the normal glucose level of the culture medium (5.55 mM) plus the same volume of mannitol as the osmotic control.

2.8. Isometric Force Measurement in Wire Myograph

Arterial segments including aortas, carotid arteries and renal arteries from the SD rats as well as aortas from C57BL/6J mice were cut to ~ 2 mm in length and were suspended in a Multi Myograph System (Danish Myo Technology, Denmark) for a functional study to detect changes in isometric tension. The blood vessel segments were stretched to optimal baseline tension and were equilibrated for 60 min, followed by induction of vasoconstriction using 60 mM KCl. Endothelium-dependent relaxations (EDRs) were studied in response to cumulative addition of acetylcholine (ACh, 3 nM-10 μM, Sigma-Aldrich, St. Louis, MO, USA) upon precontraction of phenylephrine (Phe, Sigma-Aldrich) in endothelium-intact rings. Then sodium nitroprusside (SNP, 1 Nm–10 μM, Sigma-Aldrich)-induced endothelium-independent relaxations were measured in Phe-contracted rings. Each experiment was performed on rings prepared from different experimental animals.

The baseline and Phe concentrations were at 10 mN and 1 μM for rat aortas, 5 mN and 3 μM for rat carotid arteries, 2.5 mN and 1 μM for rat renal arteries, and 3 mN and 3 μM for mouse aortas.

2.9. Western Blotting

Aortas from rats after chronic treatment were isolated and snap frozen in liquid nitrogen. They were subsequently homogenized in ice-cold RIPA lysis buffer (Beyotime Biotechnology, Shanghai, China) plus cOmplete protease inhibitor cocktail (Roche, Basel, Switzerland) and PhosSTOP phosphatase inhibitors (Roche, Basel, Switzerland). The lysates were incubated on ice for 30 min and then centrifuged at 15,000 rpm for 30 min. The supernatant was collected and the protein concentration was measured using the bicinchoninic acid (BCA) assay (Beyotime Biotechnology, Shanghai, China). The protein samples (15 μg) were electrophoresed through 10% sodium dodecyl sulfate polyacrylamide gel electrophoresis (SDS-PAGE) and transferred to PVDF membranes (Millipore, Billerica, MA, USA) using wet transfer. Non-specific binding sites were blocked with 1% BSA in 0.05% Tween-20 PBS, and then the membrane was probed overnight at 4 °C with a primary antibody against the target protein including, iNOS, COX-2, VCAM-1, ICAM-1, phosphor (p)-AMPKα at Thr172 and AMPKα (Cell Signaling Tech, Danvers, MA, USA). The blots were incubated with the corresponding HRP-coupled secondary antibody (Beyotime Biotechnology, Shanghai, China) for 1 h at room temperature. GAPDH was selected as the housekeeping protein and used to check the equal loading of each sample. Protein bands were finally developed with the American ECLTM Advanced Western Blotting Detection Kit (GE Healthcare Life Sciences, Uppsala, Sweden), and images were captured using the ChemiDoc^{TM} MP Imaging System (Bio-Rad, Hercules, CA, USA). The intensities of signals were analyzed by ImageJ software (National Institute of Health, Bethesda, MA, USA).

2.10. Statistical Analysis

All data were shown as mean ± standard error of mean (SEM) of n-independent experiments. Comparisons among groups were analyzed using one-way ANOVA followed by Bonferroni post hoc tests and Student's *t*-test in the GraphPad Prism software (GraphPad Software, San Diego, CA, USA). $p < 0.05$ was considered as a statistically significant difference.

3. Results

3.1. CRP Extract Mitigates High Blood Pressure in Diabetic Rats

Male Sprague Dawley rats were randomly assigned into three groups which had similar body weight at the start of the experiment, and rats feeding on a high-fat diet (HFD) had a moderately higher increase in body weight than the control group feeding on normal chow (Figure 1A). The food intake in terms of weight for consumption of HFD was less than that of normal chow by the control rats (Figure 1B); nevertheless, the calorie intake was similar among the groups (Figure 1C). After streptozotocin (STZ) injection at week 10, the body weight in the diabetic rats was reduced, whilst the food/calorie intake was increased. The CRP treatment (400 mg/kg/day, 4 weeks) had no effect on body weight and food/calorie intake in the diabetic rats. The blood glucose levels of the model rats were significantly higher at each time point than those of the control rats (Figure 1E,F). The systolic blood pressure (SBP) in HFD/STZ-induced diabetic rats was significantly increased, and the hypertensive condition was significantly mitigated by CRP treatment (Figure 1D). However, the blood glucose level was not altered by CRP treatment, as indicated in the results of the fasting blood glucose (FBG) determination and oral glucose tolerance test (Figure 1E,F). The elevated FBG and glucose intolerance supported that the type 2 diabetes model was successfully induced in rats.

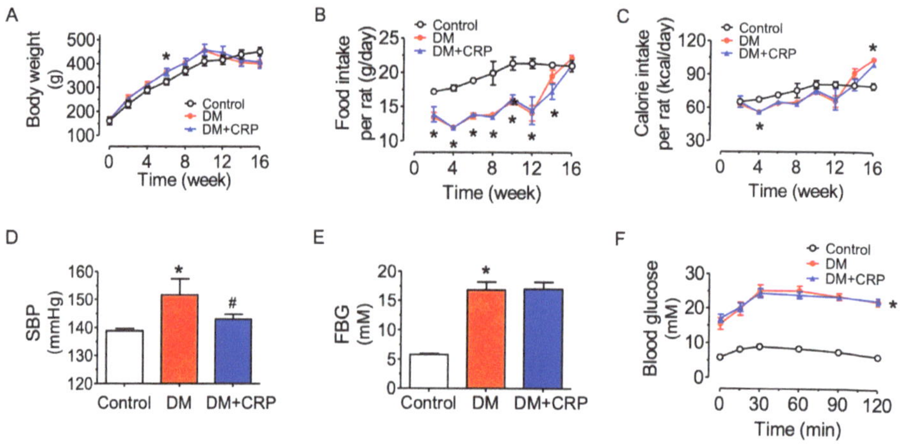

Figure 1. Effects of tangerine peel (*Citrus reticulatae Pericarpium*, CRP) extract administration on body weight, blood pressure and blood glucose level of diabetic rats. (**A**) Body weight changes. (**B**,**C**) Food and energy intake. (**D**) Systolic blood pressure (SBP) measured by tail-cuff method. (**E**) Fasting blood glucose (FBG) measured after 12 h of fasting. (**F**) Oral glucose tolerance test (OGTT) after 12 h of fasting. Data are mean ± SEM from five rats for each group. * $p < 0.05$ DM vs. Control; # $p < 0.05$ DM + CRP vs. DM.

3.2. CRP Extract Ameliorates Lipid Metabolism Dysregulation and Liver Function Injury in Diabetic Rats

At 16 weeks, the diabetic rats exhibited significant abnormalities in lipid metabolism, including significant increases in plasma levels of triglycerides, total cholesterol, low-density lipoprotein (LDL-C), and significant decreases in high-density lipoprotein (HDL-C), and such changes were remarkably reversed by four-week CRP administration (Figure 2A–D). Similarly, plasma AST and ALT activities were diminished in the CRP-treated group (Figure 2E,F), implying that the liver function damage of diabetic rats was significantly improved after chronic CRP treatment.

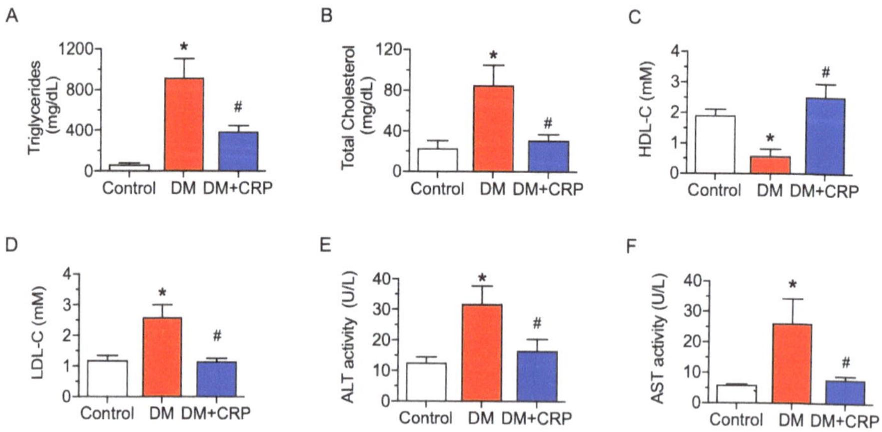

Figure 2. Effect of CRP extract on plasma lipid profile and liver function biomarkers in rats. Levels of (**A**) triglycerides, (**B**) total cholesterol, (**C**) high-density lipoprotein cholesterol (HDL-C), (**D**) low-density lipoprotein cholesterol (LDL-C), and (**E**) alanine aminotransferase (ALT) and (**F**) aspartate transaminase (AST) activities detected in plasma. Data are mean ± SEM from five rats for each group. * $p < 0.05$ DM vs. Control; # $p < 0.05$ DM + CRP vs. DM.

3.3. CRP Extract Alleviates Endothelial Dysfunction Associated with Diabetes

Aortas, carotid arteries and renal arteries of the control and diabetic rats were dissected for vascular functional assays. Compared with the arteries from the control group, the acetylcholine (ACh)-induced endothelium-dependent relaxations (EDRs) were markedly reduced in diabetic rats (Figure 3A–C). Treatment with CRP extract at 400 mg/kg/day for 4 weeks significantly reversed the impairments in aortas, carotid arteries and renal arteries. On the other hand, diabetes or CRP treatment did not affect the SNP-induced endothelium-independent relaxations (Figure 3D–F), indicating the vascular smooth muscle function responding to NO was unaltered.

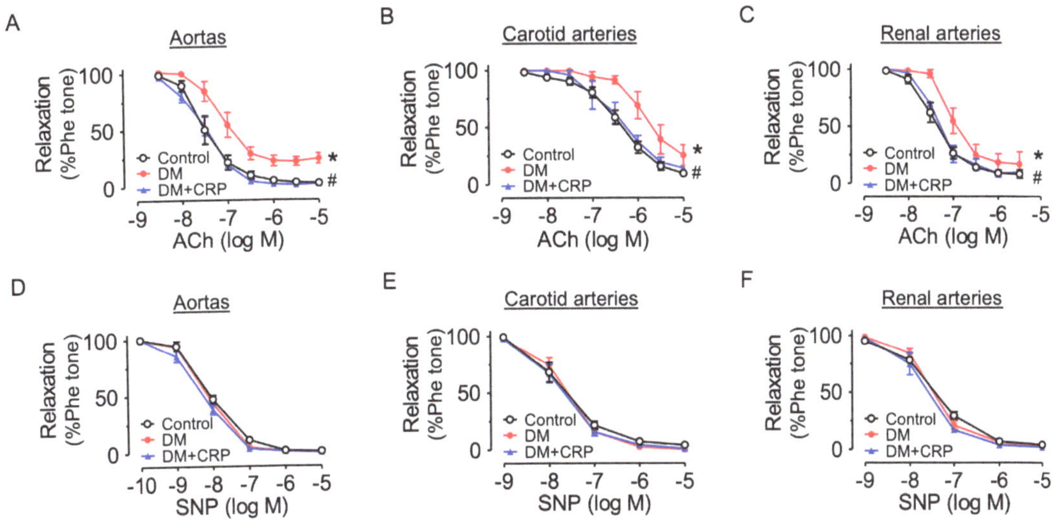

Figure 3. Vascular protective effects of CRP extract on diabetic rats. (**A–C**) Effects of oral administration of CRP (400 mg/kg/day for 4 weeks) on acetylcholine (ACh)-induced endothelium-dependent relaxations in the aortas, carotid arteries and renal arteries of diabetic rats. (**D–F**) Sodium nitroprusside (SNP)-induced endothelium-independent relaxations in the aortas, carotid arteries and renal arteries of diabetic rats. Data are mean ± SEM from five rats for each group. * $p < 0.05$ DM vs. Control; # $p < 0.05$ DM + CRP vs. DM.

3.4. CRP Extract Inhibited Inflammation in the Aortas of Diabetic Rats

The protein expressions of common inflammatory markers such as iNOS, COX-2, VCAM-1 and ICAM-1 were significantly elevated (Figure 4A–E), whereas the phosphorylated AMPKα at Thr172 was suppressed in the aortas of the diabetic rats (Figure 4A,F). These changes in protein expression were reversed after 4 weeks of CRP extract treatment. The GAPDH levels indicated equal loading of each sample in the Western blot analysis.

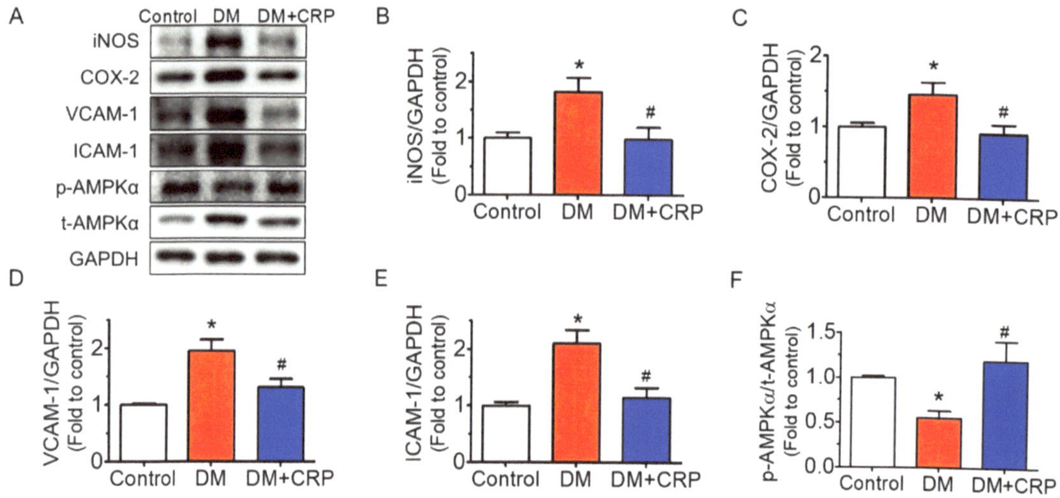

Figure 4. Effect of CRP extract on the protein expressions of inflammation markers in the rat aortas. (**A**) Representative blots and (**B–F**) summarized data showing the effect of CRP extract treatment (400 mg/kg body weight, 4 weeks) on aortic inflammation in diabetic rats, including iNOS (130 kDa), COX-2 (74 kDa), VCAM-1 (110 kDa), and ICAM-1 (90 kDa) compared with GAPDH (36 kDa) as well as the phosphorylation of AMPKα at Thr172 (p-AMPKα; 62 kDa) compared with its total protein (t-AMPKα). Data are mean ± SEM from five rats for each group. * $p < 0.05$ DM vs. Control; # $p < 0.05$ DM + CRP vs. DM.

3.5. CRP Extract Protects against Endothelial dysfunction in Hyperglycemic conditions

Furthermore, we confirmed the protective effect of CRP extract on both rat and mouse aortas ex vivo. The rat aortas were incubated with high glucose (44.4 mM, 5 h) to mimic the hyperglycemic environment in diabetes. Ex vivo high glucose exposure impaired the ACh-induced EDRs compared with the normal glucose control (11.1 mM, 5 h with mannitol added as a osmotic control), and the impairment was reversed by CRP co-treatment at 400 μg/mL (Figure 5A). Similarly, CRP extract prevented high glucose (30 mM, 48 h)-induced EDR impairment in mouse aortas (Figure 5B).

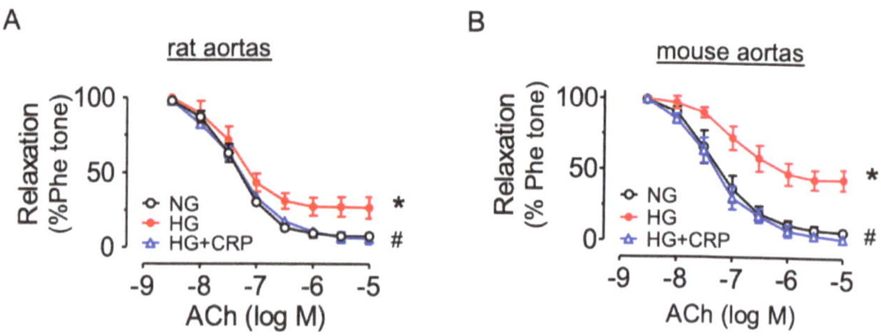

Figure 5. Vasoprotective effect of CRP in aortas from SD rats and C57BL/6J mice ex vivo. Effect of CRP extract (400 μg/mL) on acetylcholine (ACh)-induced endothelium-dependent relaxations of aortas (**A**) from rats exposed to high glucose (HG, 44.4 mM, 5 h) and (**B**) from mice (30 mM, 48 h). Data are mean ± SEM of 4 experiments. * $p < 0.05$ HG vs. NG; # $p < 0.05$ HG + CRP vs. HG.

4. Discussion

The present results suggest that the CRP extract has a significant protective effect against vascular inflammation and endothelial dysfunction associated with diabetes on

AMPK activation. We observe that HFD/STZ-induced diabetic rats have high blood pressure, increased levels of plasma lipids and liver enzymes, upregulated inflammatory markers in aortas, and impaired vasodilation in thoracic aortas, carotid arteries and renal arteries, which are reversed by chronic administration of CRP extract in vivo.

Previous studies have demonstrated that different types of citrus peel extracts exert beneficial effects on metabolic disorders, including lowered body weight, improved hyperglycemia, reduced dyslipidemia and reduced liver function abnormalities in obese and diabetic mice [24–27]. In consistence with previous studies, the current study shows that chronic consumption of CRP extract decreases plasma levels of total cholesterol, triglycerides, LDL, AST, ALT and increases plasma HDL level in diabetic rats. However, our data illustrate no significant changes in body weight and glucose tolerance with oral treatment of CRP extract, which contradicts previous studies. These contradictions may be attributed to the difference in treatment dosage, duration, and animal models used.

Fresh citrus peel has been shown to protect against atherosclerosis, reducing the fatty plaque in coronary arteries and aortas and enhancing the antioxidant capacity in rabbit plasma [28]. The polymethoxy-flavonoid nobiletin present in citrus peel significantly improves hemodynamic parameters, oxidative stress, collagen levels, vascular reactivity, cardiac hypertrophy index and myocardial fibrosis in STZ-induced diabetic rats [29]. In addition, this compound exerts endothelium-independent vasodilatory effects to attenuate phenylephrine-induced constriction in rat aortas by increasing cyclic guanosine monophosphate (cGMP) levels through guanylate cyclase activation [22]. These previous findings suggest that tangerine peel and/or its ingredients may affect vascular function. Our present study is probably the first to demonstrate that CRP has a good vascular protective effect in mitigating high blood pressure and enhancing endothelium-dependent vasodilation in conduit arteries such as the aortas, carotid arteries and renal arteries in diabetic rats in vivo. Meanwhile, the endothelial-independent relaxations induced by cumulative addition of NO donor SNP in the arteries were unaltered, implying a normal vascular smooth muscle function [30]. The same improvement in EDRs was observed in isolated aortas from both rats and mice upon exposure to high glucose ex vivo. Notably, the beneficial effect of restoring lipid profiles in diabetic rats might partially mediate the protective properties of CRP extract in vasculature. However, the ex vivo experimental evidence suggests that the contribution of lipid modulation should be minimal and that CRP extract possesses direct vasoprotective activity as ambient lipid levels are constant in the culture medium on isolated aortas.

Inflammatory processes actively participate in the development of vascular dysfunction associated with chronic metabolic diseases. Both obesity and diabetes are inflammatory conditions. Surplus concentrations of nutrients, such as glucose and free fatty acids (FFAs), are observed in obesity and type 2 diabetes [31]; in parallel, inflammation is found in various tissues including fat, liver, muscle, islets, and blood vessels [32–36]. Vascular risk factors including hyperglycemia, advanced glycation end products (AGEs), oxidized lipids found in diabetes and obesity can elevate the expressions of VCAM-1 and ICAM-1 and cause inflammation, and inflammation aggravates the progression of cardiovascular disease [37]. We have recently reported that CRP extract has anti-inflammatory effects in macrophages [19]. In addition, the main flavonoid components of CRP, consisting of hesperidin, nobiletin, and tangeretin, inhibit neuroinflammation [38]. Thus, it is reasonable to postulate that CRP might improve endothelial function through anti-inflammatory effects.

Our findings show that CRP restrained vascular inflammation, downregulating the expressions of iNOS, COX-2, ICAM-1, and VCAM-1 and activating AMPK in the aortas of HFD/STZ-induced diabetic rats. ICAM-1 and VCAM-1 are critical for macrophage activation and recruitment and strongly linked to inflammation. AMPK is a phylogenetically conserved serine/threonine protein kinase. AMPK regulates not only energy homeostasis but also vascular homeostasis. AMPK activation is widely considered to elicit anti-inflammatory effects and protect endothelial function in the vascular system [39]. AMPK can be stimulated by various natural and synthetic compounds, and AMPK modulates multiple targets. AMPK activation by adenine suppresses the inflammatory response in endothelial cells, where COX-2, ICAM-1, and VCAM-1 expressions are inhibited [40]. Activation of AMPK is reported to inhibit inflammation through phosphorylation of p300 and inactivation of p300 histone acetyltransferase, and thereby the nuclear factor kappa B (NF-κB) pathway [41]. Methotrexate alleviates inflammation in an AMPK-CREB-dependent manner [42]. Additionally, AMPK is suggested to trigger the biogenesis and maturation of microRNAs (miRNAs), thereby modulating vascular homeostasis [43,44]. Our data indicate that CRP extract ameliorates vascular inflammation and endothelial dysfunction in diabetes, possibly mediated by AMPK activation; nevertheless, the comprehensive mechanism underlying AMPK activation and the involvement of other proteins and signaling pathways need to be further explored in future study. Natural flavones are suggested to be protective against inflammation associated with diabetes [45]. CRP contains flavones such as hesperidin and other ingredients. Investigation and identification of the active ingredients present in CRP extract contributing to its vasoprotective effect should be performed in the future.

5. Conclusions

In conclusion, chronic administration of tangerine peel extract confers protection against vascular inflammation and endothelial dysfunction in arteries from diabetic rats, possibly via AMPK activation. CRP treatment also normalizes blood pressure and plasma lipid levels. These novel findings strengthen the prospects for the potential use of tangerine peel in functional foods or health supplements for patients with metabolic disorders and vascular complications.

Author Contributions: Conceptualization, M.F. and W.S.C.; methodology, X.Z. and Y.W.; formal analysis, X.Z. and Y.W.; investigation, Y.W., X.Z. and C.Z.; writing—original draft preparation, Y.W.; writing—review and editing, H.K., M.F. and W.S.C.; supervision, W.S.C. All authors have read and agreed to the published version of the manuscript.

Funding: This research was funded by the Research Committee of the University of Macau (Grant Number: SRG2019-00154-ICMS and MYRG2019-00157-ICMS), and the Science and Technology Development Fund, Macau SAR (Grant Number: 0117/2020/A and SKL-QRCM(UM)-2020–2022), the National Natural Science Foundation of China (31901713), the Natural Science Foundation of Guangdong Province (No. 2021A1515011049), the Talent Project of the Guangdong Academy of Agricultural Sciences (No. R2020PY-JX011), and the Research Group Construction Project of the Guangdong Academy of Agricultural Sciences (No. 202109TD).

Institutional Review Board Statement: The animal study protocol was conducted according to the guidelines of the Declaration of Helsinki and approved by the Animal Research Ethics Committee of the University of Macau (Protocol Code: UMARE-032-2021; approved on 17 January 2022).

Informed Consent Statement: Not applicable.

Data Availability Statement: The data presented in this study are available on request from the corresponding author.

Acknowledgments: The authors thank the Animal Research Core in the Faculty of Health Sciences, the University of Macau, for breeding and providing the mice and rats used in the present study.

Conflicts of Interest: The authors declare no conflict of interest.

References

1. Einarson, T.R.; Acs, A.; Ludwig, C.; Panton, U.H. Prevalence of cardiovascular disease in type 2 diabetes: A systematic literature review of scientific evidence from across the world in 2007–2017. *Cardiovasc. Diabetol.* **2018**, *17*, 83. [CrossRef] [PubMed]
2. Emerging Risk Factors Collaboration. Diabetes mellitus, fasting blood glucose concentration, and risk of vascular disease: A collaborative meta-analysis of 102 prospective studies. *Lancet* **2010**, *375*, 2215–2222. [CrossRef] [PubMed]
3. Bakker, W.; Eringa, E.C.; Sipkema, P.; van Hinsbergh, V.W.M. Endothelial dysfunction and diabetes: Roles of hyperglycemia, impaired insulin signaling and obesity. *Cell Tissue Res.* **2009**, *335*, 165–189. [CrossRef] [PubMed]

4. Sena, C.M.; Carrilho, F.; Seiça, R. Endothelial Dysfunction in Type 2 Diabetes: Targeting Inflammation. *Endothel. Dysfunct. Old Concepts New Chall.* **2018**, *24*, 23110.
5. Duncan, B.B.; Schmidt, M.I.; Pankow, J.S.; Ballantyne, C.M.; Couper, D.; Vigo, A.; Hoogeveen, R.; Folsom, A.R.; Heiss, G. Low-Grade Systemic Inflammation and the Development of Type 2 Diabetes: The Atherosclerosis Risk in Communities Study. *Diabetes* **2003**, *52*, 1799–1805. [CrossRef]
6. Hingorani, A.D.; Cross, J.; Kharbanda, R.K.; Mullen, M.J.; Bhagat, K.; Taylor, M.; Donald, A.E.; Palacios, M.; Griffin, G.E.; Deanfield, J.E.; et al. Acute systemic inflammation impairs endothelium-dependent dilatation in humans. *Circulation* **2000**, *102*, 994–999. [CrossRef]
7. Qi, J.; Wu, Q.; Cheng, Q.; Chen, X.; Zhu, M.; Miao, C. High Glucose Induces Endothelial COX2 and iNOS Expression via Inhibition of Monomethyltransferase SETD8 Expression. *J. Diabetes Res.* **2020**, *2020*, 2308520. [CrossRef]
8. Theofilis, P.; Sagris, M.; Oikonomou, E.; Antonopoulos, A.S.; Siasos, G.; Tsioufis, C.; Tousoulis, D. Inflammatory Mechanisms Contributing to Endothelial Dysfunction. *Biomedicines* **2021**, *9*, 781. [CrossRef]
9. Hardie, D.G. Adenosine monophosphate-activated protein kinase: A central regulator of metabolism with roles in diabetes, cancer, and viral infection. *Cold Spring Harb. Symp. Quant. Biol.* **2011**, *76*, 155–164. [CrossRef]
10. Ewart, M.-A.; Kennedy, S. AMPK and vasculoprotection. *Pharmacol. Ther.* **2011**, *131*, 242–253. [CrossRef]
11. Cheang, W.S.; Tian, X.Y.; Wong, W.T.; Lau, C.W.; Lee, S.S.; Chen, Z.Y.; Yao, X.; Wang, N.; Huang, Y. Metformin protects endothelial function in diet-induced obese mice by inhibition of endoplasmic reticulum stress through 5′ adenosine monophosphate-activated protein kinase-peroxisome proliferator-activated receptor delta pathway. *Arter. Thromb. Vasc. Biol.* **2014**, *34*, 830–836. [CrossRef] [PubMed]
12. Zhou, C.; Tan, Y.; Xu, B.; Wang, Y.; Cheang, W.S. 3,4′,5-Trimethoxy-trans-stilbene Alleviates Endothelial Dysfunction in Diabetic and Obese Mice via Activation of the AMPK/SIRT1/eNOS Pathway. *Antioxidants* **2022**, *11*, 1286. [CrossRef]
13. Commission, C.P. *Pharmacopoeia of the People's Republic of China*; China Medical Science Press: Beijing, China, 2020; p. 199.
14. Hua, L.S.L. Advance of Main Chemical Components and Quality Control of Citri Reticulatae Pericarpium. *Pharm. Today* **2020**, *30*, 861–864.
15. Wu, Y.; He, Y.; Wang, R.; Zhao, X. Preventive Effect of Flavonoid Extract from the Peel of Gonggan (Citrus reticulata Blanco Var. Gonggan) on CCl(4)-Induced Acute Liver Injury in Mice. *J. Inflamm. Res.* **2021**, *14*, 5111–5121. [CrossRef]
16. Fu, M.; Zou, B.; An, K.; Yu, Y.; Tang, D.; Wu, J.; Xu, Y.; Ti, H. Anti-asthmatic activity of alkaloid compounds from Pericarpium Citri Reticulatae (Citrus reticulata 'Chachi'). *Food Funct.* **2019**, *10*, 903–911. [CrossRef] [PubMed]
17. Wang, F.; Chen, L.; Chen, S.; Chen, H.; Liu, Y. Microbial biotransformation of Pericarpium Citri Reticulatae (PCR) by Aspergillus niger and effects on antioxidant activity. *Food Sci. Nutr.* **2021**, *9*, 855–865. [CrossRef] [PubMed]
18. Singh, B.; Singh, J.P.; Kaur, A.; Singh, N. Phenolic composition, antioxidant potential and health benefits of citrus peel. *Food Res. Int.* **2020**, *132*, 109114. [CrossRef]
19. Zhang, X.; Zhou, Y.; Cheong, M.S.; Khan, H.; Ruan, C.-C.; Fu, M.; Xiao, J.; Cheang, W.S. Citri Reticulatae Pericarpium extract and flavonoids reduce inflammation in RAW 264.7 macrophages by inactivation of MAPK and NF-κB pathways. *Food Front.* **2022**, *3*, 785–795. [CrossRef]
20. Ling, Y.; Shi, Z.; Yang, X.; Cai, Z.; Wang, L.; Wu, X.; Ye, A.; Jiang, J. Hypolipidemic effect of pure total flavonoids from peel of Citrus (PTFC) on hamsters of hyperlipidemia and its potential mechanism. *Exp. Gerontol.* **2020**, *130*, 110786. [CrossRef]
21. Mulvihill, E.E.; Burke, A.C.; Huff, M.W. Citrus Flavonoids as Regulators of Lipoprotein Metabolism and Atherosclerosis. *Annu. Rev. Nutr.* **2016**, *36*, 275–299. [CrossRef]
22. Kaneda, H.; Otomo, R.; Sasaki, N.; Omi, T.; Sato, T.; Kaneda, T. Endothelium-independent vasodilator effects of nobiletin in rat aorta. *J. Pharmacol. Sci.* **2019**, *140*, 48–53. [CrossRef] [PubMed]
23. Zhang, M.; Lv, X.Y.; Li, J.; Xu, Z.G.; Chen, L. The characterization of high-fat diet and multiple low-dose streptozotocin induced type 2 diabetes rat model. *Exp. Diabetes Res.* **2008**, *2008*, 704045. [CrossRef] [PubMed]
24. Guo, J.; Tao, H.; Cao, Y.; Ho, C.-T.; Jin, S.; Huang, Q. Prevention of Obesity and Type 2 Diabetes with Aged Citrus Peel (Chenpi) Extract. *J. Agric. Food Chem.* **2016**, *64*, 2053–2061. [CrossRef] [PubMed]
25. Lu, Y.; Xi, W.; Ding, X.; Fan, S.; Zhang, Y.; Jiang, D.; Li, Y.; Huang, C.; Zhou, Z. Citrange fruit extracts alleviate obesity-associated metabolic disorder in high-fat diet-induced obese C57BL/6 mouse. *Int. J. Mol. Sci.* **2013**, *14*, 23736–23750. [CrossRef]
26. Park, H.J.; Jung, U.J.; Cho, S.J.; Jung, H.K.; Shim, S.; Choi, M.S. Citrus unshiu peel extract ameliorates hyperglycemia and hepatic steatosis by altering inflammation and hepatic glucose- and lipid-regulating enzymes in db/db mice. *J. Nutr. Biochem.* **2013**, *24*, 419–427. [CrossRef]
27. Lee, Y.S.; Cha, B.Y.; Saito, K.; Choi, S.S.; Wang, X.X.; Choi, B.K.; Yonezawa, T.; Teruya, T.; Nagai, K.; Woo, J.T. Effects of a Citrus depressa Hayata (shiikuwasa) extract on obesity in high-fat diet-induced obese mice. *Phytomedicine* **2011**, *18*, 648–654. [CrossRef]
28. Boshtam, M.; Asgary, S.; Moshtaghian, J.; Naderi, G.; Jafari-Dinani, N. Impacts of fresh lime juice and peel on atherosclerosis progression in an animal model. *ARYA Atheroscler.* **2013**, *9*, 357–362.
29. Parkar, N.A.; Bhatt, L.K.; Addepalli, V. Efficacy of nobiletin, a citrus flavonoid, in the treatment of the cardiovascular dysfunction of diabetes in rats. *Food Funct.* **2016**, *7*, 3121–3129. [CrossRef]
30. Gagov, H.; Gribkova, I.V.; Serebryakov, V.N.; Schubert, R. Sodium Nitroprusside-Induced Activation of Vascular Smooth Muscle BK Channels Is Mediated by PKG Rather Than by a Direct Interaction with NO. *Int. J. Mol. Sci.* **2022**, *23*, 2798. [CrossRef]
31. Boden, G. Role of fatty acids in the pathogenesis of insulin resistance and NIDDM. *Diabetes* **1997**, *46*, 3–10. [CrossRef]

32. Berk, B.C.; Weintraub, W.S.; Alexander, R.W. Elevation of C-reactive protein in "active" coronary artery disease. *Am. J. Cardiol.* **1990**, *65*, 168–172. [CrossRef] [PubMed]
33. Maedler, K.; Sergeev, P.; Ris, F.; Oberholzer, J.; Joller-Jemelka, H.I.; Spinas, G.A.; Kaiser, N.; Halban, P.A.; Donath, M.Y. Glucose-induced beta cell production of IL-1beta contributes to glucotoxicity in human pancreatic Isletsj. *Clin. Investig.* **2002**, *110*, 851–860. [CrossRef]
34. Navarro-Gonzalez, J.F.; Mora-Fernandez, C. The role of inflammatory cytokines in diabetic nephropathy. *J. Am. Soc. Nephrol.* **2008**, *19*, 433–442. [CrossRef] [PubMed]
35. Ridker, P.M.; Cushman, M.; Stampfer, M.J.; Tracy, R.P.; Hennekens, C.H. Inflammation, aspirin, and the risk of cardiovascular disease in apparently healthy men. *N. Engl. J. Med.* **1997**, *336*, 973–979. [CrossRef] [PubMed]
36. Wellen, K.E.; Hotamisligil, G.S. Inflammation, stress, and diabetes. *J. Clin. Investig.* **2005**, *115*, 1111–1119. [CrossRef]
37. Steven, S.; Frenis, K.; Oelze, M.; Kalinovic, S.; Kuntic, M.; Bayo Jimenez, M.T.; Vujacic-Mirski, K.; Helmstadter, J.; Kroller-Schon, S.; Munzel, T.; et al. Vascular Inflammation and Oxidative Stress: Major Triggers for Cardiovascular Disease. *Oxid. Med. Cell Longev.* **2019**, *2019*, 7092151. [CrossRef]
38. Ho, S.-C.; Kuo, C.-T. Hesperidin, nobiletin, and tangeretin are collectively responsible for the anti-neuroinflammatory capacity of tangerine peel (Citri reticulatae pericarpium). *Food Chem. Toxicol.* **2014**, *71*, 176–182. [CrossRef]
39. Jansen, T.; Kvandova, M.; Daiber, A.; Stamm, P.; Frenis, K.; Schulz, E.; Munzel, T.; Kroller-Schon, S. The AMP-Activated Protein Kinase Plays a Role in Antioxidant Defense and Regulation of Vascular Inflammation. *Antioxidants* **2020**, *9*, 525. [CrossRef]
40. Cheng, Y.F.; Young, G.H.; Lin, J.T.; Jang, H.H.; Chen, C.C.; Nong, J.Y.; Chen, P.K.; Kuo, C.Y.; Kao, S.H.; Liang, Y.J.; et al. Activation of AMP-Activated Protein Kinase by Adenine Alleviates TNF-Alpha-Induced Inflammation in Human Umbilical Vein Endothelial Cells. *PLoS ONE* **2015**, *10*, e0142283. [CrossRef]
41. Zhang, Y.; Qiu, J.; Wang, X.; Zhang, Y.; Xia, M. AMP-activated protein kinase suppresses endothelial cell inflammation through phosphorylation of transcriptional coactivator p300. *Arter. Thromb. Vasc. Biol.* **2011**, *31*, 2897–2908. [CrossRef]
42. Thornton, C.C.; Al-Rashed, F.; Calay, D.; Birdsey, G.M.; Bauer, A.; Mylroie, H.; Morley, B.J.; Randi, A.M.; Haskard, D.O.; Boyle, J.J.; et al. Methotrexate-mediated activation of an AMPK-CREB-dependent pathway: A novel mechanism for vascular protection in chronic systemic inflammation. *Ann. Rheum. Dis.* **2016**, *75*, 439. [CrossRef] [PubMed]
43. Gelinas, R.; Mailleux, F.; Dontaine, J.; Bultot, L.; Demeulder, B.; Ginion, A.; Daskalopoulos, E.P.; Esfahani, H.; Dubois-Deruy, E.; Lauzier, B.; et al. AMPK activation counteracts cardiac hypertrophy by reducing O-GlcNAcylation. *Nat. Commun.* **2018**, *9*, 374. [CrossRef] [PubMed]
44. Cheng, C.K.; Shang, W.; Liu, J.; Cheang, W.S.; Wang, Y.; Xiang, L.; Lau, C.W.; Luo, J.Y.; Ng, C.F.; Huang, Y.; et al. Activation of AMPK/miR-181b Axis Alleviates Endothelial Dysfunction and Vascular Inflammation in Diabetic Mice. *Antioxidants* **2022**, *11*, 1137. [CrossRef] [PubMed]
45. Miao, L.; Liu, C.; Cheong, M.S.; Zhong, R.; Tan, Y.; Rengasamy, K.R.R.; Leung, S.W.S.; Cheang, W.S.; Xiao, J. Exploration of natural flavones' bioactivity and bioavailability in chronic inflammation induced-type-2 diabetes mellitus. *Crit. Rev. Food Sci. Nutr.* **2022**, 1–28. [CrossRef]

nutrients

MDPI

Article

C-Phycocyanin and *Lycium barbarum* Polysaccharides Protect against Aspirin-Induced Inflammation and Apoptosis in Gastric RGM-1 Cells

Yu-Chen Liu [1], Chun-Chao Chang [2,3,4], Hirofumi Matsui [5] and Jane C.-J. Chao [1,4,6,7,*]

1 School of Nutrition and Health Sciences, Taipei Medical University, 250 Wu-Hsing Street, Taipei 110301, Taiwan
2 Division of Gastroenterology and Hepatology, Department of Internal Medicine, Taipei Medical University Hospital, 252 Wu-Hsing Street, Taipei 110301, Taiwan
3 Division of Gastroenterology and Hepatology, Department of Internal Medicine, School of Medicine, Taipei Medical University, 250 Wu-Hsing Street, Taipei 110301, Taiwan
4 TMU Research Center for Digestive Medicine, Taipei Medical University, 250 Wu-Hsing Street, Taipei 110301, Taiwan
5 Division of Gastroenterology, Faculty of Medicine, University of Tsukuba, 1-1-1 Tennodai, Tsukuba 305-8575, Japan
6 Master Program in Global Health and Health Security, Taipei Medical University, 250 Wu-Hsing Street, Taipei 110301, Taiwan
7 Nutrition Research Center, Taipei Medical University Hospital, 252 Wu-Hsing Street, Taipei 110301, Taiwan
* Correspondence: chenjui@tmu.edu.tw; Tel.: +886-2-2736-1661 (ext. 6548); Fax: +886-2-2737-3112

Abstract: Aspirin causes gastrotoxicity and damaged epithelial defense via cyclooxygenase inhibition. C-phycocyanin (CPC) and *Lycium barbarum* polysaccharides (LBP), an active ingredient of *Spirulina platensis* and wolfberry, respectively, exerted antioxidation, anti-inflammation, and/or immunoregulation. The actions of CPC and/or LBP on gastric damage induced by aspirin were explored in rat gastric mucosal RGM-1 cells. Gastric injury was performed by 21 mM aspirin for 3 h after the pretreatment of CPC and/or LBP (100–500 μg/mL) for 24 h in RGM-1 cells. Proinflammatory, anti-inflammatory, and apoptotic markers were examined by ELISA or gel electrophoresis and Western blotting. Cell viability and interleukin 10 (IL-10) were reduced by aspirin. Increased proinflammatory markers, caspase 3 activity, and Bax protein were observed in RGM-1 cells with aspirin treatment. Aspirin elevated nuclear factor-κB (NF-κB), extracellular signal-regulated kinase (ERK), p38, and c-Jun N-terminal kinase (JNK) activation, while CPC and/or LBP increased IL-10, and attenuated proinflammatory markers, Bax protein, NF-κB, and the activation of ERK and JNK. Therefore, CPC and/or LBP possess anti-inflammation by restraining the activation of the ERK signaling pathway, and LBP decreases apoptosis by suppressing the JNK signaling pathway activation in gastric RGM-1 cells with aspirin-induced epithelial damage.

Keywords: apoptosis; aspirin; C-phycocyanin; extracellular signal-regulated kinase; inflammation; *Lycium barbarum* polysaccharides

Citation: Liu, Y.-C.; Chang, C.-C.; Matsui, H.; Chao, J.C.-J. C-Phycocyanin and *Lycium barbarum* Polysaccharides Protect against Aspirin-Induced Inflammation and Apoptosis in Gastric RGM-1 Cells. *Nutrients* **2022**, *14*, 5113. https://doi.org/10.3390/nu14235113

Academic Editors: Maria Digiacomo and Doretta Cuffaro

Received: 25 October 2022
Accepted: 28 November 2022
Published: 1 December 2022

1. Introduction

Aspirin, a non-steroidal anti-inflammatory drug (NSAID), has analgesic and anti-inflammatory characteristics. Low-dose aspirin (75–325 mg/day) has been used for cardiovascular protection and was demonstrated in clinical practice guidelines for established atherosclerotic cardiovascular disease and acute management of myocardial infarction [1]. However, long-term (>3 months) administration of low-dose aspirin was correlated with upper gastrointestinal side effects such as dyspepsia and peptic ulcer [2]. The acidity (pKa 3.5) and lipid solubility (log P = 1.15) of aspirin could impair the barrier of gastric mucosa [3]. In addition, aspirin inhibited the activity of cyclooxygenases (COX), thereby reducing the production of gastroprotective prostaglandins [3]. NSAIDs also generate free radicals and inhibit antioxidant enzymes in the process of inducing peptic ulcers [4]. The generation of free radicals could be caused by Bax, caspase-8, caspase-9, and caspase-3, which may further lead to apoptosis of gastrointestinal mucosal cells [5]. Previous studies have shown that under the induction of lipopolysaccharide, extracellular signal-regulated kinase (ERK) could be activated, and resulted in the inhibition of IκB kinase (IKK) phosphorylation and then the activation of nuclear factor-κB (NF-κB) pathway to promote the production of related inflammatory factors [6].

Phycobiliproteins, such as phycocyanin and phycocyanobilin, with similar structures to biliverdin and bilirubin [7], respectively, are biliprotein pigments of *Spirulina platensis* [8]. These proteins are divided into two categories: red phycoerythrin and blue phycocyanin, and the latter includes C-phycocyanin (CPC), R-phycocyanin, and allophycocyanin [8]. CPC is not toxic and carcinogenic and has been applied as a dye in food and cosmetics [8]. CPC also possesses antioxidant, anti-inflammatory, and free radical-eliminating characteristics [8]. The antioxidant capacity of phycocyanin was similar to biliverdin [7]. In addition, CPC has been found to stimulate cell migration into the wound site [9] and promote wound healing in animal experiments [10].

Lycium barbarum L. was also called wolfberry or goji berry and has been utilized for over 2300 years as one of the traditional Chinese herbal medicines [11]. There were 31 polysaccharides found in *Lycium barbarum*, which accounted for 5–8% of the dry weight [12]. The main monosaccharides were xylose, glucose, rhamnose, arabinose, fructose, trehalose, mannose, ribose, and galactose [12], and different compositions of *Lycium barbarum* polysaccharides (LBP) such as LBPA3, LBPB1, LBP-IV, and LpGp1 (-4) and possible structures with $(1 \rightarrow 3)$, $(1 \rightarrow 4)$, or $(1 \rightarrow 6)$ α- or β-glycosidic bond were reported previously [13]. LBP has been shown to own antioxidant and anti-cancer characteristics and protect the liver and small intestine from ischemia-reperfusion injury [14].

Aspirin has been applied to cardiovascular disease. However, chronic aspirin use could lead to gastric lesions. Although certain medications can be used to treat gastric damage caused by aspirin, they have several side effects. Our motivation was to use food ingredients to eliminate gastric damage caused by aspirin. CPC and LBP are both natural ingredients of edible food and have been proven to have antioxidant and anti-inflammatory effects. However, the actions of CPC and LBP on gastric damage caused by aspirin remained unexplored. Therefore, we studied whether CPC and/or LBP protected against gastric lesions in rat gastric mucosal RGM-1 cells damaged by aspirin.

2. Materials and Methods

2.1. Cells and Treatments

We bought rat normal gastric mucosa RGM-1 cells from RIKEN BioResource Center (Tsukuba, Ibaraki, Japan), and the morphology of RGM-1 cells was shown previously [15]. Cells were maintained in 20% fetal bovine serum, Dulbecco's modified Eagle's medium, and Ham's F-12 mixture. The preparation of aspirin (Sigma Chemical Co., St. Louis, MO, USA) was referred to in the previous study [16]. Aspirin was dissolved in dimethyl sulfoxide and then added to the medium. CPC was provided by Far East Biotechnology. The spirulina powder was mixed with water for 24-h extraction followed by centrifugation, and the supernatant was collected and lyophilized. This lyophilized spirulina extract contained 24.4% phycocyanin (CPC and isophycocyanin), 35–45% polysaccharide, 10–20% non-phycocyanin protein, 5–8% moisture, and 10–12% ash. The treatment of LBP contained 40% LBP (GojiMax® 40%, Priority Healthfood Corporation, New Taipei, Taiwan), which was analyzed by spectrophotometry. Gastric lesions were performed by aspirin (21 mM) for 3 h after the pretreatment of CPC and/or LBP (100–500 μg/mL) for 24 h in RGM-1 cells. Cells and the conditioned media were collected for further analyses. We used the colorimetric method to examine cell viability as the method of Saito et al. [17] using the MTS (3-(4,5-dimethylthiazol-2-yl)-5-(3-carboxymethoxyphenyl)-2-(4-sulfophenyl)-2H-tetrazolium, inner salt) kit (K300-2500, BioVision Inc., Milpitas, CA, USA). The viable cells were determined by colored formazan compound at 490–500 nm.

2.2. Proinflammatory and Anti-Inflammatory Markers and Regulators

The proinflammatory markers such as TNF-α (Cloud-Clone Corp., Houston, TX, USA), IL-6 (Cloud-Clone Corp.), and NF-κB (NF-κB (p65) Transcription Factor Assay Kit, Cayman Chemical Co., Ann Arbor, MI, USA) and anti-inflammatory marker such as IL-10 (Cloud-Clone Corp.) were examined by the ELISA kits. The absorbance of TNF-α, IL-6, and IL-10 in the conditioned medium was quantitated at 450 nm. The cells were washed with phosphate-buffered saline (PBS) with phosphatase inhibitors, added to the cell lysis buffer and NP-40 assay buffer, and centrifuged at 14,000× g for 30 s to precipitate the cell pellet. The cell pellet was dissolved in nuclear extraction buffer and centrifuged at 14,000× g for 10 min for NF-κB analysis. We finally measured the DNA binding activity of NF-κB in the nuclear fraction at 450 nm.

Protein expression of p-ERK, ERK, p-IκB-α, and IκB-α was determined by sodium dodecyl sulfate (SDS)-polyacrylamide gel electrophoresis (PAGE) and Western blotting. The cell suspension was applied to 10% SDS-PAGE at 40 mA for 1.5 h, and proteins were transferred to the nitrocellulose membrane by semi-dry electroblotting apparatus at 220 mA for 1 h and 20 min. The membrane was blocked in the blocking buffer (3% bovine serum albumin and Tris-buffered saline with Tween 20), incubated with primary antibody for p-ERK, ERK, p-IκB-α, or IκB-α (Santa Cruz Biotechnology Inc., Santa Cruz, CA, USA) and secondary antibody (Santa Cruz Biotechnology Inc.), and washed by Tris-buffered saline (TBS) and Tween 20 in between. The bands on the membrane were visualized using a commercial kit (Opti-ECL HRP Reagent Kit, Bioman Scientific Co., Ltd., New Taipei, Taiwan). Protein expression of β-actin was an internal control. The membrane was probed with different primary antibodies after the membrane was stripped by stripping buffer and washed by TBS and Tween 20 in between. The relative level was determined by chemiluminescence using an imaging system (UVP ChemiDoc-It 515 Imaging System Vision Works 8.18, UVP, LLC, Upland, CA, USA) and the software (Media Cybernetics, Inc., Rockville, MD, USA).

2.3. Apoptotic Markers and Regulators

Caspase 9 and caspase 3 activities were evaluated by the colorimetric kits (BioVision Inc., Milpitas, CA, USA). The cells were mixed with cell lysis buffer and centrifuged at 10,000× g for 1 min. Proteins were quantitated, added to the reaction buffer and chromophore DEVD-pNA (aspartyl-glutamyl-valyl-aspartyl p nitroanilide) at 37 °C for 2 h, and these activities were finally measured at 400 nm. The relative enzyme activity was expressed as % of the control group.

Apoptotic regulators such as Bax, Bcl-2, p-p38, p38, p-JNK, and JNK (BioVision Inc.) were isolated by 10% SDS-PAGE and determined by Western blotting mentioned above. The transferred proteins on the nitrocellulose membrane were identified by the corresponding primary antibodies (Santa Cruz Biotechnology Inc.) and incubated with the secondary antibodies and *chemiluminescence* reagent (Opti-ECL HRP Reagent Kit, Bioman Scientific Co., Ltd.). Protein expression of β-actin was an internal control. The membrane was stripped by the method described above for incubation with different primary antibodies. The image of protein bands was examined by the method described above.

2.4. Statistical Analysis

Data are expressed as mean ± standard deviation (SD). The statistical method was done by one-way analysis of variance (ANOVA), and the treatment effect was determined by Tukey's multiple comparison test using GraphPad Prism 5.0 (GraphPad Software Inc., San Diego, CA, USA). The value of $p < 0.05$ was significant statistically.

3. Results

3.1. Cell Viability

RGM-1 cells were pretreated with CPC (30–900 μg/mL) or LBP (50–1000 μg/mL) for 24 h and then induced lesions by 21 mM aspirin for 3 h. Cell viability was significantly reduced by pretreating with high doses of CPC (750 and 900 μg/mL) (Figure 1a) or LBP (800 and 1000 μg/mL) (Figure 1b). Considering not affecting cell viability by the pretreatment, 3 different doses (100, 250, and 500 μg/mL) for CPC or LBP were used in this study. Additionally, to explore whether the action of combined CPC and LBP was better, 250 μg/mL CPC and 250 μg/mL LBP (C + L) were used.

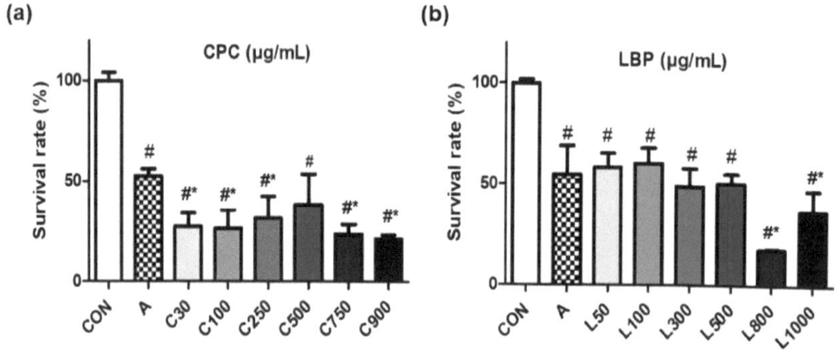

Figure 1. Effects of (**a**) C-phycocyanin (CPC) or (**b**) *Lycium barbarum* polysaccharides (LBP) on the cell survival rate of RGM-1 cells with aspirin-induced gastric damage. CON: control group, A: aspirin-induced group, C: CPC treated group, L: LBP treated group. Data are mean ± SD (n = 6). # $p < 0.05$ compared with the CON group. * $p < 0.05$ compared with the A group.

3.2. Effects of CPC and LBP on Proinflammatory and Anti-Inflammatory Markers

Aspirin significantly increased the secretion of tumor necrosis factor-α (TNF-α) (1418 ± 64 pg/mL) (Figure 2a) and interleukin-6 (IL-6) (898 ± 7 pg/mL) (Figure 2b), and NF-κB (p65) DNA binding activity ratio (93.5% ± 6.9%) (Figure 2c) in RGM-1 cells. The treatment of CPC at 100, 250, or 500 μg/mL and C + L (250 μg/mL CPC and 250 μg/mL LBP) significantly decreased TNF-α concentration (1291 ± 39 pg/mL, 1269 ± 57 pg/mL, 1289 ± 43 pg/mL, and 1263 ± 41 pg/mL, respectively) compared with the aspirin-induced group. The treatment of CPC and/or LBP at all doses significantly attenuated IL-6 secretion and NF-κB (p65) DNA binding activity ratio compared with aspirin alone. Aspirin significantly reduced IL-10 secretion compared with the control group (Figure 2d). The treatment of CPC or LBP at higher doses (250 and 500 μg/mL) and C + L significantly increased IL-10 secretion compared with the aspirin-induced group.

3.3. Effects of CPC and LBP on Inflammatory Regulators

Aspirin significantly elevated the relative ratio of the phosphorylated extracellular signal-regulated kinase (p-ERK)/ERK (Figure 3a) and phosphorylated IκB-α (p-IκB-α)/IκBα (Figure 3b). However, the treatment of 100 μg/mL CPC or 250 μg/mL and 500 μg/mL LBP decreased the relative ratio of p-ERK/ERK, and a high dose of CPC or LBP (500 μg/mL) and C + L treatments significantly decreased the relative ratio of p-IκBα/IκBα.

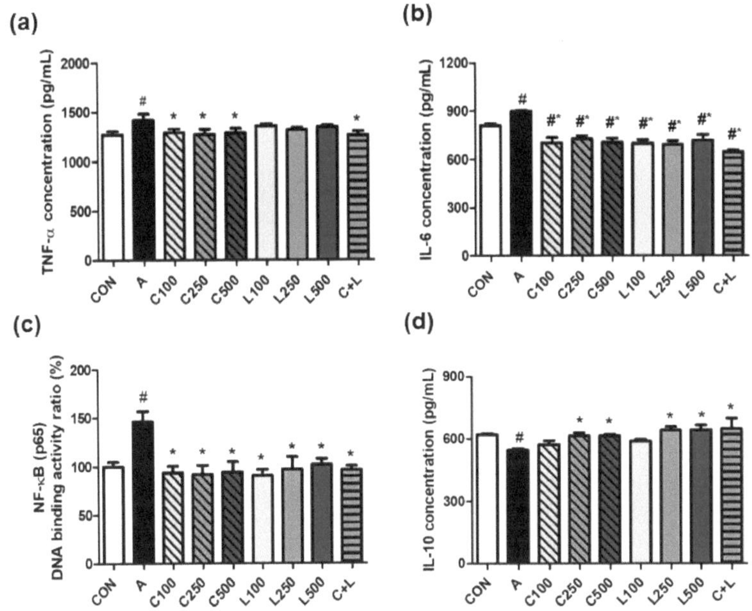

Figure 2. Effects of aspirin, CPC, and/or LBP on proinflammatory markers (**a**) TNF-α, (**b**) IL-6, (**c**) NF-κB (p65) DNA binding activity ratio, and anti-inflammatory marker (**d**) IL-10. CON: control group, A: aspirin-induced group, C: CPC treated group, L: LBP treated group. Data are mean ± SD ($n = 3$). # $p < 0.05$ compared with the CON group. * $p < 0.05$ compared with the A group.

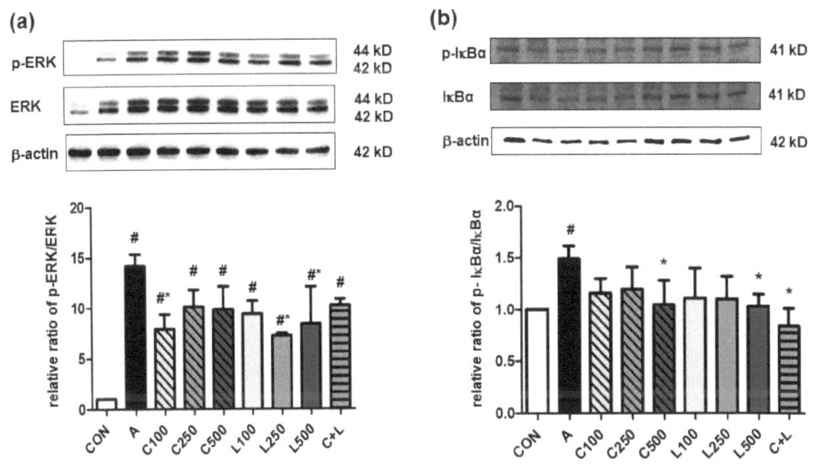

Figure 3. Effects of aspirin, CPC, and/or LBP on the relative ratio of (**a**) p-ERK/ERK ($n = 3$) and (**b**) p-IκBα/IκBα ($n = 4$). CON: control group, A: aspirin-induced group, C: CPC treated group, L: LBP treated group. Data are mean ± SD. # $p < 0.05$ compared with the CON group. * $p < 0.05$ compared with the A group.

3.4. Actions of CPC and LBP on Apoptotic Markers

Aspirin did not change caspase 9 activity (Figure 4a). A high dose of LBP (500 μg/mL) and C + L treatments significantly decreased caspase 9 activity compared with the control and aspirin-induced groups. Aspirin significantly reduced caspase 3 activity (Figure 4b). The treatment of CPC and/or LBP significantly attenuated caspase 3 activity.

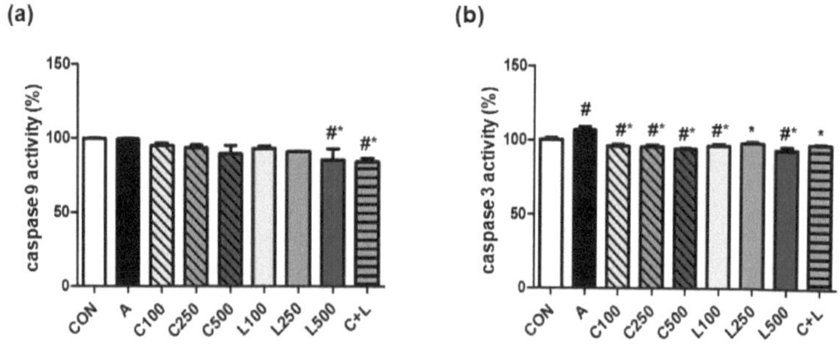

Figure 4. Effects of aspirin, CPC, and/or LBP on apoptotic markers (**a**) caspase 9 activity and (**b**) caspase 9 activity. CON: control group, A: aspirin-induced group, C: CPC treated group, L: LBP treated group. Data are mean ± SD ($n = 3$). # $p < 0.05$ compared with the CON group. * $p < 0.05$ compared with the A group.

3.5. Effects of CPC and LBP on Apoptotic Regulators

Aspirin elevated the pro-apoptotic relative ratio of p-Bax/Bax compared with the control group (Figure 5a). The treated group with LBP at 500 μg/mL or combined CPC and LBP significantly attenuated the relative ratio of p-Bax/Bax. However, no significant differences were shown in the anti-apoptotic relative ratio of p-Bcl-2/Bcl-2 between any two groups (Figure 5b). Aspirin significantly increased the relative ratios of p-p38/p38 (Figure 5c) and phosphorylated c-Jun N-terminal kinase (p-JNK)/JNK (Figure 5d). However, the treatment of CPC or LBP, either alone or a combination of both, could not alter the relative ratio of p-p38/p38 compared with the groups with or without aspirin. The treatment of LBP at all doses and C + L significantly decreased the relative ratio of p-JNK/JNK.

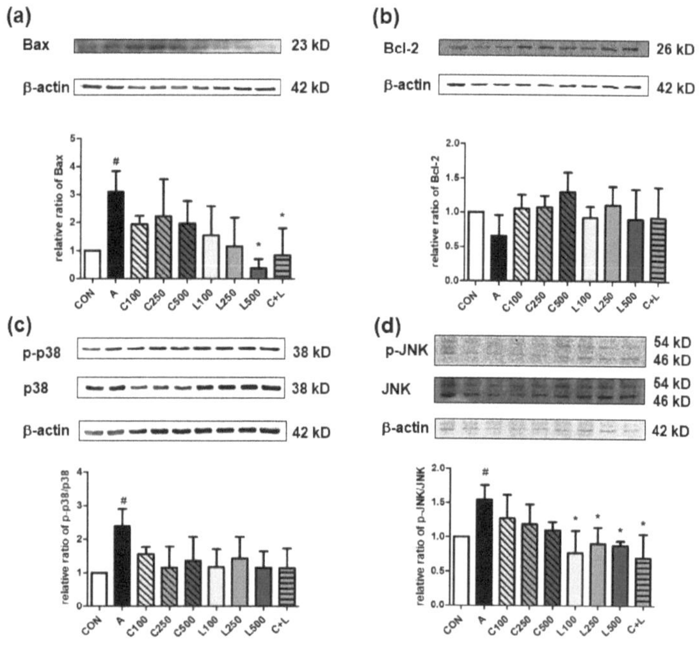

Figure 5. Effects of aspirin, CPC, and/or LBP on the relative ratio of (**a**) Bax, (**b**) Bcl-2, (**c**) phosphorylated p38 (p-p38)/p38, and (**d**) phosphorylated c-Jun N-terminal kinase (p-JNK)/JNK. CON: control group, A: aspirin-induced group, C: CPC treated group, L: LBP treated group. Data are mean ± SD ($n = 4$). # $p < 0.05$ compared with the CON group. * $p < 0.05$ compared with the A group.

4. Discussion

RGM-1 cells were induced lesions by aspirin at a dose of 21 mM for 3 h in this study, and the survival rate was 53–55%. A previous study demonstrated that the cell survival rate of RGM-1 cells treated with 5 mM or 10 mM aspirin for 3 h was less than 80% and 60%, respectively [18]. The survival rate of RGM-1 cells was significantly decreased after cells were damaged by 10 mM aspirin for 18 h [17]. In this study, the survival rate of RGM-1 cells was not altered after cells were incubated with various doses of CPC (30–900 μg/mL) or LBP (50–1000 μg/mL) for 24 h (Figure S1), indicating both CPC and LBP were safe interventions. Additionally, the randomized, double-blind, placebo-controlled human studies found that a high dose of CPC-containing extract (2.3 g/day) for 2 weeks [19] or LBP at a dose of 1.632 g/day for 30 days [20] was well tolerated in adults, and was beneficial for the regulation on certain physical functions.

We demonstrated that aspirin caused significant increases in NF-κB (p65) DNA binding activity ratio and protein expression ratio of p-ERK/ERK. Similarly, indomethacin, an NSAID, could increase p-ERK [21]. Increased p-ERK could activate the IKK-β complex, which in turn activated NF-κB and generated inducible nitric oxide synthase (iNOS), which then led to neutrophil infiltration and inflammation [22]. We speculated that aspirin might activate NF-κB and cause inflammation by activating ERK. Our study showed that the treatment of CPC (100–500 μg/mL) decreased the secretion of TNF-α and IL-6 in RGM-1 cells. Consistent with our finding, CPC at the dose of 50–250 μg/mL led to decreases in elevated TNF-α, IL-6, matrix metalloproteinase-3 (MMP-3), and NO-induced by IL-1β in canine chondrocytes [22]. The intravenous administration of CPC (30 or 50 mg/kg) in rats attenuated carrageenan-induced inflammation by reducing TNF-α and iNOS [23]. A recent in vivo study also showed that spirulina could restore gastric damage caused by aspirin via balancing oxidant and antioxidant system [24]. We showed that the intervention of 100–500 μg/mL LBP for 24 h significantly inhibited the activation of the NF-κB pathway and reduced proinflammatory cytokine IL-6 secretion but not TNF-α secretion in RGM-1 cells. A previous study found that the treatment of LBP at a dose of 600 μg/mL for 24 h followed by lipopolysaccharide-induced damage increased cell survival but suppressed NF-κB protein expression and TNF-α concentration in BV-2 microglial cells [25]. However, LBP could not inhibit TNF-α expression through the NF-κB signaling pathway, which was mediated by toll-like receptors [26]. The secretion of TNF-α in RGM-1 cells in our study was not altered, which could be due to less dosage of LBP compared with that in a previous study mentioned above. An animal study showed that oral feeding of LBP (1 or 10 mg/kg) for 10 h in mice with liver injury induced by carbon tetrachloride for 8 h after 2-h pretreatment of LBP reduced mRNA expression of TNF-α, IL-1β, and iNOS, and decreased NF-κB activity, indicating that LBP may suppress the expression of proinflammatory markers by inhibiting the activation of NF-κB [27].

Aspirin was reported to produce free radicals, which led to the activation of JNK and p38 mitogen-activated protein kinase (MAPK) pathways, altered the balance between pro-apoptotic and anti-apoptotic markers, and further stimulated the release of cytochrome c from mitochondria to activate caspase 9 and caspase 3 [28]. Previous studies showed that anti-apoptotic Bcl-2 protein expression was decreased, and caspase 9 and caspase 3 activities were activated in human gastric epithelial AGS cells damaged by 40 mM aspirin, indicating that aspirin could induce apoptosis in human gastric epithelial cells [29,30]. However, we found that aspirin increased caspase 3 activity, but did not change Bcl-2 protein expression and caspase 9 activity, probably because the dosage of aspirin and/or the incubation time was not enough. The treatment of 10 μM CPC for 4 h attenuated doxorubicin-induced apoptosis in rat cardiomyocytes by suppressing cytochrome c release, Bax protein expression, and caspase 3 activity, and increasing Bcl-2 protein expression and the ratio of Bcl-2 to Bax [31]. The other study demonstrated that CPC decreased apoptosis induced by islet amyloid polypeptide in rat insulinoma β INS-1E cells by reducing oxidative stress and regulating JNK and p38 MAPK pathways [32]. However, our study showed that even caspase 3 activity was inhibited by CPC, protein expression of Bax, Bcl-2, p-p38, and p-JNK was not changed significantly by CPC. The inconsistent results between the previous studies in rat cardiomyocytes or rat insulinoma β cells and the present study in rat normal gastric mucosal cells are probably because of different cell lines and various inducers for cell apoptosis. We found that LBP reduced both caspase 9 and caspase 3 activities by suppressing the activation of JNK. A previous study demonstrated that LBP suppressed apoptosis induced by hydrogen peroxide through increasing Bcl-2 and decreasing Bax protein expression in human lens epithelial SRA01/04 cells [33]. The animal study showed that LBP (40 mg/kg) inhibited apoptosis induced by focal ischemic injury in mice by down-regulating protein expression of cleaved poly(ADP-ribose) polymerase 1, cytochrome c, Bax, caspase 3, and caspase 9, but increasing Bcl-2 protein expression [34]. Additionally, both previous studies mentioned above pointed out that LBP suppressed apoptosis by attenuating oxidative stress [29,30]. Therefore, LBP inhibited apoptosis by modulating pro-apoptotic and anti-apoptotic factors and suppressing oxidative stress.

This study showed that the treatment of LBP at a higher dose of 500 μg/mL might exert the best effects on reducing aspirin-induced inflammation and apoptosis in RGM cells. The anti-inflammatory mechanisms of LBP could be involved in significantly reducing the activation of ERK, thereby attenuating IκBα phosphorylation, NF-κB (p65) DNA binding activity, and finally, proinflammatory IL-6 secretion, and increasing anti-inflammatory IL-10 secretion. The anti-apoptotic action of LBP could be attributed to reducing JNK activation, Bax protein expression, and caspase 3 activity. However, the combined CPC and LBP did not obviously demonstrate additional or synergistic effects on anti-inflammation or anti-apoptosis.

Other natural ingredients or extracts have been observed to exert gastroprotective effects on aspirin-induced damage. An in vitro study found that licoflavone at the dose of 6–12 mg/L had potential gastroprotective effects on antioxidation via enhancing antioxidant enzyme activities and reducing lipid peroxides and on anti-apoptosis via decreasing Bax expression and increasing Bcl-2 expression in human gastric mucosal GES-1 cells damaged by aspirin (18.5 mM) through suppressing ERK1/2 signaling pathway [35], which may have similar properties as CPC and LBP. Araloside A, an active ingredient from *Aralia elata* bark, at the dose of 10–40 mg/kg for 7 days, inhibited gastric damage by aspirin through decreasing gastric activities of H^+/K^+-ATPase, apoptotic enzymes, and protein expression of Bax, but enhancing Bcl-2 expression in mice [36]. Artesunate, a derivative from *Artemisia annua L.* as an antimalarial compound, at the dose of 50 or 150 mg/kg has been reported to decrease aspirin-induced gastric lesions in rats by reducing oxidative stress, gastric TNF-α, IL-6, COX-2, and NF-κB expression [37]. Mice given spirulina at the dose of 250–500 mg/kg for 3 days reduced gastric damage induced by aspirin via decreasing lipid peroxidation and inflammatory markers such as TNF-α and COX-2 [38]. Additionally, basil extract at the dose of 100 or 200 mg/kg for 5 days inhibited TNF-α and IL-6 but increased IL-4 and prostaglandin E_2 in mice serum to protect against gastric injury by aspirin [39]. Bambara groundnut extract is originally from Africa as a common food source and an antioxidant. Pretreatment with Bambara groundnut extract at the dose of 200 or 400 mg/kg for 3 weeks protected against rat gastric ulcer induced by aspirin and pyloric ligation through elevating superoxide dismutase and glutathione peroxidase activities and attenuating malondialdehyde lipid peroxides [40]. Another in vivo study showed that supplementation with 10% pomegranate peel powder for 30 days increased nitric oxide in plasma and decreased gastric TNF-α and COX-2 gene expression in rats with aspirin-induced gastric ulcer [41]. Overall, these results showed that natural ingredients or extracts had gastroprotective actions on aspirin-induced injury by various mechanisms, including antioxidation, anti-inflammation, antacid, anti-apoptosis, and stimulation of gastric mucosal protective factors.

The previous review found that omeprazole combined with rebamipide may be the most effective treatment for aspirin-induced gastric damage compared with several common drugs such as rabeprazole, phosphatidylcholine complex, famotidine, ranitidine bismuth citrate, and lansoprazole from 10 randomized controlled studies [42]. Past studies also compared the gastroprotective effects of natural compounds with medicine [43,44].

Rats were treated with pentagalloyl glucose (50–200 mg/kg) isolated from *Harpullia pendula* pericarp compared with those treated with melatonin, omeprazole, or famotidine (10–40 mg/kg) to explore the gastroprotective properties of the natural compound and medication against gastric injury caused by aspirin [43]. Rats were pretreated with pentagalloyl glucose, melatonin, omeprazole, or famotidine for 1 h, followed by the induction of gastric injury by aspirin for 1 h, and the results showed elevated prostaglandin E_2 and reduced COX-2 concentrations in the stomach [43]. Another research used ranitidine as a positive control to clarify the effect of *Cyperus rotundus* rhizome extract on gastric lesions caused by aspirin [43]. The results demonstrated that *C. rotundus* rhizome extract at the dose of 250–500 mg/kg suppressed gastric lipid peroxidation and exerted antioxidant action, which showed a similar effect to the anti-ulcer drug ranitidine [44]. However, we did not perform a positive control in this study, and a positive control such as omeprazole can be used to demonstrate whether CPC and/or LBP may have the same effect as the drug in the future study. In addition, the effects of combined natural compounds or extracts with anti-ulcer medicine could be further investigated to clarify whether the combination could reduce the dosage of the anti-ulcer drug in order to not only eliminate the side effects of the drug but also ameliorate aspirin-induced gastric lesions.

5. Conclusions

The present study demonstrated that CPC and/or LBP at a high dose of 500 μg/mL reversed gastric injury caused by aspirin via elevating IL-10 and attenuating proinflammatory markers, NF-κB, caspase 3 activity, Bax protein, and the activation of ERK and JNK in RGM-1 cells. Therefore, CPC and/or LBP exert anti-inflammatory effects against damaged gastric RGM-1 cells by aspirin via inhibiting the activation of the ERK signaling pathway. LBP reduces apoptosis by restraining the activation of the JNK signaling pathway.

Supplementary Materials: The following supporting information can be downloaded at: https://www.mdpi.com/article/10.3390/nu14235113/s1, Figure S1: Effects of (a) C-phycocyanin (CPC) or (b) *Lycium barbarum* polysaccharides (LBP) on the cell survival rate of RGM-1 cells.

Author Contributions: Conceptualization, Y.-C.L., C.-C.C. and J.C.-J.C.; methodology, Y.-C.L., H.M. and J.C.-J.C.; validation, C.-C.C. and H.M.; formal analysis, Y.-C.L.; writing—original draft preparation, Y.-C.L.; writing—review and editing, J.C.-J.C.; funding acquisition, J.C.-J.C. All authors have read and agreed to the published version of the manuscript.

Funding: This research was funded by the National Science and Technology Council, Taiwan, grant number MOST105-2320-B-038-036-MY3.

Institutional Review Board Statement: Not applicable.

Informed Consent Statement: Not applicable.

Data Availability Statement: Data of the present study are not publicly available. Data are available upon reasonable request and with permission from the authors.

Acknowledgments: We sincerely thank the grant funding by the National Science and Technology Council, Taiwan.

Conflicts of Interest: The authors declare no conflict of interest.

References

1. Damman, P.; van't Hof, A.W.; Ten Berg, J.M.; Jukema, J.W.; Appelman, Y.; Liem, A.H.; de Winter, R.J. 2015 ESC guidelines for the management of acute coronary syndromes in patients presenting without persistent ST-segment elevation: Comments from the Dutch ACS working group. *Neth. Heart J.* **2017**, *25*, 181–185. [CrossRef]
2. Hsu, P.I.; Tsai, T.J. Epidemiology of upper gastrointestinal damage associated with low-dose aspirin. *Curr. Pharm. Des.* **2015**, *21*, 5049–5055. [CrossRef] [PubMed]
3. Darling, R.L.; Romero, J.J.; Dial, E.J.; Akunda, J.K.; Langenbach, R.; Lichtenberger, L.M. The effects of aspirin on gastric mucosal integrity, surface hydrophobicity, and prostaglandin metabolism in cyclooxygenase knockout mice. *Gastroenterology* **2004**, *127*, 94–104. [CrossRef] [PubMed]

4. Firdaus, S.B.; Ghosh, D.; Chattyopadhyay, A.; Jana, K.; Bandyopadhyay, D. A combination of aqueous curry (*Murraya koenigii*) leaf extract and melatonin protects against piroxicam induced gastric ulcer in male albino rats: Involvement of antioxidant mechanism(s). *J. Pharm. Res.* **2014**, *8*, 428–436.
5. Musumba, C.; Pritchard, D.M.; Pirmohamed, M. Cellular and molecular mechanisms of NSAID-induced peptic ulcers. *Aliment. Pharmacol. Ther.* **2009**, *3*, 517–531. [CrossRef] [PubMed]
6. Slomiany, B.L.; Slomiany, A. Role of epidermal growth factor receptor transactivation in the amplification of *Helicobacter pylori*-elicited induction in gastric mucosal expression of cyclooxygenase-2 and inducible nitric oxide synthase. *OA Inflamm.* **2013**, *1*, 1–8. [CrossRef]
7. Zheng, J.; Inoguchi, T.; Sasaki, S.; Maeda, Y.; McCarty, M.F.; Fujii, M.; Ikeda, N.; Kobayashi, K.; Sonoda, N.; Takayanagi, R. Phycocyanin and phycocyanobilin from *Spirulina platensis* protect against diabetic nephropathy by inhibiting oxidative stress. *Am. J. Physiol. Regul. Integr. Comp. Physiol.* **2013**, *304*, R110–R120. [CrossRef]
8. Kuddus, M.; Singh, P.; Thomas, G.; Al-Hazimi, A. Recent developments in production and biotechnological applications of C-phycocyanin. *Biomed. Res. Int.* **2013**, *2013*, 742859. [CrossRef]
9. Madhyastha, H.K.; Radha, K.S.; Nakajima, Y.; Omura, S.; Maruyama, M. uPA dependent and independent mechanisms of wound healing by C-phycocyanin. *J. Cell. Mol. Med.* **2008**, *12*, 2691–2703. [CrossRef]
10. Gur, C.S.; Erdogan, D.K.; Onbasılar, I.; Atilla, P.; Cakar, N.; Gurhan, I.D. In vitro and in vivo investigations of the wound healing effect of crude Spirulina extract and C-phycocyanin. *J. Med. Plants Res.* **2013**, *7*, 425–433.
11. Xin, Y.F.; Wan, L.L.; Peng, J.L.; Guo, C. Alleviation of the acute doxorubicin-induced cardiotoxicity by *Lycium barbarum* polysaccharides through the suppression of oxidative stress. *Food Chem. Toxicol.* **2011**, *49*, 259–264. [CrossRef] [PubMed]
12. Cheng, J.; Zhou, Z.W.; Sheng, H.P.; He, L.J.; Fan, X.W.; He, Z.X.; Sun, T.; Zhang, X.; Zhao, R.J.; Gu, L.; et al. An evidence-based update on the pharmacological activities and possible molecular targets of *Lycium barbarum* polysaccharides. *Drug Des. Devel. Ther.* **2015**, *9*, 33–78. [PubMed]
13. Tian, X.; Liang, T.; Liu, Y.; Ding, G.; Zhang, F.; Ma, Z. Extraction, structural characterization, and biological functions of *Lycium barbarum* polysaccharides: A review. *Biomolecules* **2019**, *9*, 389. [CrossRef] [PubMed]
14. Cheng, D.; Kong, H. The effect of *Lycium barbarum* polysaccharide on alcohol-induced oxidative stress in rats. *Molecules* **2011**, *16*, 2542–2550. [CrossRef] [PubMed]
15. Kobayashi, I.; Kawano, S.; Tsuji, S.; Matsui, H.; Nakama, A.; Sawaoka, H.; Ohno, T. RGM1, a cell line derived from normal gastric mucosa of rat. *In Vitro Cell. Dev. Biol. Anim.* **1996**, *32*, 259–261. [CrossRef] [PubMed]
16. Qiao, L.; Hanif, R.; Sphicas, E.; Shiff, S.J.; Rigas, B. Effect of aspirin on induction of apoptosis in HT-29 human colon adenocarcinoma cells. *Biochem. Pharmacol.* **1998**, *55*, 53–64. [CrossRef] [PubMed]
17. Saito, R.; Tamura, M.; Matsui, H.; Nagano, Y.; Suzuki, H.; Kaneko, T.; Mizokami, Y.; Hyodo, I. Qing Dai attenuates nonsteroidal anti-inflammatory drug-induced mitochondrial reactive oxygen species in gastrointestinal epithelial cells. *J. Clin. Biochem. Nutr.* **2015**, *56*, 8–14. [CrossRef] [PubMed]
18. Mustonen, H.; Puolakkainen, P.; Kemppainen, E.; Kiviluoto, T.; Kivilaakso, E. Taurocholate potentiates ethanol-induced NF-κB activation and inhibits caspase-3 activity in cultured rat gastric mucosal cells. *Dig. Dis. Sci.* **2009**, *54*, 928–936. [CrossRef]
19. Jensen, G.S.; Drapeau, C.; Lenninger, M.; Benson, K.F. Clinical safety of a high dose of phycocyanin-enriched aqueous extract from arthrospira (spirulina) platensis: Results from a randomized, double-blind, placebocontrolled study with a focus on anticoagulant activity and platelet activation. *J. Med. Food* **2016**, *19*, 645–653. [CrossRef]
20. Amagase, H.; Sun, B.; Borek, C. *Lycium barbarum* (goji) juice improves in vivo antioxidant biomarkers in serum of healthy adults. *Nutr. Res.* **2009**, *29*, 19–25. [CrossRef]
21. Yadav, S.K.; Adhikary, B.; Chand, S.; Maity, B.; Bandyopadhyay, S.K.; Chattopadhyay, S. Molecular mechanism of indomethacin-induced gastropathy. *Free Radic. Biol. Med.* **2012**, *52*, 1175–1187. [CrossRef]
22. Martinez, S.E.; Chen, Y.; Ho, E.A.; Martinez, S.A.; Davies, N.M. Pharmacological effects of a C-phycocyanin-based multicomponent nutraceutical in an in-vitro canine chondrocyte model of osteoarthritis. *Can. J. Vet. Res.* **2015**, *79*, 241–249.
23. Shih, C.M.; Cheng, S.N.; Wong, C.S.; Kuo, Y.L.; Chou, T.C. Antiinflammatory and antihyperalgesic activity of C-phycocyanin. *Anesth. Analg.* **2009**, *108*, 1303–1310. [CrossRef]
24. Sherif, A.K.; Ahmed, H.B.; Hagar, Y.R. Spirulina supplements: An approach moderating aspirin persuaded histological and ultra-structural alterations in albino rats gastric mucosa. *Ultrastruct. Pathol.* **2022**, *46*, 204–216.
25. Teng, P.; Li, Y.; Cheng, W.; Zhou, L.; Shen, Y.; Wang, Y. Neuroprotective effects of *Lycium barbarum* polysaccharides in lipopolysaccharide-induced BV2 microglial cells. *Mol. Med. Rep.* **2013**, *7*, 1977–1981. [CrossRef]
26. Youchao, Q.; Guozhen, D.; Guanghui, F.; Ning, P. Effect of *Lycium barbarum* polysaccharides on cell signal transduction pathways. *Biomed. Pharmacother.* **2022**, *147*, 112620.
27. Xiao, J.; Liong, E.C.; Ching, Y.P.; Chang, R.C.C.; So, K.F.; Fung, M.L.; Tipoe, G.L. *Lycium barbarum* polysaccharides protect mice liver from carbon tetrachloride-induced oxidative stress and necroinflammation. *J. Ethnopharmacol.* **2012**, *139*, 462–470. [CrossRef]
28. Bhatt, D.L.; Scheiman, J.; Abraham, N.S.; Antman, E.M.; Chan, F.K.; Furberg, C.D.; Johnson, D.A.; Mahaffey, K.W.; Quigley, E.M.; Harrington, R.A.; et al. ACCF/ACG/AHA 2008 expert consensus document on reducing the gastrointestinal risks of antiplatelet therapy and NSAID use: A report of the American College of Cardiology Foundation Task Force on Clinical Expert Consensus Documents. *J. Am. Coll. Cardiol.* **2008**, *52*, 1502–1517. [CrossRef]
29. Power, J.J.; Dennis, M.S.; Redlak, M.J.; Miller, T.A. Aspirin-induced mucosal cell death in human gastric cells: Evidence supporting an apoptotic mechanism. *Dig. Dis. Sci.* **2004**, *49*, 1518–1525. [CrossRef]

30. Redlak, M.J.; Power, J.J.; Miller, T.A. Role of mitochondria in aspirin-induced apoptosis in human gastric epithelial cells. *Am. J. Physiol. Gastrointest. Liver Physiol.* **2005**, *289*, G731–G738. [CrossRef]
31. Khan, M.; Varadharaj, S.; Shobha, J.C.; Naidu, M.U.; Parinandi, N.L.; Kutala, V.K.; Kuppusamy, P. C-phycocyanin ameliorates doxorubicin-induced oxidative stress and apoptosis in adult rat cardiomyocytes. *J. Cardiovasc. Pharmacol.* **2006**, *47*, 9–20. [CrossRef] [PubMed]
32. Li, X.L.; Xu, G.; Chen, T.; Wong, Y.S.; Zhao, H.L.; Fan, R.R.; Chan, J.C. Phycocyanin protects INS-1E pancreatic beta cells against human islet amyloid polypeptide-induced apoptosis through attenuating oxidative stress and modulating JNK and p38 mitogen-activated protein kinase pathways. *Int. J. Biochem. Cell Biol.* **2009**, *41*, 1526–1535. [CrossRef] [PubMed]
33. Qi, B.; Ji, Q.; Wen, Y.; Liu, L.; Guo, X.; Hou, G.; Zhong, J. *Lycium barbarum* polysaccharides protect human lens epithelial cells against oxidative stress–induced apoptosis and senescence. *PLoS ONE* **2014**, *9*, e110275. [CrossRef] [PubMed]
34. Wang, T.; Li, Y.; Wang, Y.; Zhou, R.; Ma, L.; Hao, Y.; Yu, J. *Lycium barbarum* polysaccharide prevents focal cerebral ischemic injury by inhibiting neuronal apoptosis in mice. *PLoS ONE* **2014**, *9*, e90780. [CrossRef] [PubMed]
35. Li, Y.; Wan, Q. Mechanism of licoflavone in protecting human gastric mucosal epithelial cells from aspirin-induced injury. *Chin. Gen. Pract.* **2018**, *21*, 3971–3975.
36. He, H.; Li, X.; Yu, H.; Zhu, S.; He, Y.; Komatsu, K.; Guo, D.; Li, X.; Wang, J.; Luo, H.; et al. Gastroprotective effect of araloside A on ethanol- and aspirin-induced gastric ulcer in mice: Involvement of H^+/K^+-ATPase and mitochondrial-mediated signaling pathway. *J. Nat. Med.* **2019**, *73*, 339–352. [CrossRef]
37. Sneh, V.; Vijay, L.K. Artesunate affords protection against aspirin–induced gastric injury by targeting oxidative stress and proinflammatory signaling. *Pharmacol. Rep.* **2018**, *70*, 390–397.
38. Mahmoud, Y.I.; Abd El-Ghffar, E.A. Spirulina ameliorates aspirin-induced gastric ulcer in albino mice by alleviating oxidative stress and inflammation. *Biomed. Pharmacother.* **2019**, *109*, 314–321. [CrossRef]
39. Abd El-Ghffar, E.A.; Al-Sayed, E.; Safia, S.M.; Eldahshan, O.A.; Efferth, T. The protective role of *Ocimum basilicum L.* (Basil) against aspirin-induced gastric ulcer in mice: Impact on oxidative stress, inflammation, motor deficits and anxiety-like behavior. *Food Funct.* **2018**, *9*, 4457–4468.
40. Morufu, E.B.; Elizabeth, E.B.; Jacinta, N.O.; Ogochukwu, S.M.; Fankou, S.A.D. Protective roles of *Vigna subterranea* (Bambara nut) in rats with aspirin-induced gastric mucosal injury. *J. Integr. Med.* **2018**, *16*, 342–349.
41. Mohamed, M.S.; Mabrok, H.B. Protective Effect of pomegranate peel powder against gastric ulcer in rats. *Biointerface Res. Appl. Chem.* **2021**, *12*, 4888–4899.
42. Zhang, W.T.; Wang, M.R.; Hua, G.D.; Li, Q.Y.; Wang, X.J.; Lang, R.; Weng, W.L.; Xue, C.M.; Zhu, B.C. Inhibition of aspirin-induced gastrointestinal injury: Systematic review and network meta-analysis. *Front. Pharmacol.* **2021**, *12*, 730681. [CrossRef]
43. Moharram, F.A.; El Dib, R.A.; Wahman, L.F.; El-Shenawy, S.M.; El-Awdan, S. Gastro protective effect of pentagalloyl glucose on aspirin induced gastric mucosal ulcer in comparison with omeprazole, Famotidine and Melatonin. *J. Pharm. Sci. Res.* **2017**, *9*, 785–795.
44. Thomas, D.; Govindhan, S.; Baiju, E.C.; Padmavathi, G.; Kunnumakkara, A.B.; Padikkala, J. *Cyperus rotundus L.* prevents non-steroidal anti-inflammatory drug-induced gastric mucosal damage by inhibiting oxidative stress. *J. Basic Clin. Physiol. Pharmacol.* **2015**, *26*, 485–490. [CrossRef] [PubMed]

nutrients

MDPI

Article

Anti-Inflammatory and Antimicrobial Activities of Portuguese *Prunus avium* L. (Sweet Cherry) By-Products Extracts

Ana R. Nunes [1,2], José D. Flores-Félix [1], Ana C. Gonçalves [1,3], Amílcar Falcão [3,4], Gilberto Alves [1] and Luís R. Silva [1,5,*]

1 CICS-UBI—Health Sciences Research Centre, University of Beira Interior, Av. Infante D. Henrique, 6200-506 Covilhã, Portugal
2 CNC—Centre for Neuroscience and Cell Biology, Faculty of Medicine, University of Coimbra, 3004-504 Coimbra, Portugal
3 Laboratory of Pharmacology, Faculty of Pharmacy, University of Coimbra, 3000-548 Coimbra, Portugal
4 CIBIT—Coimbra Institute for Biomedical Imaging and Translational Research, University of Coimbra, 3000-548 Coimbra, Portugal
5 CPIRN-UDI-IPG—Research Unit for Inland Development, Center for Potential and Innovation of Natural Resources, Polytechnic Institute of Guarda, 6300-554 Guarda, Portugal
* Correspondence: luisfarmacognosia@gmail.com

Abstract: The bioactivity of natural by-products in food and pharmaceutical applications is the subject of numerous studies. Cherry production and processing generates large amounts of biowaste, most of which is not used. The recovery of these by-products is essential for promoting the circular economy and to improving sustainability in the food industry. In this work, we explored the anti-inflammatory and antimicrobial potential of two different extracts from stems, leaves, and flowers of Portuguese cherries. The anti-inflammatory potential was studied on lipopolysaccharide (LPS)-stimulated mouse macrophages (RAW 264.7) by evaluating the effect of by-products on cellular viability and nitric oxide (NO) production. Disc diffusion and minimum inhibitory concentration (MIC) were used to determine antimicrobial activity. The cherry by-products had no cytotoxic effect on RAW 264.7 cells, and were able to inhibit nitrite production in a dose-dependent manner. Moreover, all aqueous infusions showed good antioxidant activity against NO radicals. Moreover, leaf extracts showed the best activity against most of the strains studied. The results revealed, for the first time, interesting anti-inflammatory and antimicrobial properties of cherry by-products. This could potentially be of interest for their therapeutic use in the treatment of inflammation-related diseases or in controlling the growth of microorganisms.

Keywords: *Prunus avium* L.; by-products; anti-inflammatory activity; antibacterial activity

Citation: Nunes, A.R.; Flores-Félix, J.D.; Gonçalves, A.C.; Falcão, A.; Alves, G.; Silva, L.R. Anti-Inflammatory and Antimicrobial Activities of Portuguese *Prunus avium* L. (Sweet Cherry) By-Products Extracts. *Nutrients* **2022**, *14*, 4576. https://doi.org/10.3390/nu14214576

Academic Editors: Maria Digiacomo and Doretta Cuffaro

Received: 6 October 2022
Accepted: 27 October 2022
Published: 31 October 2022

1. Introduction

Since ancient times, plants formed the basis for the development of many therapeutic products used in traditional medicine with significant pharmacological properties [1]. Natural products consist of various bioactive compounds that have been shown to have health-promoting properties, keep the body healthy, and protect against a variety of diseases, many of which are related to oxidative stress [2]. Nowadays, consumers are looking for healthier foods/products for their daily life. In this context, food and pharmaceutical industries are increasingly interested in finding compounds with good biological properties from natural sources [3,4]. Detailed knowledge of phytochemical composition, biological properties, safety profile, and environmental toxicity is essential for the characterization of herbal products.

Prunus avium L., commonly known as sweet cherry, is a high-value fruit of great nutritional and economic importance, widely distributed throughout the world [5–7]. The literature dealing with this specie has shown that the cherries possess interesting biological properties, and are indispensable in human nutrition [8–11]. According to various studies, the presence of phytochemicals such as phenolic compounds (flavonoids and non-flavonoids), minerals, and organic volatiles is the basis for these properties [7,12].

Cherry harvesting and industrial processing (e.g., into candies, jellies, juices, and fresh-cut fruits), produces a large amount of by-products, such as leaves, stems, seeds, and pulp. In addition, cherry flowers decompose in the soil after spring, without being used or recycled in any way [7]. Although the by-products are treated as biowaste, recent studies have shown that these basic plant materials are an excellent source of bioactive compounds that can be used in value-added products [13–15]. Traditionally, cherry stems in aqueous infusion have been used as diuretics, sedatives, anti-inflammatory agents, and to promote cardiovascular health [7]. Previous works developed by our research team have described the antioxidant, antiproliferative, and antidiabetic effects of *P. avium* fruits and their by-products [8,11,13,14]. However, little is known about the anti-inflammatory and antimicrobial potential of cherry by-products.

The anti-inflammatory process is a normal response of the body to an infection or injury. Immunological biochemical reactions lead to the release of reactive oxygen species (ROS) and reactive nitrogen species (RNS) and, consequently, to an increase in oxidative stress [16]. When inflammation becomes chronic or unregulated it can lead to serious pathological conditions. Therefore, the search for new alternatives that act on inflammation became the subject of the study [17]. The anti-inflammatory effect of extracts from sweet cherries has been recently studied and showed promising results, mainly due to their richness in phenolic compounds [11].

Despite the already demonstrated biological properties of *P. avium* fruits and their by-products, their antimicrobial properties are still poorly studied. A study conducted by Afonso and co-workers [10] showed that extracts from cherry stems are able to inhibit Gram-positive bacteria, probably due to the presence of phenolic acids in this by-product. Similarly, previous studies have shown that cherry stems, leaves, and flowers have a high content of these phenolic compounds [10]. In addition, the increase multidrug-resistant bacteria associated with the decrease of molecules with antibacterial properties makes it necessary to find new compounds containing molecules with this activity [18]. In this context, secondary metabolites produced by plants, which act as defense compounds against pathogens, could be a good strategy to explore.

The present work aimed to investigate the anti-inflammatory and antimicrobial properties of leaves, stems and flowers of a Portuguese *P. avium* (cv. *Saco*) in vitro. For this purpose, the effect of these by-products was analyzed in a mouse macrophage cell line (RAW 264.7) stimulated with lipopolysaccharide (LPS), as an inflammatory model. Nitric oxide (NO) production and scavenging capacity were studied. In addition, antimicrobial activity was determined by disk diffusion and minimum inhibitory concentrations (MICs) using the broth microdilution method.

2. Materials and Methods

2.1. Chemicals and Reagents

N-(1-naphthyl)ethylenediamine dihydrochloride, sulfanilamide and sodium nitroprusside dihydrate (SNP) were acquired from Alfa Aesar (Karlsruhe, Germany). Dulbecco's modified Eagle's medium (DMEM), fetal bovine serum (FBS), penicillin, streptomycin, trypsin-ethylenediaminetetraacetic acid solution, 3-(4,5-dimethylthiazol-2-yl)-2,5-diphenyltetrazolium bromide (MTT), dimethyl sulfoxide (DMSO), and β-nicotinamide adenine dinucleotide (NADH) were from Sigma-Aldrich (St. Louis, MO, USA). Purified water was acquired using the Milli-Qplus185 system (Millipore, Billerica, MA, USA). Müeller-Hinton agar (MHA) was purchased in VWR (Prolabo Chemicals, Radnor, PA, USA).

2.2. Plant Material

P. avium by-products, namely leaves, stems, and flowers were collected between April and June 2018 in the Fundão region (Portugal), as previously described [13]. Samples were frozen, freeze-dried in a freeze dryer lab equipment (ScanVac CoolSafe, LaboGene APS, Allerød, Denmark), reduced to a dried powder, and stored at −20 °C until analysis [13].

2.3. Extracts Preparation

Hydroethanolic extracts and aqueous infusions of *P. avium* by-products were prepared according to the method previously described [13]. Firstly, leaves, stems, and flowers were reduced in a dried powder. For the hydroethanolic extracts, 1 g of the dried powder of each sample was dissolved in ethanol/water (50:50, *v*/*v*), and posteriorly sonicated for 30 min. After that, the hydroethanolic solutions were shaken at room temperature for 2 h and then again sonicated for 30 min. Finally, the extracts were filtered using a 0.45 µm membrane (Millipore, Bedford, MA, USA) in a vacuum system, evaporated under reduced pressure, freeze-dried, and stored at −20 °C until further use.

Regarding to aqueous infusions, 1 g of dried powder of each sample was subjected to an infusion (100 mL of water) at 100 °C for 3 min, according to the manufacturer's instructions for daily herbal infusions. Then, infusions were filtered as described above to hydroethanolic extracts, freeze-dried, and stored at −20 °C until further analysis. The extraction yields were reported in a previous work [13].

2.4. Cell Culture

RAW 264.7 cell line, was cultured in a DMEM high glucose medium, supplemented with 10% (*v*/*v*) of non-inactivated FBS, 1% penicillin/streptomycin, and 1.5 g/L sodium bicarbonate, at 37 °C in a humidified atmosphere of 95% air and 5% CO_2. The cells were passaged using a cell scrapper, before reaching confluence and according to ATCC recommendations.

2.5. Anti-Inflammatory Potential

2.5.1. Effect of Cherry By-Products Extracts on Cellular Viability

To determine the effects of hydroethanolic extracts and aqueous infusions of *P. avium* by-products on cellular viability, the 3-(4,5-dimethyl-thiazol-2-yl)-2,5-diphenyl tetrazolium bromide (MTT) assay was performed, as previously described [19]. Briefly, 2.5×10^4 cells per well were seeded in a 96-well plate and incubated with different concentrations (50, 100, 200, 400, and 800 µg/mL) of the extracts, for 24 h, according to the described method, with minor adjustments [19]. Then, the medium was then removed and the MTT solution was added to each well (1 mg mL^{-1}) and the plates were incubated, in the dark at 37 °C for 4 h. The absorbance of formazan crystals was measured at 570 nm using an XmarckTM microplate absorbance spectrophotometer (Bio-Rad, Hercules, CA, USA). Six independent experiments were performed in triplicate.

2.5.2. Nitric Oxide (NO) Radicals Production

For nitrite accumulation in culture medium, cells were seeded at a density of 1×10^5 per well in a 96-well plate for 24 h, according to the method described by Oliveira and co-workers [20]. After that, the medium was removed, and the cells were incubated for 2 h with different dilutions, as described in the previous section. Subsequently, 1 µg mL^{-1} LPS was used to stimulate the cells, for 22 h. Production of NO radicals was determined by measuring nitrite accumulation in the supernatant, as previously described [20]. The Griess reagent (1% (*w*/*v*) sulfanilamide in 5% (*w*/*v*) phosphoric acid and 0.1% (*w*/*v*) N-(1-naphthyl)-ethylenediamine dihydrochloride) was mixed into the supernatants, and incubated at room temperature for 30 min. Absorbance was measured at 550 nm.

2.5.3. Determination of Radicals NO Scavenging Activity

NO radicals scavenging activity was evaluated according to the method explained by Jesus and co-workers [21]. Briefly, the same concentrations used in the cells were dissolved in potassium phosphate buffer (100 mM, pH 7.4) and mixed with SNP (20 mM), as described by Gonçalves and co-workers [11]. Plates were then incubated for 1 h at room temperature, under light. Griess reagent, composed by 1% sulfanilamide and 0.1% naphthylethylenediamine in 2% phosphoric acid (H_3PO_4), was added to the wells, and the multiwell plates were incubated for 10 min, in the dark. Then, the absorbance was measured at 560 nm. The radical scavenging activity NO was determined by comparing the absorbance values between the extracts and the control and corresponded to the mean ± standard deviations of three independent experiments, performed in triplicate.

2.6. *Antimicrobial Potential*

2.6.1. Bacterial Strains and Growth Conditions

Thirteen bacterial strains were used in this work acquired from American Type Culture Collection (ATCC, Manassas, VT, USA), BCCM/LMG Bacteria Collection (Belgian Co-Ordinated Collections of Micro-organisms, Gent, Belgium), and the Spanish Type Culture Collection (Valencia, Spain). Six Gram-positive bacteria (*Micrococcus luteus* (CECT 243), *Enterococcus faecalis* (ATCC 29212), *Bacillus cereus* (ATCC 11778), *Listeria monocytogenes* LMG 16779, *Staphylococcus aureus* ATCC 25923, and *Bacillus subtilis* (CECT 35)), and seven Gram-negative bacteria (*Salmonella typhimurium* (ATCC 13311), *Pseudomonas aeruginosa* ATCC 27853, *Escherichia coli* ATCC 25922, *Klebsiella pneumoniae* (ATCC 13883), *Proteus mirabilis* (CECT 170), *Serratia marcescens* (CECT 159) and *Acinetobacter baumannii* (LMG 1025)).

2.6.2. Disk Diffusion Method

The disk diffusion test was performed according to CLSI M02-A12 (2015) for bacteria [22]. Sterile Petri dishes (90 mm Ø) were used using 25 mL of medium. A 0.5 McFarland suspensions were made from bacterial cultures in sterile saline solution. Antimicrobial disks (Filtar Lab, 6 mm Ø) were impregnated with 10 µL each of a hydroethanolic extract or an aqueous infusion of *P. avium* leaves, stems, and flowers, following to the method previously described [23], with small modifications. Gentamycin was used as a positive control. MHA and with tryptic soy agar (TSA) medium were used for this experiment. The plates were incubated at 37 °C for 24 h. Posteriorly, the inhibition zone was visualized and measured in millimeters. The obtained results are reported as means ± standard deviations of three independent experiments.

2.6.3. Evaluation of the Minimum Inhibitory Concentration (MIC)

The susceptibility of bacteria to the hydroethanolic extracts and aqueous infusions of cherry leaves, stems, and flowers was determined using the broth microdilution method, as previously described [24]. Briefly, inoculums were prepared by suspension in NaCl 0.85% (*w*/*v*), and turbidity was adjusted to 0.5 McFarland to get a final concentration of approximately 5 × 106 CFU mL^{-1}. Experiments were performed in 96-well plate with a range of concentrations between 2 and 0.015 mg mL^{-1}, posteriorly incubated at 37 °C for 24 h. Then, 30 µL of a resazurin solution (0.01%) was added to each well and incubated for 2 h at 37 °C. The assay was performed in triplicate, and results were reported as modal values.

2.7. *Evaluation of P. avium by-Products Biocompatibility*

The cytotoxicity of hydroethanolic extracts and aqueous infusions of sweet cherry leaves, stems, and flowers was investigated using the normal human dermal fibroblasts (NHDF) cell line. This cell line was obtained from the American Type Culture Collection (ATCC; Manassas, VA, USA) and grown in high-glucose DMEM medium supplemented with 10% FBS and 1% antibiotic/antimycotic and 25 µg mL^{-1} amphotericin B (Sigma-Aldrich, Inc., St. Louis, MO, USA). 1×10^4 cells per well were seeded in 96-well plate and incubated at 37 °C in a humidified atmosphere of 95% air and 5% CO_2 for 24 h. After, the cells were incubated with different concentrations of extracts of *P. avium* by-products (50, 100, 200, 400, and 800 µg mL^{-1}), according to the method described previously [13]. After 24 h, the MTT assay described in Section 2.5.1 was performed. The absorbance was measured at 570 nm using an XmarckTM Microplate Absorbance Spectrophotometer (Bio-Rad, Hercules, CA, USA). Six independent experiments were performed in triplicate.

2.8. Statistical Analysis

The results were analyzed using GraphPad Prism 8 (GraphPad Software, San Diego, CA, USA). Data were expressed as mean ± standard deviations or mean ± standard error of the mean. One-way analysis variance (ANOVA) with Tukey's multiple comparison test or Student's *t*-test were performed to compare the effect of each concentration of *P. avium* by-products extract with controls. A p value < 0.05 was considered statistically significant.

3. Results and Discussion

3.1. Anti-Inflammatory Potential

The anti-inflammatory process is the body's natural mechanism in response to infection or injury. However, if the inflammatory state is overestimated, it can contribute to the onset and development of numerous diseases, such as cancer, cardiovascular and neurological diseases [25]. Plant extracts consist of various phytochemicals, mainly phenolic compounds [6,7]. According to the literature, these compounds possess a plethora of health benefits, including anti-inflammatory properties [11,20].

Macrophages play a crucial function in the innate immune response, so in this work, we used RAW 264.7 cells as an in vitro inflammatory model to investigate the possible anti-inflammatory activity of *P. avium* stems, leaves, and flowers.

3.1.1. Effect of *P. avium* By-Products on the Viability of the Raw 264.7 Cell Line

To determine whether the hydroethanolic extracts and aqueous infusions of sweet cherry leaves, stems, and flowers affect macrophages viability of, the MTT assay was performed after 24 h of exposure. The concentration of the extracts was adjusted to 50, 100, 200, 400 and 800 μg mL^{-1}. As shown in Figure 1, measurement of cell viability after treatment with all tested concentrations of the two extracts from *P. avium* stems and flowers aqueous infusion resulted in cell viability ≥90% (Figure 1C,D,F). The hydroethanolic extract and aqueous infusion both from leaves showed a cell viability ≥90% in a concentrations range of 50 to 400 μg mL^{-1} and 50 to 200 μg mL^{-1}, respectively (Figure 1A,B). The hydroethanolic extract of flowers showed cell viability between 80% and 90% (Figure 1E). Similar results were obtained by Orabona and co-workers [26] for *Crocus sativus* L. by-products. Based on these results, we can assume that the concentrations used are safe and non-toxic for further studies.

The biological properties of secondary plant metabolites have been extensively studied [7,27,28]. According to previous studies about phytochemical composition of *P. avium* by-products, it is known that cherry leaves, stems, and flowers are composed by several phenolic compounds, namely hydroxycinnamic acid, flavonols, flavan-3-ols, and flavanones [13,21]. Among several phenolics present in these by-products, the 3-caffeoylquinic acid cis, 5-caffeoylquinic acid trans, quercetin 3-*O*-rutinoside, quercetin 3-*O*-glucoside, 3-coumaroyl-5-caffeoylquinic acid, and 3-coumaroyl-4-caffeoylquinic acid are the most abundant [13]. The maintenance of RAW 264.7 cells viability when exposed to several concentrations and different extracts may be related to the presence of these hydroxycinnamic acids and flavonoids. Furthermore, these phenolic compounds are described as being able to reduce inflammation [29]. In addition, they are molecules with low cytotoxicity [29,30].

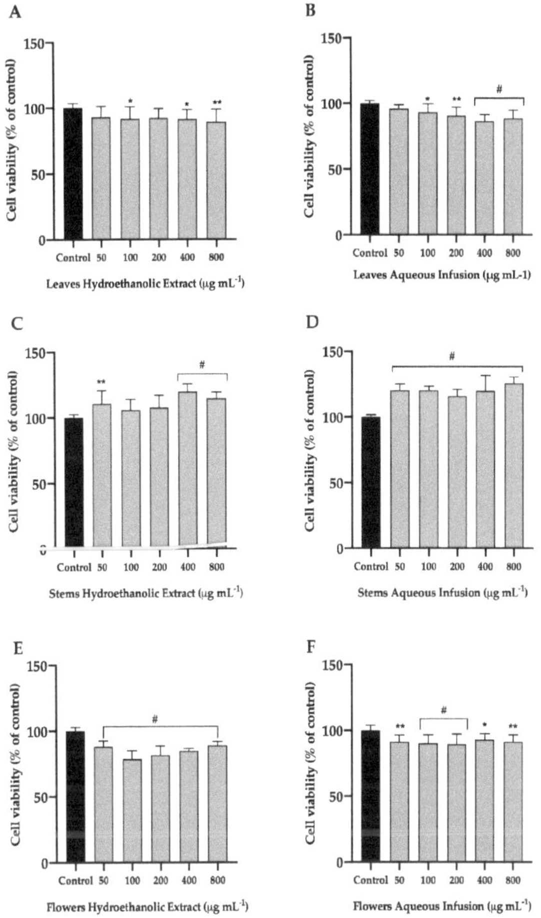

Figure 1. Effect of hydroethanolic extract (**A**,**C**,**E**) and aqueous infusion (**B**,**D**,**F**) of *Prunus avium* L. leaves, stems, and flowers on RAW 264.7 macrophages. Cells were plated and submitted to increasing concentrations of extracts for 24 h. 3-(4,5-dimethylthiazol-2-yl)-2,5-diphenyltetrazolium bromide (MTT) assay was performed to assess cell viability. The presented data correspond to the means ± standard deviation of three independent assays and are represented as percentage (%) of control cells. Statistical analysis: One-way ANOVA was performed for each concentration compared to control; * p value < 0.05, ** $p < 0.01$, and # $p < 0.001$ were considered significant.

3.1.2. Effect of *P. avium* By-Products on NO Production

Numerous inflammatory mediators are released as part of the inflammatory process. In addition, macrophages produce NO in large quantities, and it is known that natural products have shown a good response to this increase [31]. Lipopolysaccharide are extracellular elements of Gram-negative bacteria and possess potent stimuli for monocytes and macrophages. When these cells are activated by a stimulus induced by LPS, they produce inflammatory mediators, including NO, and release them via the regulation of pro-inflammatory factors [32]. Therefore, in this work, we investigated the anti-inflammatory potential of hydroethanolic extracts and aqueous infusions of cherry leaves, stems, and flowers through their ability to inhibit the production of NO in stimulated RAW 264.7 macrophages. An inflammatory stimulus was elicited by LPS (1 μg mL^{-1}), to simulate an inflammatory state. The production of nitrites (NO stable metabolites), was measured after cell exposure to *P. avium* by-products extracts by the colorimetric Griess reaction.

As shown in Figure 2, all extracts of cherry by-products were able to decrease the production of NO, in LPS-stimulated RAW 264.7 cells, in a dose-dependent manner. In particular, the hydroethanolic extracts of leaves and stems and the aqueous infusions of flowers were able to decrease NO to below 50% at extract concentrations of 400 μg mL^{-1} or more (Figure 2A,C–F). Interestingly, both extracts of flowers were promising, and showed a very significant reduction in NO, at extract concentrations of 200 μg mL^{-1} or greater (Figure 2E,F). To our knowledge, no similar studies have been performed with sweet cherry by-products. However, similar profiles were obtained for sweet cherry phenolic-rich fractions of the same variety of by-products [11].

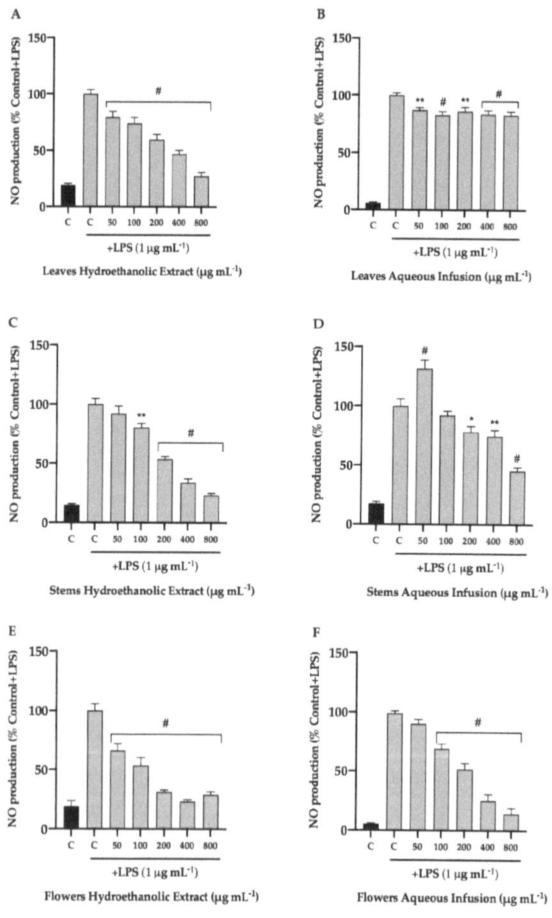

Figure 2. Effect of hydroethanolic extract (**A**,**C**,**E**) and aqueous infusion (**B**,**D**,**F**) of *Prunus avium* L. leaves, stems, and flowers on macrophage nitric oxide (NO) production upon an inflammatory stimulus. Cells were plated and exposed to Ctrl medium exposed to increasing concentrations of extracts for 24 h, in the presence of lipopolysaccharide (LPS) (1 μg mL^{-1}). Griess assay was performed to assess nitrites levels in the supernatant. The presented data correspond to the means ± standard error of the mean of three independent assays and are represented as percentage (%) of control cells (C) cells exposed to LPS (Ctrl LPS). Statistical analysis: One-way ANOVA was performed for each concentration compared to control with LPS; * p value < 0.05, ** $p < 0.01$, and # $p < 0.001$ were considered significant.

3.1.3. Effect of *P. avium* By-Products on NO Scavenging Activity

To determine whether the decrease in production of NO was due to the scavenging activity of extracts from *P. avium* leaves, stems, and flowers, the NO scavenging activity was considered. As shown in Figure 3, all hydroethanolic extracts and aqueous infusions of cherry by-products showed significant scavenging activity in a concentration-dependent manner. The aqueous infusions of cherry stems, flowers, and leaves were the most active (Figure 3B,D,F) showing a scavenging activity greater than 50% (IC_{50} values of 111.4 ± 0.31, 149.3 ± 0.94, and 183.5 ± 0.72 μg mL^{-1}, respectively). Hydroethanolic extracts and aqueous infusions of leaves, stems, and flowers demonstrated an excellent scavenging activity against NO radicals compared to ascorbic acid (IC_{50} = 767.67 ± 4.04 μg mL^{-1}) (data not shown). The results obtained are in agreement with previous studies [21].

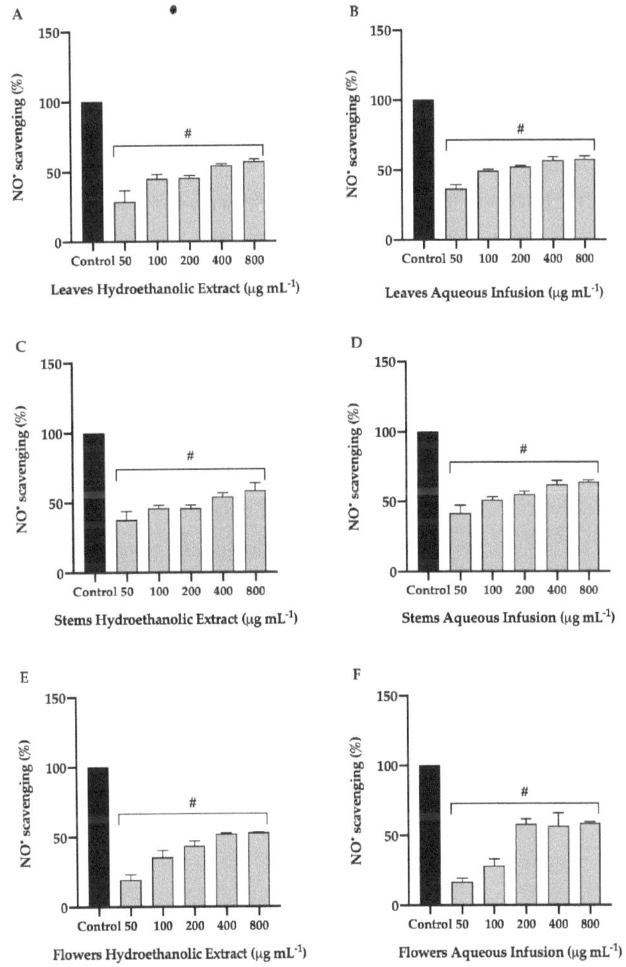

Figure 3. Effect of hydroethanolic extract (**A**,**C**,**E**) and aqueous infusion (**B**,**D**,**F**) of *Prunus avium* L. leaves, stems, and flowers on NO scavenging activity. Griess assay was performed to assess NO scavenging in supernatant. The presented data correspond to the means ± standard deviation of three independent assays and are represented as percentage (%) of control. Statistical analysis: One-way ANOVA was performed for each concentration compared to control; # $p < 0.001$ was considered significant.

The phenolic compounds are natural molecules with a relevant antioxidant potential Oxidative stress can contribute to the inflammatory process through the activation of several anti-inflammatory mediators. In this context, phenolics appear as quite useful compounds in the inhibition of oxidative stress. The antioxidant properties of *P. avium* by-products previously studied [13,21] are associated with the capacity to capture free radicals The chemical structure of phenolic compounds in sweet cherry leaves, stems, and flowers is capable of transferring hydrogen atoms to radicals, thus reducing their concentration [11] In addition, the NO scavenging activity of the extracts may help to break the chain of reactions triggered by excessive formation of NO and, thus, prevent the development of diseases, which also favors their use in the treatment of inflammatory disorders.

3.2. Antimicrobial Potential

Screening of antimicrobial activity of hydroethanolic extracts and aqueous infusions of *P. avium* leaves, stems, and flowers was evaluated using several Gram-positive and Gram-negative bacteria. For this purpose, the disk-diffusion method was first performed, which measures the influence of the *P. avium* extracts on the normal growth of bacteria on agar plates. The larger the diameter without growing bacteria around the paper disc impregnated with the extract, the higher the growth inhibitory properties of the extract. The diameters of the inhibition zones are shown in Table 1. Thus, higher antimicrobial activity against *B. cereus* (ATCC 11778) was observed with the hydroethanolic extract of stems, followed by stems aqueous infusion (inhibition halo of 7.24 ± 0.14 mm and 4.06 ± 0.2 mm, respectively) (Table 1). In addition, the hydroethanolic extract of stems was also able to inhibit the growth of *S. aureus* (ATCC 25923), with an inhibition halo of 3.04 ± 0.16 mm. According to the study conducted by Afonso and co-workers [10], extracts from cherry stems were also found to be more effective, for Gram-positive bacteria. The strain *P. mirabilis* (CECT 17) was the most resistant to the hydroethanolic extract from leaves, with an inhibition halo of 0.32 ± 0.15 mm. In general, the hydroethanolic extracts of *P. avium* by-products were those that showed the highest microbial activity against Gram-positive bacteria. These by-products showed almost no antimicrobial activity against Gram-negative bacteria (Table 1).

The antimicrobial activity of *P. avium* by-products has been associated with the content of phenolic compounds. Indeed, several authors reported that phenolics, such as *p*-coumaric acid, *p*-coumaroylquinic acid derivative, sakuranetin, neochlorogenic and chlorogenic acids, and catechin are some compounds involved in the antimicrobial properties of plant extracts [11,38–40]. Aromatic hydroxyl groups present in many phenolics have high affinity for bacterial membranes, interfering with membrane and cytoplasmic organelles, which may lead to bacterial death [10,41,42].

According to visual MIC, high activity of the hydroethanolic extract of leaves followed by the aqueous infusion leaves was observed, with MIC values ranging from 0.016 to 2 mg mL^{-1} for most of the bacterial strains studied (Table 2). The obtained results show that the by-products of *P. avium* have inhibitory activity against more than half of the tested strains, namely, *S. tiphymurium*, *E. faecalis*, *S. marcescens*, and *B. cereus* (Table 2). It can be concluded that the chemical composition of sweet cherry leaves, stems, and flowers contains a wide range of phytochemicals with antimicrobial properties. Many mechanisms may lead to stronger antimicrobial activity, such as the chemical composition of the by-products and the method of extraction of the bioactive compounds. According to Gullon and collaborators [43], the antimicrobial activity can be associated with the combination of several bioactive compounds.

Table 1. Diameters (mm) of inhibition zones for disk-diffusion method of hydroethanolic extracts and aqueous infusions from *P. avium* by-products.

	Hydroethanolic Extracts			Aqueous Infusions			Control
	Leaves	Stems	Flowers	Leaves	Stems	Flowers	Gentamycin
Gram-positive							
M. luteus CECT 243	0.00 ± 0.00	1.86 ± 0.25	1.00 ± 0.08	0.96 ± 0.05	1.16 ± 0.19	0.00 ± 0.00	15.83 ± 0.17
E. faecalis ATCC 29212	0.00 ± 0.00	2.85 ± 0.23	0.00 ± 0.00	0.00 ± 0.00	0.80 ± 0.21	0.00 ± 0.00	11.05 ± 0.38
B. cereus ATCC 11778	1.37 ± 0.17	7.24 ± 0.14	1.33 ± 0.17	1.14 ± 0.03	4.06 ± 0.2	1.20 ± 0.05	15.60 ± 0.82
L. monocytogenes LMG 16779	0.00 ± 0.00	1.94 ± 0.19	0.00 ± 0.00	0.00 ± 0.00	0.99 ± 0.01	0.00 ± 0.00	15.93 ± 0.13
S. aureus ATCC 25923	0.00 ± 0.00	3.04 ± 0.16	0.00 ± 0.00	0.00 ± 0.00	0.86 ± 0.09	0.00 ± 0.00	12.93 ± 0.43
B. subtilis CECT 35	0.00 ± 0.00	0.95 ± 0.06	0.76 ± 0.13	0.00 ± 0.00	0.64 ± 0.16	0.00 ± 0.00	29.70 ± 0.88
Gram-negative							
S. tiphymurium ATCC 13311	0.00 ± 0.00	0.00 ± 0.00	0.00 ± 0.00	0.00 ± 0.00	0.00 ± 0.00	0.00 ± 0.00	13.30 ± 0.65
P. aeruginosa ATCC 27853	0.00 ± 0.00	0.00 ± 0.00	0.00 ± 0.00	0.00 ± 0.00	0.00 ± 0.00	0.00 ± 0.00	18.03 ± 0.12
E. coli ATCC 25922	0.00 ± 0.00	0.00 ± 0.00	0.00 ± 0.00	0.00 ± 0.00	0.00 ± 0.00	0.00 ± 0.00	12.57 ± 0.22
K. pneumoniae ATCC 13883	0.00 ± 0.00	0.00 ± 0.00	0.00 ± 0.00	0.00 ± 0.00	0.00 ± 0.00	0.00 ± 0.00	10.30 ± 0.08
P. mirabilis CECT 17	0.32 ± 0.15	0.00 ± 0.00	0.00 ± 0.00	0.00 ± 0.00	0.00 ± 0.00	0.00 ± 0.00	17.73 ± 0.18
S. marcescens CECT 159	0.00 ± 0.00	0.00 ± 0.00	0.00 ± 0.00	0.00 ± 0.00	0.00 ± 0.00	0.00 ± 0.00	18.83 ± 0.09
A. baumannii LMG 1025	0.00 ± 0.00	0.00 ± 0.00	0.00 ± 0.00	0.00 ± 0.00	0.00 ± 0.00	0.00 ± 0.00	19.33 ± 0.27

Values for inhibition zone are presented as means ± standard deviations; ATCC—American Type Culture Collection; LMG—Laboratory for Microbiology of the Faculty of Sciences of the Ghent University; CECT—Spanish Type Culture Collection.

Table 2. Minimum inhibitory concentration (MIC) values (mg mL^{-1}) of hydroethanolic extracts and aqueous infusions from *Prunus avium* L. by-products.

	Hydroethanolic Extracts			Aqueous Infusions			Negative Control	Positive Control
	Leaves	Stems	Flowers	Leaves	Stems	Flowers	DMSO	Gentamicin
Gram-positive								
M. luteus CECT 243	0.5	2	2	2	1	>2	>2	0.016
E. faecalis ATCC 29212	0.016	0.5	2	0.016	1	>2	>2	0.016
B. cereus ATCC 11778	0.016	1	2	0.016	1	2	>2	0.016
L. monocytogenes LMG 16779	0.016	1	0.031	0.5	0.5	>2	>2	0.016
S. aureus ATCC 25923	0.5	2	2	2	2	2	>2	0.016
B. subtilis CECT 35	1	2	>2	2	2	>2	>2	0.016
Gram-negative								
S. tiphymurium ATCC 13311	0.016	0.031	2	0.016	0.031	1	>2	0.016
P. aeruginosa ATCC 27853	1	0.5	0.5	2	0.5	2	>2	0.016
E. coli ATCC 25922	0.5	1	1	2	1	2	>2	0.016
K. pneumoniae ATCC 13883	1	1	2	0.016	1	>2	>2	0.016
P. mirabilis CECT 17	0.016	2	>2	1	1	>2	>2	0.016
S. marcescens CECT 159	0.016	1	2	0.016	1	>2	>2	0.016
A. baumannii LMG 1025	0.016	1	>2	0.25	1	2	>2	0.016

3.3. Effect of *P. avium* By-Products on NHDF Cell Line

To understand the biocompatibility of *P. avium* by-product extracts with human cells, the effects of hydroethanolic extracts and aqueous infusions of leaves, stems, and flowers were studied in NHDF cells (Figure 4). The results show that, both extracts from stems increased significantly the cellular viability of these cells (Figure 4C,D). Although aqueous infusions of leaves and flowers slightly reduced the viability of NDHD, and considering ISO 1093:5-2009, it is possible that the concentrations used did not present cytotoxicity when compared to untreated cells (Control) (Figure 4B,F).

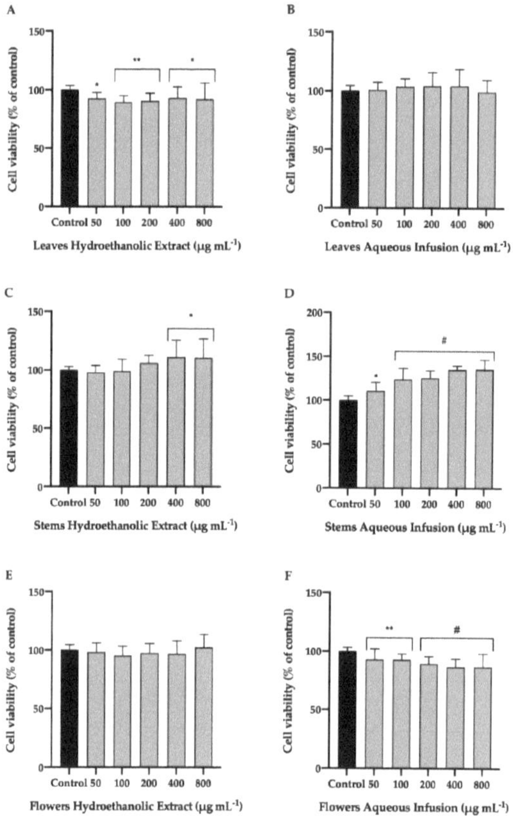

Figure 4. Effect of hydroethanolic extract (**A**,**C**,**E**) and aqueous infusion (**B**,**D**,**F**) of *Prunus avium* L. leaves, stems, and flowers on the normal human dermal fibroblast (NHDF) cell line. Cells were plated and exposed to increasing concentrations of extracts for 24 h. 3-(4,5-dimethylthiazol-2-yl)-2,5-diphenyltetrazolium bromide (MTT) assay was performed to assess cell viability. The presented data correspond to the means ± standard deviation of three independent assays and are represented as percentage (%) of control cells. Statistical analysis: One-way ANOVA was performed for each concentration compared to control; * p value < 0.05, ** $p < 0.01$, and # $p < 0.001$ were considered significant.

The biological properties of the extracts are related to their composition of phenolic compounds. In the literature, several studies have demonstrated that these phytochemicals, isolated or in synergy, contribute to the elimination of ROS and, consequently to the reduction of oxidative stress, promote the maintenance of healthy skin through collagen production [44–47]. Detailed knowledge of the chemical composition and biological properties of *P. avium* by-products in conjugation with its biosafety profile, biocompatibility, and use is essential for biomedical applications [48].

4. Conclusions

This work is the first to report the anti-inflammatory and antimicrobial activity of leaves, stems, and flowers of *P. avium* from the Fundão region (Portugal). Both sweet cherry by-products extracts were found to be safe and non-cytotoxic to RAW 264.7 and NDHF cell lines. The hydroethanolic extracts of leaves and stems and the aqueous infusion of flowers showed promising results suppressing inflammation by reducing the production of NO by LPS-activated RAW 264.7 macrophages. In addition, all aqueous infusions of the by-products, especially cherry flowers, showed excellent scavenging activity against NO radicals. In addition, cherry leaves, stems, and flowers were found to be more effective against Gram-positive bacteria. Moreover, the hydroethanolic extract of the leaves and the aqueous infusion of the leaves showed great antimicrobial capacity for most of the tested bacteria. Our results demonstrate, for the first time, that *P. avium* by-products are a good source of phenolics with relevant anti-inflammatory and antimicrobial potential, highlighting the importance of using these biowastes as bioactive compounds for therapeutic use in inflammation and control the microbial growth.

Author Contributions: Conceptualization, A.R.N., L.R.S. and J.D.F.-F.; methodology, A.R.N., A.C.G. and J.D.F.-F.; software, A.R.N. and J.D.F.-F.; validation, A.R.N., J.D.F.-F., A.C.G., A.F., G.A. and L.R.S.; investigation, A.R.N. and J.D.F.-F.; resources, J.D.F.-F., A.F., G.A. and L.R.S.; data curation, A.R.N. and J.D.F.-F.; writing—original draft preparation, A.R.N.; writing—review and editing, A.R.N., J.D.F.-F., A.C.G., A.F., G.A. and L.R.S.; visualization, A.R.N., J.D.F.-F., A.C.G., A.F., G.A. and L.R.S.; supervision, A.F., G.A. and L.R.S.; project administration, G.A.; funding acquisition, J.D.F.-F., A.F., G.A. and L.R.S. All authors have read and agreed to the published version of the manuscript.

Funding: This work was supported by the Foundation for Science and Technology (FCT), through funds from State Budget, and by the European Regional Development Fund (ERDF), under the Portugal 2020 Program, through the Regional Operational Program of the Center (Centro2020), through the Projects with the reference (UIDB/00709/2020). Financial support was also provided by FCT through PhD fellowships (SFRH/BD/139137/2018) awarded to Ana R. Nunes and (2020.04947.BD) awarded to Ana C. Gonçalves. José D. Flores-Félix was supported by the Marie Sklodowska-Curie grant agreement No. 101003373.

Institutional Review Board Statement: Not applicable.

Informed Consent Statement: Not applicable.

Data Availability Statement: Not applicable.

Acknowledgments: The authors acknowledge Susana Ferreira and Alexandra Coimbra from the Microbiology Laboratory I from Health Sciences Research Centre, University of Beira Interior (CICS-UBI) for their laboratorial and scientific support.

Conflicts of Interest: The authors declare no conflict of interest.

References

1. Poswal, F.S.; Russell, G.; Mackonochie, M.; MacLennan, E.; Adukwu, E.C.; Rolfe, V. Herbal Teas and their Health Benefits: A Scoping Review. *Plant Foods Hum. Nutr.* **2019**, *74*, 266–276. [CrossRef]
2. Ahnen, R.T., Jonnalagadda, S.S.; Slavin, J.L. Role of plant protein in nutrition, wellness, and health. *Nutr. Rev.* **2019**, *77*, 735–747. [CrossRef]
3. Barreira, J.C.M.; Ferreira, I.C.F.R.; Oliveira, M.B.P.P.; Pereira, J.A. Effects of different phenols extraction conditions on antioxidant activity of almond (*prunus dulcis*) fruits. *J. Food Biochem.* **2009**, *33*, 763–776. [CrossRef]
4. Ferreira-Santos, P.; Zanuso, E.; Genisheva, Z.; Rocha, C.M.R.; Teixeira, J.A. Green and Sustainable Valorization of Bioactive Phenolic Compounds from Pinus By-Products. *Molecules* **2020**, *25*, 2931. [CrossRef]
5. Chockchaisawasdee, S.; Golding, J.B.; Vuong, Q.V.; Papoutsis, K.; Stathopoulos, C.E. Sweet cherry: Composition, postharvest preservation, processing and trends for its future use. *Trends Food Sci. Technol.* **2016**, *55*, 72–83. [CrossRef]
6. Gonçalves, A.C.; Bento, C.; Jesus, F.; Alves, G.; Silva, L.R. Sweet cherry phenolic compounds: Identification, characterization, and health benefits. In *Studies in Natural Products Chemistry*; Atta-ur-Rahman, F., Ed.; Science Publishers: Amesterdam, The Netherlands, 2018; pp. 31–78. ISBN 9780444641793.
7. Nunes, A.R.; Gonçalves, A.C.; Falcão, A.; Alves, G.; Silva, L.R. *Prunus avium* L. (Sweet Cherry) By-Products: A Source of Phenolic Compounds with Antioxidant and Anti-Hyperglycemic Properties—A Review. *Appl. Sci.* **2021**, *11*, 8516. [CrossRef]
8. Gonçalves, A.C.; Rodrigues, M.; Flores-Félix, J.D.; Nunes, A.R.; Ribeiro, A.B.; Alves, G. Sweet cherry phenolics revealed to be promising agents in inhibiting P-glycoprotein activity and increasing cellular viability under oxidative stress conditions: In vitro and in silico study. *J. Food Sci.* **2021**, *87*, 450–465. [CrossRef]
9. Gonçalves, A.C.; Campos, G.; Alves, G.; García-Viguera, C.; Moreno, D.A.; Silva, L.R. Physical and phytochemical compositon of 23 Portuguese sweet cherries as conditioned by variety (or genotype). *Food Chem.* **2021**, *335*, 127637. [CrossRef]
10. Afonso, S.; Oliveira, I.V.; Meyer, A.S.; Aires, A.; Saavedra, M.J.; Gonçalves, B. Phenolic Profile and Bioactive Potential of Stems and Seed Kernels of Sweet Cherry Fruit. *Antioxidants* **2020**, *9*, 1295. [CrossRef]

11. Gonçalves, A.C.; Costa, A.R.; Flores-Félix, J.D.; Falcão, A.; Alves, G.; Silva, L.R. Anti-Inflammatory and Antiproliferative Properties of Sweet Cherry Phenolic-Rich Extracts. *Molecules* **2022**, *27*, 268. [CrossRef] [PubMed]
12. Gonçalves, A.C.; Campos, G.; Pinto, E.; Oliveira, A.S.; Almeida, A.; de Pinho, P.G.; Alves, G.; Silva, L.R. Essential and non-essential elements, and volatile organic compounds for the discrimination of twenty-three sweet cherry cultivars from Fundão, Portugal. *Food Chem.* **2022**, *367*, 130503. [CrossRef] [PubMed]
13. Nunes, A.R.; Gonçalves, A.C.; Alves, G.; Falcão, A.; García-Viguera, C.; Moreno, D.A.; Silva, L.R. Valorisation of *Prunus avium* L. By-Products: Phenolic Composition and Effect on Caco-2 Cells Viability. *Foods* **2021**, *10*, 1185. [CrossRef]
14. Nunes, A.R.; Gonçalves, A.C.; Pinto, E.; Amaro, F.; Flores-Félix, J.D.; Almeida, A.; de Pinho, P.G.; Falcão, A.; Alves, G.; Silva, L.R. Mineral Content and Volatile Profiling of *Prunus avium* L. (Sweet Cherry) By-Products from Fundão Region (Portugal). *Foods* **2022**, *11*, 751. [CrossRef]
15. Bastos, C.; Barros, L.; Duenas, M.; Calhelha, R.C.; Queiroz, M.J.R.P.; Santos-Buelga, C.; Ferreira, I.C.F.R. Chemical characterisation and bioactive properties of *Prunus avium* L.: The widely studied fruits and the unexplored stems. *Food Chem.* **2015**, *173*, 1045–1053. [CrossRef] [PubMed]
16. Hussain, T.; Tan, B.; Yin, Y.; Blachier, F.; Tossou, M.C.B.; Rahu, N. Oxidative Stress and Inflammation: What Polyphenols Can Do for Us? *Oxid. Med. Cell. Longev.* **2016**, *2016*, 7432797. [CrossRef]
17. Shahidi, F.; Yeo, J. Bioactivities of Phenolics by Focusing on Suppression of Chronic Diseases: A Review. *Int. J. Mol. Sci.* **2018**, *19*, 1573. [CrossRef]
18. Abedini, A.; Colin, M.; Hubert, J.; Charpentier, E.; Angelis, A.; Bounasri, H.; Bertaux, B.; Kotland, A.; Reffuveille, F.; Nuzillard, J.-M.; et al. Abundant Extractable Metabolites from Temperate Tree Barks: The Specific Antimicrobial Activity of *Prunus Avium* Extracts. *Antibiotics* **2020**, *9*, 111. [CrossRef] [PubMed]
19. Gonçalves, A.C.; Rodrigues, M.; Santos, A.; Alves, G.; Silva, L.R. Antioxidant Status, Antidiabetic Properties and Effects on Caco-2 Cells of Colored and Non-Colored Enriched Extracts of Sweet Cherry Fruits. *Nutrients* **2018**, *10*, 1688. [CrossRef] [PubMed]
20. Oliveira, A.S.; Rolo, J.; Gaspar, C.; Cavaleiro, C.; Salgueiro, L.; Palmeira-de-Oliveira, R.; Ferraz, C.; Coelho, S.; Pastorinho, M.R.; Sousa, A.C.; et al. Chemical characterization and bioactive potential of Thymus×citriodorus (Pers.) Schreb. preparations for anti-acne applications: Antimicrobial, anti-biofilm, anti-inflammatory and safety profiles. *J. Ethnopharmacol.* **2022**, *287*, 114935. [CrossRef]
21. Jesus, F.; Gonçalves, A.C.; Alves, G.; Silva, L.R. Exploring the phenolic profile, antioxidant, antidiabetic and anti-hemolytic potential of *Prunus avium* vegetal parts. *Food Res. Int.* **2019**, *116*, 600–610. [CrossRef]
22. Martins, J.M.; Scheffer, M.C.; de Melo Machado, H.; Schörner, M.A.; Golfetto, L.; dos Santos, T.M.; Barazzetti, F.H.; de Albuquerque, V.C.B.; Bazzo, M.L. Spectinomycin, gentamicin, and routine disc diffusion testing: An alternative for the treatment and monitoring of multidrug-resistant *Neisseria gonorrhoeae*? *J. Microbiol. Methods* **2022**, *197*, 106480. [CrossRef] [PubMed]
23. Coimbra, A.; Miguel, S.; Ribeiro, M.; Coutinho, P.; Silva, L.; Duarte, A.P.; Ferreira, S. Thymus zygis Essential Oil: Phytochemical Characterization, Bioactivity Evaluation and Synergistic Effect with Antibiotics against *Staphylococcus aureus*. *Antibiotics* **2022**, *11*, 146. [CrossRef] [PubMed]
24. Coimbra, A.T.; Luís, Â.; Batista, M.T.; Ferreira, S.; Duarte, A.P. Phytochemical Characterization, Bioactivities Evaluation and Synergistic Effect of *Arbutus unedo* and *Crataegus monogyna* Extracts with Amphotericin B. *Curr. Microbiol.* **2020**, *77*, 2143–2154. [CrossRef]
25. Chen, L.; Deng, H.; Cui, H.; Fang, J.; Zuo, Z.; Deng, J.; Li, Y.; Wang, X.; Zhao, L. Inflammatory responses and inflammation-associated diseases in organs. *Oncotarget* **2018**, *9*, 7204–7218. [CrossRef]
26. Orabona, C.; Orecchini, E.; Volpi, C.; Bacaloni, F.; Panfili, E.; Pagano, C.; Perioli, L.; Belladonna, M.L. Crocus sativus L. Petal Extract Inhibits Inflammation and Osteoclastogenesis in RAW 264.7 Cell Model. *Pharmaceutics* **2022**, *14*, 1290. [CrossRef] [PubMed]
27. Zaynab, M.; Fatima, M.; Abbas, S.; Sharif, Y.; Umair, M.; Zafar, M.H.; Bahadar, K. Role of secondary metabolites in plant defense against pathogens. *Microb. Pathog.* **2018**, *124*, 198–202. [CrossRef]
28. Ferretti, G.; Bacchetti, T.; Belleggia, A.; Neri, D. Cherry antioxidants: From farm to table. *Molecules* **2010**, *15*, 6993–7005. [CrossRef]
29. Maleki, S.J.; Crespo, J.F.; Cabanillas, B. Anti-inflammatory effects of flavonoids. *Food Chem.* **2019**, *299*, 125124. [CrossRef]
30. Chang, S.K.; Alasalvar, C.; Shahidi, F. Superfruits: Phytochemicals, antioxidant efficacies, and health effects—A comprehensive review. *Crit. Rev. Food Sci. Nutr.* **2019**, *59*, 1580–1604. [CrossRef]
31. Afonso, A.F.; Pereira, O.R.; Fernandes, Â.S.F.; Calhelha, R.C.; Silva, A.M.S.; Ferreira, I.C.F.; Cardoso, S.M. The Health-Benefits and Phytochemical Profile of *Salvia apiana* and *Salvia farinacea* var. Victoria Blue Decoctions. *Antioxidants* **2019**, *8*, 241. [CrossRef]
32. Xu, X.; Yin, P.; Wan, C.; Chong, X.; Liu, M.; Cheng, P.; Chen, J.; Liu, F.; Xu, J. Punicalagin inhibits inflammation in LPS-induced RAW264.7 macrophages via the suppression of TLR4-mediated MAPKs and NF-κB activation. *Inflammation* **2014**, *37*, 956–965. [CrossRef] [PubMed]
33. Al-Khayri, J.M.; Sahana, G.R.; Nagella, P.; Joseph, B.V.; Alessa, F.M.; Al-Mssallem, M.Q. Flavonoids as Potential Anti-Inflammatory Molecules: A Review. *Molecules* **2022**, *27*, 2901. [CrossRef] [PubMed]
34. Seeram, N.P.; Momin, R.A.; Nair, M.G.; Bourquin, L.D. Cyclooxygenase inhibitory and antioxidant cyanidin glycosides in cherries and berries. *Phytomedicine* **2001**, *8*, 362–369. [CrossRef] [PubMed]
35. Kim, E.O.; Min, K.J.; Kwon, T.K.; Um, B.H.; Moreau, R.A.; Choi, S.W. Anti-inflammatory activity of hydroxycinnamic acid derivatives isolated from corn bran in lipopolysaccharide-stimulated Raw 264.7 macrophages. *Food Chem. Toxicol.* **2012**, *50*, 1309–1316. [CrossRef] [PubMed]

36. Jung, H.; Kwak, H.-K.; Hwang, K.T. Antioxidant and antiinflammatory activities of cyanidin-3-glucoside and cyanidin-3-rutinoside in hydrogen peroxide and lipopolysaccharide-treated RAW264.7 cells. *Food Sci. Biotechnol.* **2014**, *23*, 2053–2062. [CrossRef]
37. Tang, J.; Diao, P.; Shu, X.; Li, L.; Xiong, L. Quercetin and Quercitrin Attenuates the Inflammatory Response and Oxidative Stress in LPS-Induced RAW264.7 Cells: In Vitro Assessment and a Theoretical Model. *Biomed Res. Int.* **2019**, *2019*, 7039802. [CrossRef]
38. Smailagić, A.; Ristivojević, P.; Dimkić, I.; Pavlović, T.; Dabić Zagorac, D.; Veljović, S.; Fotirić Akšić, M.; Meland, M.; Natić, M. Radical Scavenging and Antimicrobial Properties of Polyphenol Rich Waste Wood Extracts. *Foods* **2020**, *9*, 319. [CrossRef]
39. Aires, A.; Dias, C.; Carvalho, R.; Saavedra, M.J. Analysis of glycosylated flavonoids extracted from sweet-cherry stems, as antibacterial agents against pathogenic Escherichia coli isolates. *Acta Biochim. Pol.* **2017**, *64*, 265–271. [CrossRef]
40. Liu, J.; Du, C.; Beaman, H.T.; Monroe, M.B.B. Characterization of Phenolic Acid Antimicrobial and Antioxidant Structure-Property Relationships. *Pharmaceutics* **2020**, *12*, 419. [CrossRef]
41. Lou, Z.; Wang, H.; Zhu, S.; Ma, C.; Wang, Z. Antibacterial activity and mechanism of action of chlorogenic acid. *J. Food Sci.* **2011**, *76*, M398–M403. [CrossRef]
42. Nunes, R.; Pasko, P.; Tyszka-Czochara, M.; Szewczyk, A.; Szlosarczyk, M.; Carvalho, I.S. Antibacterial, antioxidant and anti-proliferative properties and zinc content of five south Portugal herbs. *Pharm. Biol.* **2017**, *55*, 114–123. [CrossRef] [PubMed]
43. Gullon, B.; Pintado, M.E.; Pérez-Álvarez, J.A.; Viuda-Martos, M. Assessment of polyphenolic profile and antibacterial activity of pomegranate peel (*Punica granatum*) flour obtained from co-product of juice extraction. *Food Control* **2016**, *59*, 94–98. [CrossRef]
44. Cuelho, C.H.F.; Alves, G.A.D.; Lovatto, M.O.; Bonilha, I.F.; Barbisan, F.; da Cruz, I.B.M.; Oliveira, S.M.; Fachinetto, R.; do Canto, G.S.; Manfron, M.P. Topical formulation containing *Ilex Paraguariensis* extract increases metalloproteinases and myeloperoxidase activities in mice exposed to UVB radiation. *J. Photochem. Photobiol. B.* **2018**, *189*, 95–103. [CrossRef] [PubMed]
45. Kim, J.; Lee, C.-W.; Kim, E.K.; Lee, S.-J.; Park, N.-H.; Kim, H.-S.; Kim, H.-K.; Char, K.; Jang, Y.P.; Kim, J.-W. Inhibition effect of *Gynura procumbens* extract on UV-B-induced matrix-metalloproteinase expression in human dermal fibroblasts. *J. Ethnopharmacol.* **2011**, *137*, 427–433. [CrossRef]
46. Kurt-Celep, İ.; Celep, E.; Akyüz, S.; İnan, Y.; Barak, T.H.; Akaydın, G.; Telci, D.; Yesilada, E. *Hypericum olympicum* L. recovers DNA damage and prevents MMP-9 activation induced by UVB in human dermal fibroblasts. *J. Ethnopharmacol.* **2020**, *246*, 112202. [CrossRef]
47. Wu, P.-Y.; Huang, C.-C.; Chu, Y.; Huang, Y.-H.; Lin, P.; Liu, Y.-H.; Wen, K.-C.; Lin, C.-Y.; Hsu, M.-C.; Chiang, H.-M. Alleviation of Ultraviolet B-Induced Photodamage by *Coffea arabica* Extract in Human Skin Fibroblasts and Hairless Mouse Skin. *Int. J. Mol. Sci.* **2017**, *18*, 782. [CrossRef]
48. Faustino, M.; Veiga, M.; Sousa, P.; Costa, E.M.; Silva, S.; Pintado, M. Agro-Food Byproducts as a New Source of Natural Food Additives. *Molecules* **2019**, *24*, 1056. [CrossRef]

nutrients

MDPI

Article

Ameliorative Effects of *Daphnopsis costaricensis* Extract against Oxazolone-Induced Atopic Dermatitis-like Lesions in BALB/c Mice

Yunji Bae [1,†], Taeyoung Kim [1,†], Nojune Park [2,†], Sangho Choi [3], Dongkeun Yi [3], Silvia Soto [4], Nelson Zamora [4], Sunam Kim [2,*] and Minhye Yang [1,*]

1 College of Pharmacy, Pusan National University, Busan 46241, Korea
2 Natural Products Research Center, Korea Institute of Science and Technology, Gangneung 25451, Korea
3 International Biological Material Research Center, Korea Research Institute of Bioscience and Biotechnology, 111 Gwahak-ro, Yuseong-gu, Daejeon 34141, Korea
4 Bioprospecting Research Unit, National Biodiversity Institute, Heredia 22-3100, Costa Rica
* Correspondence: snkim@kist.re.kr (S.K.); mhyang@pusan.ac.kr (M.Y.); Tel.: +82-33-650-3419 (S.K.); +82-51-513-6754 (M.Y.)
† These authors contributed equally to this work.

Abstract: The genus *Daphnopsis* has been traditionally used as a purgative, diuretic, stimulant, and psoriasis treatment. In this study, the anti-AD (atopic dermatitis) activities of the *Daphnopsis costaricensis* EtOH extract (DCE) were investigated in an oxazolone (OX)-induced mouse model of AD, and the anti-inflammatory effects of its active compounds were confirmed in PI-sensitized or IgE/DNP-BSA-sensitized RBL-2H3 cells. DCE improved the symptoms of OX-induced inflammatory dermatitis (swelling, erythema, and increased ear thickening) in OX-induced BALB/c mice ears and reduced epidermal thickness and mast cell infiltration. Eleven flavonoid compounds were isolated from DCE, and two compounds (7,8-dimethoxyflavone and 7,2′-dimethoxyflavone) significantly inhibited IL-4 overexpression in PI-induced RBL-2H3 cells and mast cell degranulation in IgE + DNP-BSA-induced RBL-2H3 cells. Our study indicates that DCE and two compounds (7,8-dimethoxyflavone and 7,2′-dimethoxyflavone) might effectively improve inflammatory and atopic skin symptoms.

Keywords: *Daphnopsis costariensis*; atopic dermatitis; anti-inflammatory; 7,8-dimethoxyflavone; 7,2′-dimethoxyflavone; interleukin 4; β-hexosaminidase

Citation: Bae, Y.; Kim, T.; Park, N.; Choi, S.; Yi, D.; Soto, S.; Zamora, N.; Kim, S.; Yang, M. Ameliorative Effects of *Daphnopsis costaricensis* Extract against Oxazolone-Induced Atopic Dermatitis-like Lesions in BALB/c Mice. *Nutrients* **2022**, *14*, 4521. https://doi.org/10.3390/nu14214521

Academic Editors: Maria Digiacomo and Doretta Cuffaro

Received: 5 October 2022
Accepted: 26 October 2022
Published: 27 October 2022

1. Introduction

Atopic dermatitis (AD) is a chronic skin disease caused by multiple factors. Although its main pathogenesis has not been clarified, it is known to be associated with genetic, immunological, environmental, and psychological factors [1]. Characteristic symptoms of AD are lichenification, dry skin, pruritus, eczema, and erythema [1], In particular, itching is the major diagnostic criterion of AD, and can have severe impacts on physical and emotional life [2]. Immunologically, activation of serum immunoglobulin E (IgE) production due to Th2 cytokine (IL-4 and IL-13) overexpression is the main cause of skin lesion formation [3] because it mediates inflammation by inducing the release of histamine and prostaglandin E from various inflammatory factors and mast cells [4,5]. In addition, it also damages the skin barrier, and, thus, facilitates the entry of inflammatory mediators into skin and repeats a vicious cycle of inflammation [6]. Currently, topical corticosteroids, topical calcineurin inhibitors, and antihistamines along with moisturizers are used to suppress these immune reactions in AD [7,8]. In addition, the monoclonal antibody dupilumab, which targets IL-4Rα and inhibits the biological actions of both IL-4 and IL-13, has recently been approved to treat adult patients with AD [9]. However, long-term treatment raises safety concerns of potential side effects (skin atrophy, increased percutaneous absorption, burning, and lethargy) [10,11].

Researchers are focusing on natural products to reduce the risks of conventional AD treatments [12], and various natural extracts have been reported to have the potential to treat AD [13,14]. Those described to date usually contain phenolics, such as flavonoids and coumarins, which are considered strong inhibitors of inflammatory and allergic reactions [15]. In particular, they potently inhibit the activities of inflammatory mediators such as IL-4, IgE, and proinflammatory cytokines in inflammatory animal models and cells [16,17]. Furthermore, many studies have demonstrated that natural products are safer and more effective compared to conventional chemical products [12,18]. Thus, natural products offer attractive alternatives for the treatment of AD.

Daphnopsis costaricensis is a small shrub of the family Thymelaeaceae indigenous to Costa Rica [19], and members of this genus have been used as diuretics, radical purgatives, and to treat psoriasis [20]. Previous phytochemical studies have identified secondary metabolites, including flavonoids, terpenoids, and terpenes, in this genus [20,21], but no such study has been conducted on *D. costaricensis*. Therefore, in the present study, to propose a potential natural anti-AD and anti-inflammatory substance, we investigated the effects of *D. costaricensis* EtOH extract (DCE) on AD-like lesions in an oxazolone (OX)-induced mouse model. In addition, we isolated and identified eleven compounds and examined their anti-inflammatory and anti-allergic activities on PI or IgE + DNP-BSA-induced RBL-2H3 cells.

2. Materials and Methods

2.1. Plant Material

The aerial parts of *Daphnopsis costaricensis* (Thymelaeaceae) were collected in the Osa Conservation Area, Golfo Dulce Forest Reserve, Costa Rica, in July 2013 and identified by Nelson Zamora (National Institute of Biodiversity, INBio). Voucher specimens (KRIB 0051617) are preserved at the International Biological Material Research Center (IBMRC) at the Korea Research Institute of Bioscience and Biotechnology, Daejeon, Republic of Korea.

2.2. General Experimental Procedures

JEOL 400 MHz (JEOL, Tokyo, Japan), Bruker 500 MHz (Bruker, Billerica, MA, USA), and Agilent Technologies 600 MHz instruments (Santa Clara, CA, USA) were used to obtain ^{1}H, ^{13}C, HMQC and HMBC NMR spectra. Sephadex LH-20 (25–100 μm; Pharmacia, Stockholm, Sweden) and silica gel (230–400 mesh; Merck, Darmstadt, Germany) were used to perform column chromatography, and Merck precoated silica gel 60 F_{254} Art. 5715 (Merck, Germany) plates were used to perform thin-layer chromatography (TLC). HR ESI mass spectra were obtained using an Agilent Technologies 6530 Accurate-Mass Q-TOF LC/MS. Quantitative analysis was performed using a Shimadzu HPLC system (Tokyo, Japan) equipped with an SPD-20A UV/VIS detector, two LC-20AT pumps, and a CBM-20A HPLC system controller.

2.3. Extraction and Isolation

Air-dried aerial parts of *D. costaricensis* (4.4 kg) were subjected to three cycles of reflux extraction (95% EtOH (44 L) for 90 min at 30 °C and then held for 12 h at room temperature (RT)), filtered, and filtrates were combined and evaporated in vacuo at 40 °C to give DCE (225.3 g). The DCE obtained was suspended in distilled water (2 L) and partition extracted sequentially versus 4 L of *n*-hexane, ethyl acetate (EtOAc), and *n*-butanol (*n*-BuOH) to obtain 15.9 g, 13.2 g, and 32.3 g, respectively, of the corresponding extracts.

The EtOAc extract (13.2 g) was loaded to gradient silica gel column chromatography using Hexane:EtOAc (5:1 → 100% MeOH) as eluent to obtain 9 fractions (DCE1~DCE9). Fraction DCE4 (45 mg) was fractionated into three subfractions (DCE4-1~DCE4-3) using Sephadex LH-20 with MeOH. Subfraction DCE4-2 was subjected to RP HPLC (Watchers 120 ODS-BP, S-10 μm, 150 × 10 mm) at a flow rate of 2 mL/min using a UV detector (365 nm) and MeOH:H_2O (80:20) as eluant to yield 6 subfractions (DCE4-2-1~DCE4-2-6), which afforded compound **1** (0.7 mg, t_R 15 min), and compound **2** (0.8 mg, t_R 20 min).

Subfraction DCE4-2-5 was loaded to gradient silica gel column chromatography using Hexane:EtOAc (5:1 → 100% EtOAc) and afforded compounds **3** (1.9 mg) and **4** (1 mg). Fraction DCE6 (83.6 mg) was loaded to gradient silica gel column chromatography using Hexane:EtOAc (5:1 → 100% MeOH) to yield 11 fractions (DCE6-1~DCE6-11). Compound **5** (1.6 mg, t_R 40 min) and compound **6** (1.8 mg, t_R 55 min) were afford from subfraction DCE6-7 (18 mg) by RP HPLC (Watchers 120 ODS-BP, S-10 μm, 150 × 10 mm) using a flow rate of 2 mL/min, a 365 nm UV detector, and isocratic elution with MeOH:H_2O (60:40) as eluent. Fraction DCE8 (577.1 mg) was fractionated into 12 fractions (DCE8-1~DCE8-12) using gradient silica gel column chromatography with Hexane:EtOAc (3:1 → 100% MeOH). Subfraction DCE8-7 (103.3 mg) was subjected to Sephadex LH-20 using MeOH to afford compound **7** (1.6 mg). Subfraction DCE8-9 (93.4 mg) was loaded to gradient silica gel column chromatography using Hexane:EtOAc (7:1 → 100% MeOH) to yield 8 fractions (DCE8-9-1~DCE8-9-8). DCE8-9-4 (16.1 mg) was conducted to RP HPLC (Watchers 120 ODS-BP, S-10 μm,150 × 10 mm) at a flow rate of 2 mL/min using a UV detector 330 nm by isocratic elution with 0.1% formic acid in ACN:0.1% formic acid in H_2O (42:58) as eluant to afford compounds **8** (0.6 mg, t_R 40 min), **9** (1.5 mg, t_R 52 min), and **10** (0.8 mg, t_R 69 min). Subfraction DCE8-9-5 (23 mg) was conducted to RP HPLC (Watchers 120 ODS-BP, S-10 μm, 150 × 10 mm) using a flow rate of 2 mL/min, a 330 nm UV detector by isocratic elution with 0.1% formic acid in ACN:0.1% formic acid in H_2O (36:64) to afford compound **11** (2.5 mg, t_R 31 min).

2.4. Animals

Female BALB/c mice of six-week-old were purchased from Orient Bio, Inc. (Seongnam, Korea), housed in a controlled environment (23 ± 3 °C and 55 ± 5% RH under a 12 h light/dark cycle), and allowed standard laboratory food and water ad libitum. All experimental procedures complied with the Guide for the Care and Use of Laboratory Animals issued by the National Institutes of Health (NIH publication No. 85-23, revised 2011) and were approved beforehand by the Institutional Animal Care and Use Committee of KIST (Certification No. KIST-2016-011).

2.5. Oxazolone-Induced Atopic Dermatitis BALB/c Mice

BALB/c mice were sensitized by applying 20 μL of 1% oxazolone in a mixture of acetone and olive oil (4:1) to ear surfaces once daily for 7 days. Subsequently, AD was induced by applying 20 μL of 0.1% oxazolone for 3 weeks every 2 days (the Oxazolone-group). During the AD induction period, mice in the DCE and Dexa groups were administered 1% DCE or 0.1% dexamethasone twice daily. Mice were divided into four groups (n = 4), that is, the OX, DCE, Dexa, and control (CON) groups. Animals in the CON group were administered distilled water instead of oxazolone throughout the sensitization and induction periods. On the final application day (day 28), mice were sacrificed, and samples were collected.

2.6. Histological Examination

Ear skins from BALB/c mice were fixed in 10% formaldehyde (Sigma) for 24 h, embedded in paraffin wax, serially sectioned at 4 μm, stained with hematoxylin and eosin (H&E) or toluidine blue for general morphology and mast cell infiltration. Epidermal thickness was measured using HKBasic software (KOPTIC, Seoul, Korea), and the number of mast cells infiltrating into the dermal layer was counted by randomly selecting three sections in toluidine blue-stained tissue. Histopathological changes (×200 magnification) were evaluated using the ProgRes® CapturePro application software (JENOPTIK laser, Jena, Germany).

2.7. Cell Culture

RBL-2H3 (a rat basophilic leukemia cell line) cells were purchased from the American Type Culture Collection (ATCC, Rockville, MD, USA). RBL-2H3 cell were maintained in 150 cm^2 cell culture dish with DMEM (Dulbecco's modified essential medium, HyClone, Logan, UT, USA) containing 10% FBS (fetal bovine serum), 100 U/mL penicillin, and 100 μg/mL streptomycin (HyClone) at 37 °C in a humidified atmosphere containing 5% CO_2.

2.8. Measurement of IL-4 mRNA Expression

RBL-2H3 cells were injected with DMSO or compounds isolated from DCE (10 μM) for 1 h, and inflammation was triggered by adding PI (PMA (phorbol 12-myristate 13-acetate, Sigma-Aldrich, St. Louis, MO, USA) at 50 ng/mL and ionomycin at 1 μM (Sigma-Aldrich), which induced states comparable to AD [22]. The control group was injected with DMSO without PI. After treatment for 20 h, total mRNA was harvested to synthesize cDNA, and IL-4 mRNA levels were calculated by quantitative real-time PCR (qPCR). Total RNA extraction was accomplished with the RNeasy mini kit (Qiagen, Valencia, CA, USA), and cDNA synthesis was conducted with the RevertAid First Strand cDNA Synthesis Kit (Thermo Fisher, Waltham, MA, USA). qPCR was performed using the QuantStudio™ 6 Pro Real-Time PCR System (Applied Biosystems, Foster City, MA, USA) and Fast SYBR® Green Master Mix (Applied Biosystems). qPCR analysis was repeated twice and designed by duplication per sample. Expression levels of cytokines in exposed cells were compared to those in control cells at predetermined time points using the comparative cycle threshold (Ct) method. The sequences of the primers used were IL-4 forward: 5′-ACC TTG CTG TCA CCC TGT TC-3′; IL-4 reverse: 5′-TTG TGA GCG TGG ACTCAT TC-3′; β-actin forward: 5′-TCA TCA CCA TCG GCA ACG-3′, β-actin reverse: 5′-TTC CT GAT GTC CAC GTC GC-3′. mRNA expressions were normalized versus β-actin.

2.9. Measurement of β-Hexosaminidase Release

RBL-2H3 cells were seeded in 24-well plates, cultured overnight, sensitized with 100 ng/mL IgE for 4 h at 37 °C in 5% CO_2. The IgE-sensitized cells were incubated with each of the eleven compounds (10 μM) isolated from DCE for 1 h, followed by 30 min incubation with 10 μg/mL DNP-BSA to stimulate degranulation. To assess β-hexosaminidase activity, the culture medium was supplemented with 10 mM poly-N-acetyl glucosamine in 0.1 M sodium citrate buffer (pH 4.5) in 96-well plates and incubated for 1 h at 37 °C. Absorbances at 405 nm were read using a microplate reader (Tecan Infinite M1000 Microplate Reader, Männedorf, Switzerland) [23].

2.10. Statistical Analysis

In vitro data are shown as means ± standard deviations (SDs) and in vivo data as mean ± standard errors of means (SEMs) (n = 4). Statistical analysis was performed using one-way analysis of variance (ANOVA). p values of < 0.05 were considered statistically significant.

3. Results

3.1. Effects of D. costaricensis EtOH Extract (DCE) on Oxazolone-Induced AD-Like Lesions

To verify the anti-AD effect of DCE on mouse ear skin, BALB/c mice were administered OX for four weeks. OX-induced mice showed AD-like skin symptoms, which included increases in the ear and epidermal thickness, tumefaction, erythema, and dry skin, but it was improved in the DCE group (Figure 1A). On the final application day (day 28), the animals were sacrificed and ear thickness was determined. Ear thickness on the last day of the experiment was thicker in the OX group (0.32 mm) than in the CON group (0.19 mm) (Figure 1B). However, DCE significantly reduced ear thickness (to 0.24 mm) versus the OX group. The Dexa group (0.16 mm), used as positive control, also showed reduced ear thickness versus the OX group.

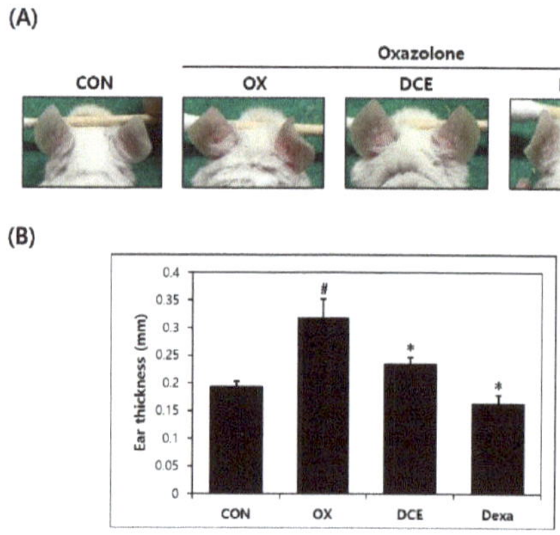

Figure 1. Effect *D. costaricensis* EtOH extract (DCE) on OX-induced AD-like lesions in the BALB/c mouse model. (**A**) Clinical features of AD-like symptoms. (**B**) Ear thickness. CON group: vehicle controls, OX group: oxazolone-treated controls, DCE group: oxazolone plus 1% DCE-treated, and Dexa group: oxazolone plus 0.1% dexamethasone-treated. Data values are presented as means ± SEMs. # $p < 0.05$ vs. the CON group, * $p < 0.05$ vs. the OX group.

3.2. Histopathologic Effects of DCE on Oxazolone-Induced AD-Like Lesions

To examine the anti-AD effect of DCE on histopathological features, the epidermal tissues of mouse ears were stained with H&E or toluidine blue to determine epidermal thicknesses and degrees of mast cell infiltration. Microscopic examination revealed epidermal thickness was thinner in the DCE group than in the OX group (Figure 2A). Mast cell counts were also lower in the DCE group (Figure 3A). Oxazolone treatment increased epidermal thickness and mast cell infiltration by 2- and 4.5-fold, respectively, versus the CON group (Figures 2B and 3B). However, in the DCE group, epidermal thickness was reduced by 22%, and mast cell infiltration was significantly reduced by 45% versus the OX group. In the Dexa group, the epidermal thickness and mast cell infiltration were reduced by 65% and 63%, respectively, versus the OX group.

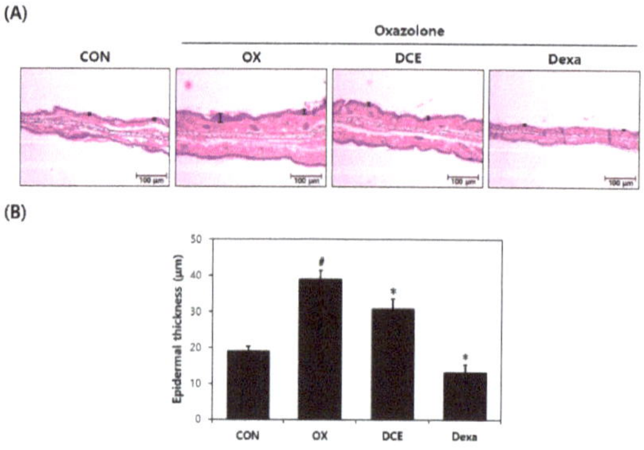

Figure 2. Effect of *D. costaricensis* EtOH extract (DCE) on H&E findings and epidermal thickness. (**A**) H&E staining results. (**B**) Epidermal thicknesses of mouse ears. CON group: vehicle controls, OX group: oxazolone-treated controls, DCE group: oxazolone plus 1% DCE-treated, and Dexa group: oxazolone plus 0.1% dexamethasone-treated. Data values are expressed as the means ± SEMs. # $p < 0.05$ vs. the CON group, * $p < 0.05$ vs. the OX group.

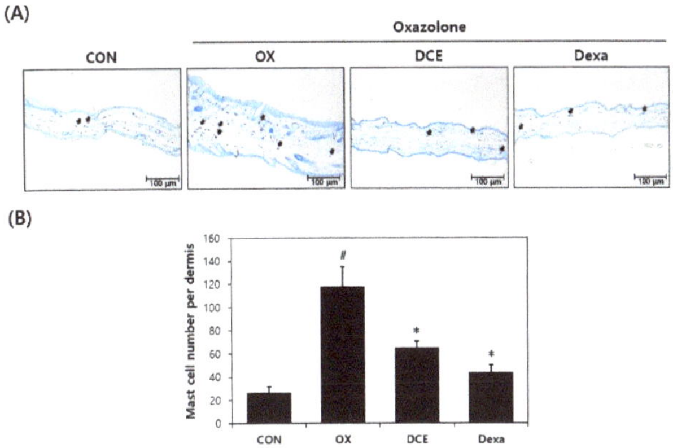

Figure 3. Effect of *D. costaricensis* EtOH extract (DCE) on toluidine blue staining findings and mast cell numbers in dermal tissues. (**A**) Toluidine blue staining results. (**B**) Mast cell numbers in dermal tissue. CON group: vehicle controls, OX group: oxazolone-treated controls, DCE group: oxazolone plus 1% DCE-treated, and Dexa group: oxazolone plus 0.1% dexamethasone-treated. Data values are expressed as means ± SEMs. # $p < 0.05$ vs. the CON group, * $p < 0.05$ vs. the OX group.

3.3. Isolation of Compounds from DCE

Figure 4 provides details of the eleven compounds isolated from DCE as identified by 1D and 2D NMR and HR-MS data, and comparisons with the literature. The isolated compounds were identified as follows: 5,2′-dimethoxyflavone (**1**) [24], 6-methoxyflavone (**2**) [24], 6′-methoxyflavone (**3**) [25], 7-methoxyflavone (**4**) [24], 7,8-dimethoxyflavone (**5**) [26], 7,2′-dimethoxyflavone (**6**) [27], 7-hydroxyflavone (**7**) [28], 6,7-dimethoxyflavone (**8**) [29], 7,8,6′-trimethoxyflavone (**9**) [30], 5,7,8,2′-tetramethoxyflavone (**10**) [31], and 7-hydroxy-6′-methoxyflavone (**11**) [32].

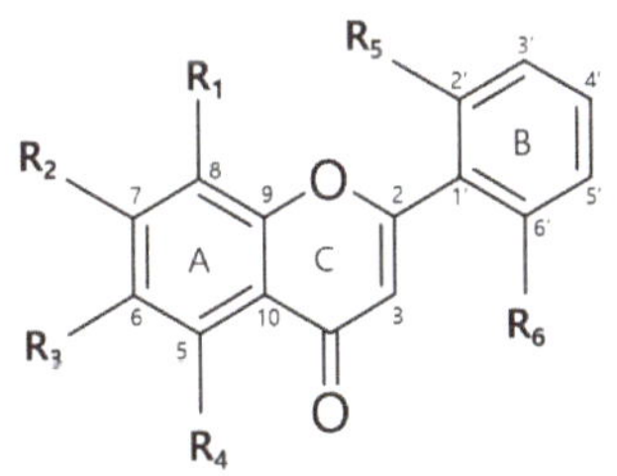

1 R_1 = H, R_2 = H, R_3 = H, R_4 = OCH_3, R_5 = OCH_3, R_6 = H
2 R_1 = H, R_2 = H, R_3 = OCH_3, R_4 = H, R_5 = H, R_6 = H
3 R_1 = H, R_2 = H, R_3 = H, R_4 = H, R_5 = H, R_6 = OCH_3
4 R_1 = H, R_2 = OCH_3, R_3 = H, R_4 = H, R_5 = H, R_6 = H
5 R_1 = OCH_3, R_2 = OCH_3, R_3 = H, R_4 = H, R_5 = H, R_6 = H
6 R_1 = H, R_2 = OCH_3, R_3 = H, R_4 = H, R_5 = OCH_3, R_6 = H
7 R_1 = H, R_2 = OH, R_3 = H, R_4 = H, R_5 = H, R_6 = H
8 R_1 = H, R_2 = OCH_3, R_3 = OCH_3, R_4 = H, R_5 = H, R_6 = H
9 R_1 = OCH_3, R_2 = OCH_3, R_3 = H, R_4 = H, R_5 = H, R_6 = OCH_3
10 R_1 = OCH_3, R_2 = OCH_3, R_3 = H, R_4 = OCH_3, R_5 = OCH_3, R_6 = H
11 R_1 = H, R_2 = OH, R_3 = H, R_4 = H, R_5 = H, R_6 = OCH_3

Figure 4. Chemical structures of the compounds isolated from DCE.

The active compound **5** exhibits a peak at m/z 283.0974 $[M+H]^+$ in HREISMS, corresponding to its molecular formula as $C_{17}H_{14}O_4$. The ^1H NMR spectrum of **5** indicated five aromatic protons of B-ring at δ_H 8.08 (m, 2H, H-2′, H-6′), 7.61 (m, 3H, H-3′, H-4′, H-5′), and *ortho*-coupling protons of A-ring at δ_H 7.79 (d, J = 8.9 Hz, 1H, H-5), 7.29 (d, J = 8.9 Hz,

1H, H-6). The observation of a sharp singlet at δ_H 6.98 (s, 1H) with the above data and a carbonyl group at δc 176.6 (C-4) showed that compound **5** was a flavone. In addition, the positions of two methoxy groups were determined by correlations between δ_H 3.96 (d, 6H, OCH_3-7, OCH_3-8) and δc 156.5 (C-7), 136.4 (C-8) in the HMBC spectrum. Based on this spectral data, the structure of **5** was identified as 7,8-dimethoxyflavone.

The active compound **6** has a molecular formula $C_{17}H_{14}O_4$ as a peak at m/z 283.0968 $[M+H]^+$ in HREISMS. The ^1H NMR spectrum of **6** identified 1′, 2′-substituted aromatic protons of B-ring at δ_H 7.94 (m, 1H, H-6′), 7.58 (td, J = 8.49, 7.4, 1.74 Hz, 1H, H-4′), 7.26 (m, 1H, H-3′), 7.16 (td, J = 7.4, 7.2, 2.4 Hz, 1H, H-5′), and *ortho*-coupling protons [δ_H 7.94 (d, J = 8.0 Hz, 1H, H-5), 7.07 (dd, J = 8.8, 2.3 Hz, 1H, H-6)] and a singlet [δ_H 7.26 (s, 1H, H-8)] of A-ring. Moreover, a singlet at δ_H 6.86 (s, 1H, H-3) and a carbonyl group at δc 176.4 (C-4) proved that compound **6** was a flavone. The positions of two methoxy groups were determined by correlations between δ_H 3.93 (s, 3H, OCH_3-2′) and δc 157.5 (C-2′), and δ_H 3.91 (s, 3H, OCH_3-7) and δc 163.8 (C-7), respectively, in the HMBC spectrum. On the basis of these spectral data, the structure of **6** was determined as 7,2′-dimethoxyflavone.

3.4. Compounds Isolated from DCE and Their Effects on RBL-2H3 Cells

The anti-inflammatory and anti-allergic effects of the eleven compounds isolated from DCE were assessed using IL-4 levels and β-hexosaminidase degranulation. PI-stimulation significantly increased the gene expression of IL-4 in RBL-2H3 cells. Pretreatment with 7,8-dimethoxyflavone (**5**), 7,2′-dimethoxyflavone (**6**), or 6,7-dimethoxyflavone (**8**) inhibited PI-induced increases in IL-4 levels by 46%, 35%, and 35%, respectively (Figure 5A). Furthermore, DNP-BSA-stimulation induced the release of β-hexosaminidase in IgE-sensitized RBL-2H3 cells, and 6′-methoxyflavone (**3**), 7-methoxyflavone (**4**), 7,8-dimethoxyflavone (**5**), or 7,2′-dimethoxyflavone (**6**) pretreatment inhibited β-hexosaminidase release by 17%, 16%, 27%, and 30%, respectively, versus IgE + DNP-BSA-stimulated RBL-2H3 cells (Figure 5B).

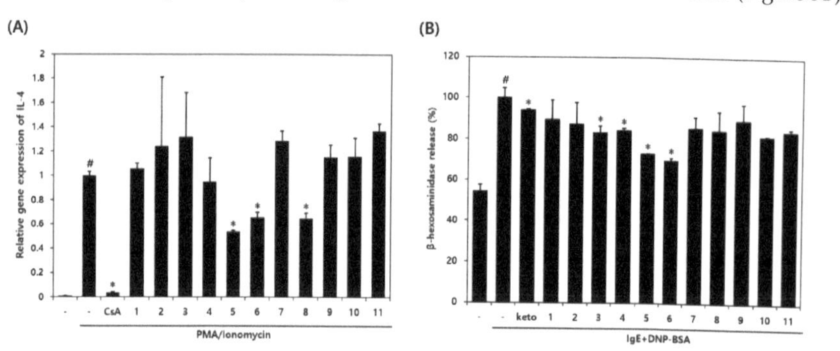

Figure 5. Anti-inflammatory and anti-allergic effects of compounds isolated from DCE on IL-4 levels and β-hexosaminidase release. (**A**) Expressions of IL-4. (**B**) β-hexosaminidase release. CsA: PI plus cyclosporin A-treated group, keto: IgE + DNP-BSA plus ketotifen-treated group. Data values are presented as the means ± SDs. # $p < 0.05$ vs. vehicle controls, * $p < 0.05$ vs. the PI or IgE + DNP-BSA-stimulated cells.

4. Discussion

Flavonoids are extensively produced by plants and present in edible plants, fruits, legumes, and tea [33]. The number of studies performed on plant flavonoids is increasing as they have been shown to have beneficial protective effects on human health without any side effects [34]. Structurally, plant-derived flavonoids have a C6-C3-C6 carbon skeleton with different substitution patterns, as demonstrated by luteolin, apigenin, diosmetin, and quercetin [34,35]. Plant flavonoids have been shown to act as anti-inflammatory agents in epidemiologic, clinical, and animal studies [35,36]. In particular, they have been shown to regulate various inflammation-related enzyme systems and transcription factors [37], and some have been reported to have affirmative therapeutic effects on chronic inflammatory skin diseases (atopic dermatitis, urticaria, and psoriasis) [37,38]. Accordingly, the discovery of novel flavonoids is attracting attention as a means of identifying prospective candidate anti-inflammatory drugs.

The Thymelaeaceae family has a broad range of biological activities, which include anti-inflammatory, anti-cancer, and antibacterial activities, and has been used to treat human diseases for centuries [20,21]. Plant extracts of the Thymelaeceae family commonly contain highly diverse phenolics and possess potent anti-inflammatory properties [21]. Representatively, *Daphne* species inhibited acute and chronic inflammation in an inflammation-induced mouse model [39,40]. *Wikstroemia* species ameliorated AD-like skin lesions and decreased serum IL-4 and IgE levels in mice [41], and the genus *Daphnopsis* has been used in diuretic, laxative, and psoriasis remedies [20]. Although bioactive compounds, including flavonoids and terpenoids, with anti-inflammatory activities have been reported in the genus *Daphnopsis*, studies on the treatment of inflammatory diseases in this genus are still rare. For this reason, we undertook to characterize the anti-AD effect of DCE in an AD-induced mouse model and to examine the anti-allergic and anti-inflammatory effects of bioactive compounds isolated from DCE on inflammatory mediator degranulation in vitro.

AD is a representative allergic inflammatory skin disease. Patients with allergic inflammatory skin diseases typically exhibit symptoms such as keratinization, swelling, and erythema [42]. These inflammatory hypersensitivity reactions increase mast cell infiltration in the epidermis and thicken the epidermis [43]. Anti-AD examination of DCE via a representative oxazolone-induced BALB/c mouse model has been performed, and mice with repeated application of oxazolone exhibited erythema and edematous dermatitis along with immunological inflammatory response [44]. As a result, DCE improved AD-like lesions, severe scratching behavior, and ear thickness induced by oxazolone treatment. In the histological study, H&E and toluidine blue staining showed DCE significantly reduced mast cell infiltration and epidermal thickening. These results show that DCE has the potential to suppress allergic inflammatory symptoms in skin.

Inflammatory hypersensitivity reactions promote the differentiation of T cells and the release of various inflammatory cytokines, including IL-4, by Th2 cells [43]. Excessive IL-4 secretion degranulates immune cells, exacerbates epidermal barrier dysfunction, and causes itching [6,45]. For this reason, many authors considered IL-4 inhibitors key markers for the development of anti-inflammatory and allergy treatments [46]. Therefore, we examined the expression of mRNA IL-4 and degranulation of mast cells in PI or IgE/DNP-BSA pretreated RBL-2H3 cells subsequently treated with each of the eleven isolated flavonoids from DCE. RBL-2H3 cells have commonly been reported to be a mucosal mast cell line, and extensively used to study IgE FcεRI interactions [47,48]. We found that 7,8-dimethoxyflavone (**5**), 7,2′-dimethoxyflavone (**6**), and 6,7-dimethoxyflavone (**8**) significantly decreased the expression of IL-4 mRNA. In addition, 6′-methoxyflavone (**3**), 7-methoxyflavone (**4**), 7,8-dimethoxyflavone (**5**), and 7,2′-dimethoxyflavone (**6**) showed strong inhibitory effects in the β-hexosaminidase release assay which is considered as a biomarker of mast cell degranulation. These observations support previous studies that methoxylation of the flavone A-ring 7-position enhances the anti-inflammatory activity and that multiple substituents have little effect on anti-inflammatory activity [49]. In addition, our results support earlier studies that hydroxylation of the flavone A-ring-7 position attenuates the anti-inflammatory activity, while hydroxyl groups on the flavone B-ring increase their inhibitory action [50]. Our results suggest that DCE containing flavonoids with anti-allergic and anti-inflammatory effects might be useful for the development of prophylactic and therapeutic agents for AD.

5. Conclusions

In summary, this study shows that DCE ameliorated AD-like pathology by decreasing ear epidermal thicknesses and mast cell infiltration in an oxazolone-induced BALB/c mouse model. Eleven active compounds were isolated from DCE, and 7,8-dimethoxyflavone and 7,2′-dimethoxyflavone were found to inhibit IL-4 overproduction and mast cell degranulation in vitro. Accordingly, our results provide that DCE has potential use as a natural treatment for AD and chronic skin disease.

Author Contributions: Conceptualization, S.K. and M.Y.; investigation, Y.B., T.K. and N.P.; data curation, S.C., D.Y., S.S. and N.Z.; writing—original draft preparation, Y.B. and T.K.; writing—review and editing, S.K. and M.Y. All authors have read and agreed to the published version of the manuscript.

Funding: This research was supported by a National Research Foundation of Korea (NRF) grant funded by the Ministry of Science, ICT, and Future Planning (NRF-2019M3A9I3080265, NRF-2019M3A9I3080266, and NRF-2022R1A2C1003088).

Institutional Review Board Statement: The study was conducted according to the guidelines issued by the Institutional Animal Care and Use Committee (IACUC) of the Korea Institute of Science and Technology (Certification No. KIST-2016-011, 2016) and in compliance with the Guide for the Care and Use of Laboratory Animals of the National Institutes of Health (2013).

Informed Consent Statement: Not applicable.

Data Availability Statement: Data are contained within the article.

Conflicts of Interest: The authors declare no conflict of interest.

References

1. Leung, D.Y.M. Pathogenesis of atopic dermatitis. *J. Allergy Clin. Immunol.* **1999**, *104*, 99–108. [CrossRef]
2. Koblenzer, C. Itching and the atopic skin. *J. Allergy Clin. Immunol.* **1999**, *104*, 109–113. [CrossRef]
3. Deo, S.S.; Mistry, K.J.; Kakade, A.M.; Niphadkar, P.V. Role played by Th2 type cytokines in IgE mediated allergy and asthma. *Lung India* **2010**, *27*, 66–71. [CrossRef] [PubMed]
4. Leung, D.Y.M. Immunopathology of atopic dermatitis. *Springer Semin. Immunopathol.* **1992**, *13*, 427–440. [CrossRef] [PubMed]
5. Branco, A.C.C.C.; Yoshikawa, F.S.Y.; Pietrobon, A.J.; Sato, M.N. Role of Histamine in Modulating the Immune Response and Inflammation. *Mediat. Inflamm.* **2018**, *2018*, 952407. [CrossRef] [PubMed]
6. Furue, M. Regulation of Skin Barrier Function via Competition between AHR Axis versus IL-13/IL-4-JAK-STAT6/STAT3 Axis: Pathogenic and Therapeutic Implications in Atopic Dermatitis. *J. Clin. Med.* **2020**, *9*, 3741. [CrossRef]
7. Giam, Y.C.; Hebert, A.A.; Dizon, M.V.; Van Bever, H.; Tiongco-Recto, M.; Kim, K.-H.; Soebono, H.; Munasir, Z.; Diana, I.A.; Luk, D.C.K. A review on the role of moisturizers for atopic dermatitis. *Asia Pac. Allergy* **2016**, *6*, 120–128. [CrossRef]
8. Simpson, E.L. Atopic dermatitis: A review of topical treatment options. *Curr. Med. Res. Opin.* **2010**, *26*, 633–640. [CrossRef]
9. Albader, S.S.; Alharbi, A.A.; Alenezi, R.F.; Alsaif, F.M. Dupilumab side effect in a patient with atopic dermatitis: A case report study. *Biologics* **2019**, *13*, 79–82. [CrossRef]
10. Kim, H.S.; Cho, S.H. Treatment for atopic dermatitis. *J. Korean Med. Assoc.* **2014**, *57*, 226–233. [CrossRef]
11. Berger, T.G.; Duvic, M.; Van Voorhees, A.S.; Frieden, I.J. The use of topical calcineurin inhibitors in dermatology: Safety concerns. Report of the American Academy of Dermatology Association Task Force. *J. Am. Acad. Dermatol.* **2006**, *54*, 818–823. [CrossRef] [PubMed]
12. Wu, S.; Pang, Y.; He, Y.; Zhang, X.; Peng, L.; Guo, J.; Zeng, J. A comprehensive review of natural products against atopic dermatitis: Flavonoids, alkaloids, terpenes, glycosides and other compounds. *Biomed. Pharmacother.* **2021**, *140*, 111741. [CrossRef] [PubMed]
13. Dohil, M.A. Natural ingredients in atopic dermatitis and other inflammatory skin disease. *J. Drugs Dermatol.* **2013**, *12*, 128–132.
14. Dawid-Pać, R. Medicinal plants used in treatment of inflammatory skin diseases. *Postep. Dermatol. Alergol.* **2013**, *30*, 170–177. [CrossRef]
15. Ambriz-Pérez, D.L.; Leyva-López, N.; Gutierrez-Grijalva, E.P.; Heredia, J.B. Phenolic compounds: Natural alternative in inflammation treatment. A Review. *Cogent Food Agric.* **2016**, *2*, 1131412. [CrossRef]
16. Shi, X.; Niu, L.; Zhao, L.; Wang, B.; Jin, Y.; Li, X. The anti-allergic activity of flavonoids extracted from *Citri Reticulatae* Pericarpium. *J. Food Process. Preserv.* **2018**, *42*, 13588. [CrossRef]
17. Li, Y.-J.; Guo, Y.; Yang, Q.; Weng, X.-G.; Yang, L.; Wang, Y.-J.; Chen, Y.; Zhang, D.; Li, Q.; Liu, X.-C.; et al. Flavonoids casticin and chrysosplenol D from *Artemisia annua* L. inhibit inflammation in vitro and in vivo. *Toxicol. Appl. Pharmacol.* **2015**, *286*, 151–158. [CrossRef]
18. Ullah, A.; Munir, S.; Badshah, S.L.; Khan, N.; Ghani, L.; Poulson, B.G.; Emwas, A.-H.; Jaremko, M. Important Flavonoids and Their Role as a Therapeutic Agent. *Molecules* **2020**, *25*, 5243. [CrossRef]
19. Daphnopsis Cotaricensis Barringer & Grayum | International Plant Names Index. Available online: https://www.ipni.org/n/275663-2 (accessed on 8 October 1984).
20. Blaskó, G.; Xun, L.; Cordell, G.A. Studies in the Thymelaeaceae, V. 2′-Hydroxyflavone from *Daphnopsis Sellowiana*: Isolation and Synthesis. *J. Nat. Prod.* **1988**, *51*, 60–65. [CrossRef]

21. Borris, R.P.; Blaskó, G.; Cordell, G.A. Ethnopharmacologic and Phytochemical Studies of the Thymelaeaceae. *J. Ethnopharmacol.* **1988**, *24*, 41–91. [CrossRef]
22. Jin, M.; Yoon, S.J.; Pyo, M.Y. Down-regulation of T helper 2-associated cytokine expression and selective transcription factors by fisetin. *J. Appl. Biol. Chem.* **2011**, *54*, 949–958. [CrossRef]
23. Lee, E.-J.; Yu, M.-H.; Garcia, C.V.; Jhee, K.-H.; Yang, S.-A. Inhibitory effect of *Zizania latifolia* chloroform fraction on allergy-related mediator production in RBL-2H3 cells. *Food Sci. Biotechnol.* **2017**, *26*, 481–487. [CrossRef] [PubMed]
24. Lee, S.H.; Moon, B.H.; Park, Y.H.; Lee, E.J.; Hong, S.W.; Lim, Y.H. Methyl Substitution Effects on ^{1}H and ^{13}C NMR Data of Methoxyflavones. *Bull. Korean Chem. Soc.* **2008**, *29*, 1793–1796. [CrossRef]
25. McKendall, M.; Smith, T.; Anh, K.; Ellis, J.; McGee, T.; Foroozesh, M.; Zhu, N.; Stevens, C.L.K. Methoxyflavone Inhibitors of Cytochrome P450. *J. Chem. Crystallogr.* **2008**, *38*, 231–237. [CrossRef]
26. Herath, W.; Mikell, J.R.; Khan, I.A. Microbial metabolism. Part 10: Metabolites of 7,8-dimethoxyflavone and 5-methoxyflavone. *Nat. Prod. Res.* **2009**, *23*, 1231–1239. [CrossRef] [PubMed]
27. Lee, J.-I.; Son, H.-S.; Park, H. An efficient synthesis of flavones from 2-hydroxybenzoic acids. *Bull. Korean Chem. Soc.* **2004**, *25*, 1945–1947. [CrossRef]
28. Aksnes, D.W.; Standnes, A.; Andersen, Ø.M. Complete Assignment of the ^{1}H and ^{13}C NMR Spectra of Flavone and Its A-Ring Hydroxyl Derivatives. *Magn. Reson. Chem.* **1996**, *34*, 820–823. [CrossRef]
29. Yoon, H.; Eom, S.L.; Hyun, J.Y.; Jo, G.H.; Hwang, D.S.; Lee, S.H.; Yong, Y.J.; Park, J.C.; L:ee, Y.H.; Lim, Y.H. ^{1}H and ^{13}C NMR Data on Hydroxy/Methoxy Flavonoids and the Effects of Substituents on Chemical Shifts. *Bull. Korean Chem. Soc.* **2011**, *32*, 2101–2104. [CrossRef]
30. Kim, Y.S.; Keyser, S.G.L.; Schneekloth, J.S. Synthesis of 2′,3′,4′-trihydroxyflavone (2-D08), an inhibitor of protein sumoylation. *Bioorg. Med. Chem. Lett.* **2014**, *24*, 1094–1097. [CrossRef]
31. Bhardwaj, D.K.; Gupta, A.K.; Chand, R.; Jain, K. Syntheses of *Andrographis paniculata* flavones. *Curr. Sci.* **1981**, *50*, 750–753.
32. Laget, M.; De Méo, M.; Wallet, J.C.; Gaydou, E.M.; Guiraud, H.; Duménil, G. Antimutagenic Activities of 24 Synthetic Flavones with the *Salmonella* Microsomal Assay. *Arch. Pharm. Res.* **1995**, *18*, 415–422. [CrossRef]
33. Panche, A.N.; Diwan, A.D.; Chandra, S.R. Flavonoids: An overview. *J. Nutr. Sci.* **2016**, *5*, 47. [CrossRef] [PubMed]
34. Karak, P. Biological activities of flavonoids: An overview. *Int. J. Pharm. Sci. Res.* **2019**, *10*, 1567–1574.
35. Singh, M.; Kaur, M.; Silakari, O. Flavones: An important scaffold for medicinal chemistry. *Eur. J. Med. Chem.* **2014**, *84*, 206–239. [CrossRef]
36. Lee, D.; Park, J.; Choi, J.; Jang, H.; Seol, J. Anti-inflammatory effects of natural flavonoid diosmetin in IL-4 and LPS-induced macrophage activation and atopic dermatitis model. *Int. Immunopharmacol.* **2020**, *89*, 107046. [CrossRef]
37. Maleki, S.J.; Crespo, J.F.; Cabanillas, B. Anti-inflammatory effects of flavonoids. *Food Chem.* **2019**, *299*, 125124. [CrossRef] [PubMed]
38. Gębka, N.; Adamczyk, J.; Gębka-Kępińska, B.; Mizgała-Izworska, E. The role of flavonoids in prevention and treatment of selected skin diseases. *J. Pre Clin. Clin. Res.* **2022**, *16*, 99–107. [CrossRef]
39. Moshiashvili, G.; Tabatadze, N.; Mshvildadze, V. The genus *Daphne*: A review of its traditional uses, phytochemistry and pharmacology. *Fitoterapia* **2020**, *143*, 104540. [CrossRef]
40. Yeşilada, E.; Taninaka, H.; Takaishi, Y.; Honda, G.; Sezik, E.; Momota, H.; Ohmoto, Y.; Taki, T. In vitro inhibitory effects of ssp. on inflammatory cytokines and activity-guided isolation of active constituents. *Cytokine* **2001**, *13*, 359–364. [CrossRef]
41. Jegal, J.; Park, N.-J.; Kim, T.-Y.; Choi, S.; Lee, S.W.; Hang, J.; Kim, S.-N.; Yang, M.H. Effect of Topically Applied *Wikstroemia Dolichantha* Diels on the Development of Atopic Dermatitis-Like Skin Symptoms in Mice. *Nutrients* **2019**, *11*, 914. [CrossRef]
42. Guttman-Yassky, E.; Nograles, K.E.; Krueger, J.G. Contrasting pathogenesis of atopic dermatitis and psoriasis—Part I: Clinical and pathologic concepts. *J. Allergy Clin. Immunol.* **2011**, *127*, 1110–1118. [CrossRef] [PubMed]
43. Hong, S.-W.; Kim, M.-R.; Lee, E.-Y.; Kim, J.H.; Kim, Y.-S.; Jeon, S.G.; Yang, J.-M.; Lee, B.-J.; Pyun, B.-Y.; Gho, Y.S.; et al. Extracellular vesicles derived from *Staphylococcus aureus* induce atopic dermatitis-like skin inflammation. *Allergy* **2011**, *66*, 351–359. [CrossRef] [PubMed]
44. Jin, H.; He, R.; Oyoshi, M.; Geha, R.S. Animal models of atopic dermatitis. *J. Investig. Dermatol.* **2009**, *129*, 31–40. [CrossRef]
45. Fujii, M. Current Understanding of Pathophysiological Mechanisms of Atopic Dermatitis: Interactions among Skin Barrier Dysfunction, Immune Abnormalities and Pruritus. *Biol. Pharm. Bull.* **2020**, *43*, 12–19. [CrossRef] [PubMed]
46. Tanaka, T.; Takahashi, R. Flavonoids and asthma. *Nutrients* **2013**, *5*, 2128–2143. [CrossRef]
47. Kinet, J.-P.; Blank, U.; Brini, A.; Jouvin, M.-H.; Küster, H.; Mejan, O.; Ra, C. The high-affinity receptor for immunoglobulin E: A target for therapy of allergic diseases. *Int. Arch. Allergy Appl. Imm.* **1991**, *94*, 51–55. [CrossRef]
48. Oliver, J.M.; Seagrave, J.; Stump, R.F.; Pfeiffer, J.R.; Deanin, G.G. Signal transduction and cellular response in RBL-2H3 mast cells. *Chem. Immunol. Allergy* **1988**, *42*, 185–245.
49. Wang, X.; Cao, Y.; Chen, S.; Lin, J.; Bian, J.; Huang, D. Anti-inflammation Activity of Flavones and Their Structure–Activity Relationship. *J. Agric. Food Chem.* **2021**, *69*, 7285–7302. [CrossRef]
50. Matsuda, H.; Morikawa, T.; Ando, S.; Toguchida, I.; Yosikawa, M. Structural requirements of flavonoids for nitric oxide production inhibitory activity and mechanism of action. *Bioorg. Med. Chem.* **2003**, *11*, 1995–2000. [CrossRef]

nutrients

Article

Is It Possible to Improve the Bioavailability of Resveratrol and Polydatin Derived from *Polygoni cuspidati* Radix as a Result of Preparing Electrospun Nanofibers Based on Polyvinylpyrrolidone/Cyclodextrin?

Magdalena Paczkowska-Walendowska [1,*], Andrzej Miklaszewski [2] and Judyta Cielecka-Piontek [1]

[1] Department of Pharmacognosy, Poznan University of Medical Sciences, Rokietnicka 3, 60-806 Poznan, Poland
[2] Institute of Materials Science and Engineering, Faculty of Materials Engineering and Technical Physics, Poznan University of Technology, Jana Pawła II 24, 61-138 Poznan, Poland
* Correspondence: mpaczkowska@ump.edu.pl; Tel.: +48-61-641-83-99

Abstract: The low bioavailability of resveratrol and polydatin obtained from *Polygoni cuspidati* extract limits the application of their pro-health properties. While nanofibers have attracted increasing attention in nutrition delivery due to their special properties, including an increase in the dissolution and permeability, which affects the bioavailability. Therefore, it is justified to obtain nanofibers from *Polygoni cuspidati* extract, which showed antioxidant and anti-inflammatory properties as a result of a presence of stilbene analogs in the *Polygoni cuspidati* extract (especially resveratrol and polydatin). In the first stage of the work, using the Design of Experiment (DoE) approach, the *Polygoni cuspidati* extract (70% of methanol, temperature 70 °C and 4 cycles) was obtained, which showed the best antioxidant and anti-inflammatory properties. Using the *Polygoni cuspidati* extract as a substrate, nanofibers were obtained by electrospinning. The identification of nanofibers was confirmed on the basis of the analysis of changes in XRPD diffractograms, SEM picture and FTIR-ATR spectra. Obtaining nanofibers from the *Polygoni cuspidati* extract significantly improved the solubility of resveratrol and polydatin (approx. 6-fold comparing to pure substance). As a consequence, the penetration coefficients of both tested resveratrol and polydatin also increased. The proposed strategy for the preparation of nanofibers from the *Polygoni cuspidati* extract is an innovative approach to better use the synergy of biological action of active compounds present in extracts. It is especially during the development of nutraceuticals based on the use of selected stilbenes.

Keywords: *Polygoni cuspidate* extract; nanofibers; resveratrol; polydatin; bioavailability

Citation: Paczkowska-Walendowska, M.; Miklaszewski, A.; Cielecka-Piontek, J. Is It Possible to Improve the Bioavailability of Resveratrol and Polydatin Derived from *Polygoni cuspidati* Radix as a Result of Preparing Electrospun Nanofibers Based on Polyvinylpyrrolidone/Cyclodextrin? *Nutrients* **2022**, *14*, 3897. https://doi.org/10.3390/nu14193897

Academic Editors: Maria Digiacomo and Doretta Cuffaro

Received: 29 August 2022
Accepted: 15 September 2022
Published: 21 September 2022

1. Introduction

Epidemiological data indicate that civilization diseases, such as cardiovascular disease, diabetes, overweight, obesity and cancer, are significantly increasing [1]. Key factors important in the development of civilization diseases are a polluted environment, low physical activity and improper diet, as well as an increase in nervous tension and stressful situations. Many of these factors increase the level of free radicals, the presence of which induces the development of many civilization diseases. The reduction of the production of free radicals and their neutralization are the main assumptions of the prevention of the development of many diet-related diseases [2]. Therefore, the formation of ROS is conditioned by ultraviolet radiation, alcohol consumption, smoking and a diet with an insufficient content of compounds with redox properties [3].

Diseases develop if there is no homeostasis in the human body, which is determined by the oxidation–antioxidant balance, and free radicals are not neutralized by antioxidants. Therefore, it is extremely important to take antioxidants, whose task is to maintain the above-mentioned oxidative–antioxidant balance and to prevent the development of civilization diseases. Among the valuable natural antioxidants it is worth pointing out polyphenols, which are secondary specialized plant metabolites [4]. An interesting group of polyphenols is stilbenes, the main representative of which is resveratrol, occurring, inter alia, in wine, grapes and berries. It protects the cardiovascular system by preventing damage to the walls of the arteries. Resveratrol inhibits the growth of cancer cells and protects protein thiol groups against oxidative oxidation. It has antitumor activity by influencing the main stages of carcinogenesis by modulating signaling pathways that control cell division and growth, apoptosis, inflammation, angiogenesis and metastasis [5].

Commercially available resveratrol is often obtained by chemical synthesis. That is why it is so valuable to take plant materials rich in this polyphenol to be able to use the potential of naturally occurring stilbenes. One of the raw materials particularly rich in silbenes is *Polygoni cuspidati rhizoma et radix*. It is worth noting that other secondary metabolites also present in the metabolite pathway of resveratrol formation have valuable pro-health properties. Additionally, *Polygonum cuspidatum,* also known as Hu Zhang in China, has the rich composition of active compounds—epicatechin, resveratrol and polydatin—as well as hydrophobic components, including emodin, physcion, torachrysone and their glycosides [6]. Hu Zhang is also used to treat a wide range of diseases due to its antiviral, antibacterial, anti-inflammatory, neuroprotective and cardioprotective effects [7].

Despite the broad spectrum of health-promoting properties, the active compounds (in particular polydatin and resveratrol) are characterized by low water solubility [8]. Additionally, when administered orally, resveratrol is readily absorbed from the small intestine. Despite rapid systemic absorption, resveratrol achieves low bioavailability, less than 1%, due to its high first-pass metabolism, mainly in enterocytes and the liver [9]. One of the methods of increasing the solubility of active compounds while bypassing enterohepatic metabolism is the production of electrospun fibers with bioactive substances for buccal administration. Solid dispersion is one of the methods that can improve active substances' dissolution and enhance their bioavailability [10]. One technique that can improve the solubility and, consequently, the bioavailability of active compounds is electrospinning. Due to the large contact surface, the nanofibers formed as a result of this process show the possibility of adhesion to the epithelium of the oral cavity, which makes it possible to bypass the hepatic metabolism in the case of active compounds absorbed within the oral cavity [11]. There are literature data on the production of electrospun nanofibers to improve the properties of polyphenols, and in particular of stilbene derivatives—namely, resveratrol. PVP-based nanofibers improved resveratrol solubility [12,13], while nanoencapsulation using poly (caprolactone) (PCL) has been reported for prolonged release of resveratrol [11] for tissue engineering and wound healing application [14]. Further, electrospun nanofibers with zein protected resveratrol from the adverse pH condition of the stomach and were released at a controlled rate in the intestinal [15]. Interestingly, there are individual reports in the literature on the use of electrospun nanofibers to increase the bioavailability of active substances [16,17]; however, there are no data of electrospun nanofibers' production with a herbal extract rich with resveratrol.

Therefore, the aim of this study is to develop an innovative preparation of nanofibers containing an extract rich in stilbenes and their associated compounds. Due to the complexity of the extract matrix, this goal was achieved in the next stages: (1) preparation of the extract with the highest contents of compounds (polydatin and resveratrol) as well as biological activity; and (2) preparation of electrospun nanofibers with increased solubility and permeability of active compounds present in the extract with the best antioxidant activity as well as the ability to inhibit hyaluronidase activity.

2. Materials and Methods

2.1. Plant Material

Plant raw material, *Polygonum cuspidatum* rhizome and root, was purchased from Herbapol Cracow (Cracow, Poland) (Lot No. 010918).

2.2. Chemicals and Reagents

Resveratrol (≥99%, HPLC), polydatin (≥95%, HPLC), emodin (phyproof® Reference Substance) and physcion (phyproof® Reference Substance) were obtained from Sigma-Aldrich (Poznan, Poland). Excipients, such as (2-Hydroxypropyl)-β-cyclodextrin (HPβCD) average Mw ~1460, were supplied from Sigma-Aldrich (Poznan, Poland), and polyvinylpyrrolidone (PVP) as Kollidon® 30 was supplied from BASF Pharma (Burgbernheim, Germany). Reagents for activity assays include the following: 2,2-Diphenyl-1-picrylhydrazyl (DPPH), potassium persulfate ($K_2S_2O_8$), 2,2′-Azino-bis(3-ethylbenzothiazoline-6-sulfonic acid) diammonium salt (ABTS, $C_{18}H_{24}N_6O_6S_4$), neocuproine, ammonium acetate, copper(II) chloride ($CuCl_2 \cdot H_2O$), sodium acetate trihydrate ($CH_3COONa \cdot 3H_2O$), 2,4,6-tris(2-pyridyl)-1,3,5-triazine (TPTZ, $C_{18}H_{12}N_6$), iron(III) chloride hexahydrate ($FeCl_3 \cdot 6H_2O$), sodium chloride, bovine serum, hexadecyltrimethylammonium bromide (CTAB), hyaluronic acid (HA); for dissolution studies, they include the following: potassium chloride, sodium chloride, di-potassium hydrogen orthophosphate, magnesium chloride, calcium chloride and xylitol; and for mucoadhesive tests, they include the following: mucin from porcine stomach were obtained from Sigma-Aldrich (Poznan, Poland). Prisma™ HT buffer, Acceptor Sink Buffer and GIT lipid solution were obtained from Pion Inc., whereas HPLC grade acetonitrile and water were obtained from Merck. High-quality pure water and ultra-high-quality pure water were prepared using a Direct-Q 3 UV Merck Millipore purification system.

2.3. Obtaining and Charakteristion of Biological Activity of Polygoni Cuspidati Extract

2.3.1. Plant Extraction Using Design of Experiment (DoE)

Using the Design of Experiments (DoE) approach, a factor experiment plan was developed for three independent variables, which were assigned three levels of values (3^2 full factorial design). As independent factors were selected, the content of extraction mixture, its temperature and the number of repetitions of the cycles were identified (Table 1).

Table 1. Factorial Extraction Process Experiment Plan.

No.	% of Methanol in the Extraction Mixture	Temperature	Number of Cycles
W1	0	30	3
W2	0	50	5
W3	0	70	4
W4	35	30	5
W5	35	50	4
W6	35	70	3
W7	70	30	4
W8	70	50	3
W9	70	70	5

The following parameters used to assess extraction efficiency were choosen: sum of content of active components, total content of phenolic compounds and antioxidant (DPPH scavenging assay) as well as antiinflammation activities (inhibition of hyaluronidase activity).

2.3.2. Determination of Selected Active Components Content and Total Phenolic Content (TPC)

The contents of main active compounds (polydatin, resveratrol, emodin and parietin) were determined by using the modified HPLC-Diode-Array Detection method described previously by Paczkowska-Walendowska et al. [8].

The total content of phenolic components was determined by using method described previously [18].

2.3.3. Determination of Biological Activity

Antioxidant Activity

Antioxidant activity was determined by using an assay with 2,2-Diphenyl-1-picrylhydrazyl (DPPH), 2,2-Azino-bis(3-ethylbenzothiazoline-6-sulfonic Acid) (ABTS) Radical Cation-Based Assays, Cupric Ion-Reducing Antioxidant Capacity (CUPRAC) Assay and Ferric Ion-Reducing Antioxidant Parameter (FRAP) Assay. All procedures were described previously [18].

Anti-Hyaluronidase Activity

The procedure of hyaluronidase inhibition was determined by the turbidimetric method described previously [18].

2.4. Obtaining of Electrospun Nanofibers Containing Polygoni Cuspidati Extract

Electrospun Nanofibers' Preparation Using Design of Experiment (DoE)

The electrospinning procedure was performed using NS + NanoSpinner Plus Electrospinning Equipment (Inovenso Ltd., Istanbul, Turkey). The amount of *Polygoni cuspidati* extract, PVP and HPβCD used to prepare nanofibers was selected on the basis of the Design of Experiment (DoE) data and 3^2 full factorial design experimental plan and is presented in Table 1. The parameters used to assess electrospinning efficiency were as follows: determination of active components content, total amount of released drug, permeability of active compounds and bioadhesion properties of systems.

Firstly, *Polygoni cuspidati* extract (W10—optimized extract prepared under the conditions: 70% of methanol in the extraction mixture, temperature 70 °C and 4 cycles) was mixed with 10 ml of ethanol, and then the appropriate amount of HPβCD was added (Table 2) and stirred with a magnetic stirrer until completely dissolved. Then, the appropriate amount of PVP was added (Table 2) and stirred with a magnetic stirrer until completely dissolved. The evenly mixed solution was transferred to a syringe and subjected to the electrospinning process with the following parameters: voltage of 25 kV, flow rate of 2 mL/min and distance of 12 cm. The nanofibers were collected in a rotary collector covered with aluminum foil. Conditions were optimized based on the preliminary trials.

Table 2. Factorial Electrospinning Process Experiment Plan.

No.	Content		
	W10 Extract (g)	**HPβCD (g)**	**PVP (g)**
F1	1	1	1
F2	1	2	3
F3	1	3	2
F4	2	1	3
F5	2	2	2
F6	2	3	1
F7	3	1	2
F8	3	2	1
F9	3	3	3

2.5. The Identification of Optimized Electrospun Nanofibers

2.5.1. Scanning Electron Microscopy (SEM)

The surface morphology of the nanofiber was visualized using SEM. The nanofibers were sputter coated with gold-palladium and then visualized by a scanning electron microscope Quanta 250 FEG (FEI, Eindhoven, The Netherlands).

2.5.2. XRPD

The crystallographic structure of the samples was analyzed by an X-ray diffraction (XRD, Panalytical Empyrean, Almelo, Netherlands) equipment with the copper anode

(CuKα—1.54 Å) at a Brag-Brentano reflection mode configuration with 45 kV and 40 mA parameters. The measurement parameters were set up for 3–60° with a 45 s per step 0.05° in all cases.

2.5.3. Fourier Transform Infrared Spectroscopy with Attenuated Total Reflectance (FTIR-ATR)

The FTIR-ATR spectra were measured between 400 cm^{-1} and 4000 cm^{-1}, with a resolution set to 1 cm^{-1}, with a Shimadzu IRTracer-100 spectrometer equipped with a QATR-10 single bounce, diamond extended range and LabSolutions IR software (Warsaw, Poland).

2.6. Characterisation of Electrospun Nanofibers

2.6.1. Determination of Active Components Content

Contents of polydatin and resveratrol were determined by using method described in Section 2.3.2.

2.6.2. Dissolution Studies

Dissolution studies of electrospun nanofibers were performed using an Agilent 708-DS dissolution apparatus. A standard basket method was used at 37 ± 0.5 °C with a stirring speed of 50 rpm. Nanofibers were placed in 300 mL of artificial saliva solution at pH 6.8 (potassium chloride (1.20 g), sodium chloride (0.85 g), di-potassium hydrogen orthophosphate (0.35 g), magnesium chloride (0.05 g), calcium chloride (0.20 g), xylitol (20.0 g) and water up to 1 L; pH was adjusted to 6.8 by 1 M HCl). The liquid samples were collected at specified time intervals, and an equal volume of temperature-equilibrated media was replaced. The samples were filtered through a 0.45 μm nylon membrane filter. The concentrations of polydatin and resveratrol in the filtered acceptor solutions were determined by the HPLC method described above. Sink conditions were preserved in the studies.

2.6.3. Permeability Studies

Permeability of active compounds enclosed in nanofibers through artificial biological membranes was investigated by using the PAMPA™ (parallel artificial membrane permeability assay) gastrointestinal tract (GIT) assay (Pion Inc., Billerica, MA, USA). Nanofibers were dissolved in donor solutions (artificial saliva solution at pH 6.8). The acceptor plates were loaded with acceptor Prisma buffer with pH 7.4. The plates were put together and incubated under conditions: temperature 37 °C and 15 min with continuous stirring 50 rpm. Each experiment was repeated at least three times. The amount of permeated active compounds was determined using the HPLC method described above.

The apparent permeability coefficients (P_{app}) were calculated from the following equation:

$$P_{app} = \frac{-ln\left(1 - \frac{C_A}{C_{equilibrium}}\right)}{S \times \left(\frac{1}{V_D} + \frac{1}{V_A}\right) \times t}$$

where V_D is the donor volume, V_A is the acceptor volume, $C_{equilibrium}$ is the equilibrium concentration $C_{equilibrium} = \frac{C_D \times V_D + C_A \times V_A}{V_D + V_A}$, C_D is the donor concentration, C_A is the acceptor concentration, S is the membrane area and t is the incubation time (in seconds).

2.6.4. In Vitro Assessment of Mucin–Biopolymer Bioadhesive Bond Strength

A viscometric method was used to quantify mucin–polymer bioadhesive bond strength. Assessment was performed according to the method described previously [8].

2.6.5. Antioxidant Activity

Antioxidant activity of electrospun nanofibers was conducted by DPPH assay described in Section 2.3.3.

2.7. Statistical Analysis

Statistical analysis was carried out with Statistica 13.3 software (TIBCO Software Inc., Palo Alto, CA, USA). The normality of the results was checked using the Shapiro–Wilk test. The differences among the mean values were tested using the ANOVA test with post hoc Tukey's range test for multiple comparisons. Differences between groups were considered to be significant at $p < 0.05$.

3. Results and Discussion

Worldwide interest in using complementary and herbal medicines for the treatment and prevention of various illnesses has grown in recent years. The actual health advantages of diverse herbal products, however, appear to be overshadowed by issues with quality (lack of consistency, safety and efficacy). Extracts from the herbal matrix must be standardized and characterized for use in modern phytopharmaceuticals. The extraction conditions determine the quality and the yield of the individual constituents; therefore, the choice of the extraction method becomes one of the most important stages in the development of modern phytopharmaceuticals [19]. One of the approaches worth implementing for the preparation of herbal extracts is the design of experiment approach, which assesses the influence of input factors (e.g., temperature, the composition of the extraction mixture and extraction method) on the properties of the extract (e.g., content of active compounds and biological activity) [19,20]. In this work, a full factorial design model was created to assess the effectiveness of the extraction process (Table 1).

Firstly, the assessment of active compounds, such as the polydatin, resveratrol, emodin and parietin, was evaluated (Figure 1, Table 3). The HPLC method was validated according to ICH guidelines, and validation parameters are collected in Table S1 (Supplementary Materials).

The content of active compounds is collected in Table 4. All extracts were also tested in regards to the total polyphenol content (Table 4).

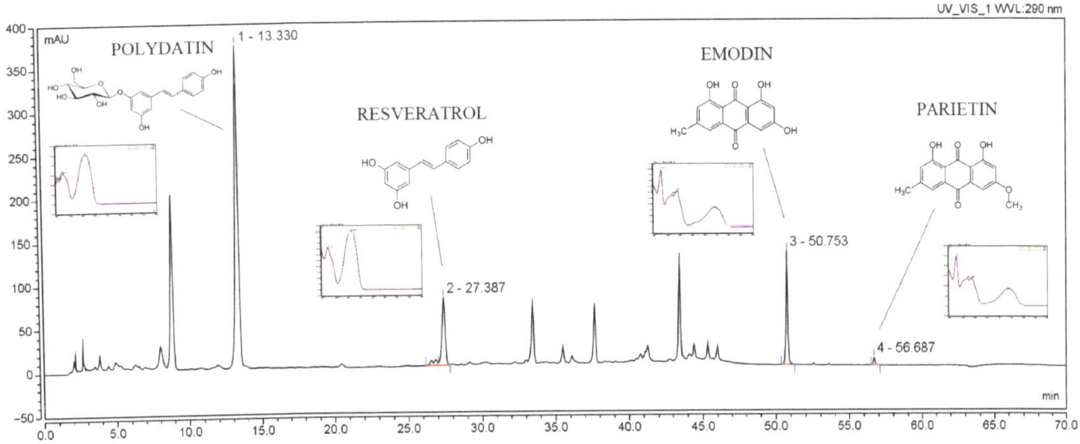

Figure 1. Chromatogram of extract W9.

Table 3. Content of Active Compounds in the Extracts.

No.	Content (μg/1 g Plant Material) Mean ± SD				Sum of Active Compounds (μg/1 g Plant Material) Mean	TPC (mg GAE/1 g Plant Material) Mean ± SD
	Polydatin	Resveratrol	Emodin	Parietin		
W1	174.29 ± 0.68	185.85 ± 10.75	7.85 ± 0.41	0.08 ± 0.02	452.61	7.85 ± 0.41
W2	288.76 ± 22.64	206.69 ± 3.99	11.97 ± 0.62	0.23 ± 0.02	729.16	11.97 ± 0.62
W3	1384.68 ± 6.51	307.38 ± 1.57	12.95 ± 1.95	0.18 ± 0.01	1892.28	12.95 ± 1.95
W4	1060.43 ± 8.74	324.95 ± 15.49	20.35 ± 0.72	0.19 ± 0.08	1631.68	20.35 ± 0.72
W5	1388.60 ± 66.33	353.33 ± 12.53	21.96 ± 0.88	0.20 ± 0.01	2001.47	21.96 ± 0.88
W6	2891.96 ± 26.31	369.26 ± 3.88	23.11 ± 1.06	0.15 ± 0.02	3738.75	23.11 ± 1.06
W7	2730.72 ± 30.61	414.12 ± 5.30	23.09 ± 2.22	0.29 ± 0.02	3931.77	23.09 ± 2.22
W8	4179.04 ± 29.44	449.50 ± 1.24	30.75 ± 2.14	0.13 ± 0.08	4925.62	30.75 ± 2.14
W9	4199.43 ± 68.10	652.67 ± 17.44	35.01 ± 1.77	0.14 ± 0.02	5241.30	35.01 ± 1.77

Table 4. Antioxidant and Anti-Hialuronidase Activities.

No.	DPPH IC_{50} (mg/mL)	ABTS IC_{50} (mg/mL)	CUPRAC IC_{50} (mg/mL)	FRAP IC_{50} (mg/mL)	Hyaluronidase Inhibition IC_{50} (mg/mL)
	Mean ± SD				
W1	0.80 ± 0.02	0.66 ± 0.08	2.16 ± 0.02	0.49 ± 0.03	73.52 ± 2.37
W2	0.65 ± 0.01	0.49 ± 0.04	1.63 ± 0.03	0.30 ± 0.01	44.02 ± 1.48
W3	0.50 ± 0.01	0.42 ± 0.04	1.33 ± 0.30	0.29 ± 0.01	33.21 ± 2.40
W4	0.45 ± 0.01	0.44 ± 0.04	0.45 ± 0.02	0.23 ± 0.01	25.45 ± 0.87
W5	0.31 ± 0.02	0.38 ± 0.03	0.30 ± 0.02	0.18 ± 0.01	19.98 ± 1.67
W6	0.25 ± 0.01	0.31 ± 0.02	0.27 ± 0.03	0.13 ± 0.01	11.34 ± 2.45
W7	0.22 ± 0.03	0.21 ± 0.04	0.19 ± 0.01	0.14 ± 0.01	10.43 ± 1.33
W8	0.16 ± 0.02	0.18 ± 0.02	0.13 ± 0.01	0.11 ± 0.01	4.69 ± 0.33
W9	0.13 ± 0.01	0.14 ± 0.01	0.11 ± 0.01	0.09 ± 0.01	4.35 ± 0.28
	IC_{50} (μg/mL)	IC_{50} (μg/mL)	IC_{50} (μg/mL)	IC_{50} (μg/mL)	
Resveratrol	22.32 ± 0.20	10.84 ± 0.23	20.82 ± 1.29	9.17 ± 0.68	
Polydatin	35.06 ± 1.11	25.21 ± 2.32	68.73 ± 0.48	12.90 ± 0.29	

Then, the sum of the content of active compounds was analyzed. Based on the Pareto diagram (Figure S1, Supplementary Materials), it can be indicated that the percentage of methanol in the extraction mixture and the temperature are statistically significant factors affecting the content of the sum of active compounds. Both effects have a positive sign, i.e., with the increase in the percentage of methanol in the extraction mixture and with the increase in temperature, the sum of active compounds increases.

The TPC was the one output parameter in the DoE model. Based on the Pareto diagram for the TPC (Figure S2, Supplementary Materials), it can be indicated that the percentage of methanol in the extraction mixture and the temperature are statistically significant factors affecting the content of the TPC. Both effects have a positive sign, i.e., with the increase in the percentage of methanol in the extraction mixture and with the rise in temperature, the TPC increases.

In assessing the impact of extraction parameters on its efficiency, it is crucial to determine the final physical properties and biological activity of the prepared extracts. For this purpose, antioxidant activity (measured by four methods: DPPH, ABTS, CUPRAC and FRAP) and anti-inflammatory activity measured as hyaluronidase enzyme inhibition were assessed. All results are presented in Table 4.

In the case of antioxidant activity (Figure S3, Supplementary Materials), the percentage of methanol in the extraction mixture and the temperature are statistically significant factors affecting the IC_{50}. As expected, in the case of anti-hyaluronidase activity (Figure S4, Supplementary Materials), only the temperature has a statistically significant effect. In that case, effects have a negative sign, that is, as the starting values increase, the IC_{50} decreases.

The antioxidant properties of *P. cuspidati* extract have been widely described in literature data [8]. In the study mentioned above, we indicated that the extract was characterized by the best activity, where 1% *m/w* of β-cyclodextrin was added to the extraction mixture (ethanol: water (5:5 *v:v*)). The antioxidant activity was IC_{50} = 0.16 mg/mL, which is a higher value than the best extract in this study (for W9 IC_{50} = 0.16 mg/mL). Due to the method described in this study, it was possible to obtain a content of resveratrol that is 2 times higher than in the cited study [8]. It is worth noting that both resveratrol and polydatin are characterized by high antioxidant activity [21]; hence, this is shown by the results of the action of the entire extract. For example, Li et al. noted that polydatin was superior to resveratrol in inhibiting malondialdehyde (MDA) production, indicating that polydatin may better protect cells from damage. Therefore, resveratrol and polydatin treatment could be effective for attenuating AAPH-induced oxidative stress in HepG2 cells. Recent results suggest that resveratrol and polydatin were able to alleviate oxidative stress and increase the expression of related antioxidant factors [21]. In the light of these reports, one can explain the higher activity of the W9 extract and the higher content of polydatin. Moreover, it is the justification that the use of extracts shows synergy of active compounds' action.

In addition to the antioxidant activity of the extracts, there are reports on their anti-inflammatory activity, expressed by inhibiting cyclooxygenase [8,22]. In this study, a different mechanism of anti-inflammatory action was used, i.e., the ability to inhibit the hyaluronidase enzyme. Such activity is shown by the prepared extracts (Table 4). Inhibition of hyaluronidase activity by polyphenolic compounds is related to, among others, the presence of hydroxyl groups. Interestingly, it has been proven that glycones are more potent inhibitors than their corresponding glycosides [23], which indicates the need to monitor the resveratrol content in the prepared extracts.

Based on the research results and statistical analyses, it was possible to indicate the technical parameters of the extraction process, leading to an extract with the best properties and the highest activity. The significance of input factors was analyzed, including those with the positive sign (sum of active compounds, TPC) (Figure 2a) and with a negative sign (antioxidant and anti-hyaluronidase activities) (Figure 2b). Based on the utility contour profiles model, it was possible to predict the model and indicate optimized parameters of the extraction process, which are as follows: 70% of methanol in the extraction mixture, temperature 70 °C and 4 cycles (statistically insignificant parameter).

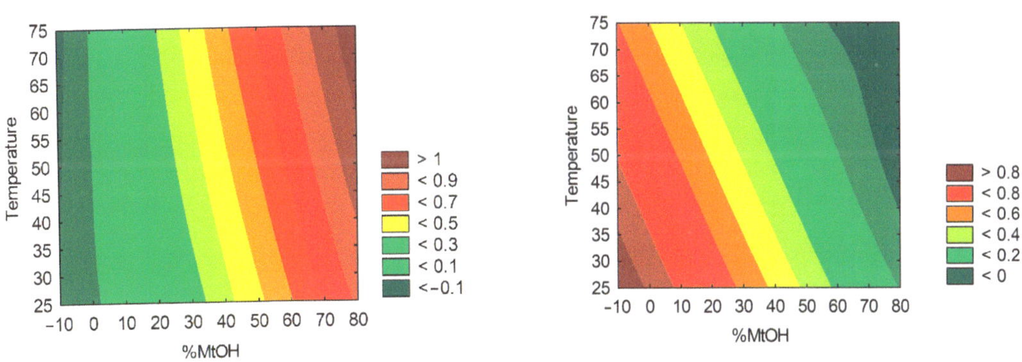

(a) (b)

Figure 2. Model utility contour profiles for effect with a positive sign (**a**) and negative sign (**b**) for extract optimalization.

Optimized process parameters were used to prepare the W10 extract, the activity of which was determined to validate the model. The W10 extract was used in the second part of the experimental work, i.e., for the preparation of electrospun nanofibers. Electrospun nanofibers have been formulated to provide a rapidly dissolving pharmaceutical form providing an optimized extract.

Optimized process parameters allowed for the creation of guidelines for the preparation of an extract with the best biological activities (W10 extract). The W10 extract was used in the second part of the experimental work during the preparation of electrospun nanofibers [24]. In this case, the content of extract, PVP and HPβCD was also analyzed (Table 2). A proper solvent mixture was selected to dissolve all compounds and enable nanofibers' production. The optimization of the electrospinning conditions with respect to the concentration of the polymer, applied voltage, flow rate and distance between the collector and the needle tip was performed, thus producing a continuous stretch of fibers.

The obtained electrospun nanofibers were characterized in terms of their morphology (SEM), structure (XRPD) and possible intermolecular chemical bond formation (FTIR-ATR).

Scanning electron microscopy for systems no. 2–6 and 9 confirmed the nanofiber structure. The systems (1, 7 and 8) did not create the structure of nanofibers, and it was not possible to remove them from the aluminum foil; therefore, they were not used for further research. The presented SEM image of nanofibers no. 5 (Figure 3) showed that obtained procedure produced the nanostructures with a diameter below 500 nm with homogeneous composition and no defects.

Figure 3. SEM image for nanofibers no. 5.

The X-ray diffractograms showed that starting materials (lyophilized extract, PVP, HPβCD) display a high broadening of the diffraction peaks, and this with a low intensity indicates the amorphous structure of these systems (Figure 4) [13]. Moreover, background nanofibers produced from PVP and HPβCD, as well as final nanofibers no. 5, are also characterized by amorphousness.

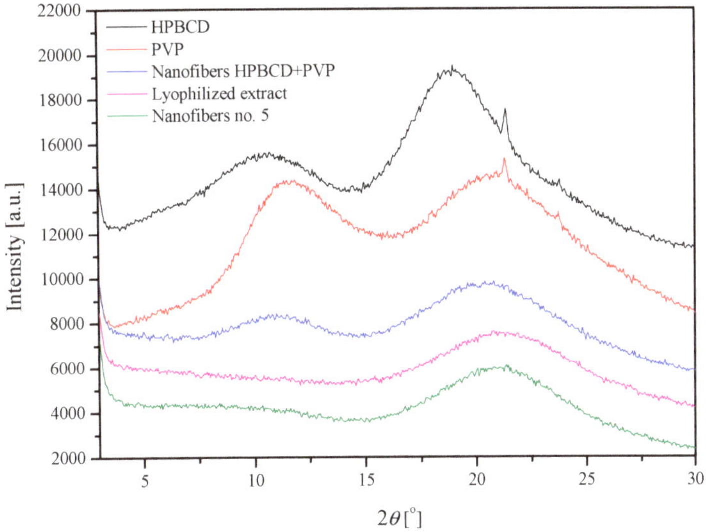

Figure 4. Diffractograms for powder systems and nanofibers no. 5.

Additionally, it was noted that nanofiber samples spectra show both a polymer matrix and excipient extract presence through the main peak position movement, according to Table 5. The polymer matrix peak position is shifted to a lower 2θ angle and above a structural displacement (network expansion), it remains present by the embed excipient and extract. As we may observe for nanofiber HPβCD-PVP vs. nanofiber no. 5 sample comparison, some of polymer matrix micro-strain could be raveled by adding the extract; however, others will grow by the basic structural orientation of the system components. Further analysis for changeable composition and shares of the system may allow for using structural methods for simple predictions of the reaction (a micro-strain shear, network deformation direction and displacement factor) that take place at the nearly amorphous state of the sample.

Table 5. XRPD Signals' Positions.

Sample	PVP	HPβCD	Lyophilized Extract	Nanofiber HPBCD-PVP	Nanofiber no. 5
(1) Peak position [2θ]	11.45	10.26	-	11.11	9.48
(2) Peak position [2θ]	21.28	18.72	21.37	20.61	21.16
Matrix peak position displacement [2θ]	-	-	-	(1) −0.34 (2) −0.67	(1) −1.97 (2) −0.12
Matrix peak position displacement [Å]			-	(1) 0.24 (2) 0.13	(1) 1.76 (2) 0.02

At last, the intramolecular interactions between the extract, PVP and HPβCD were analyzed by FTIR-ATR (Figure 5). To determine whether new interactions between the extract and excipients during the electrospinning process occurred, spectra of the lyophilized extract, PVP, HPβCD, background nanofibers produced from PVP and HPβCD, as well as nanofiber no. 5, were taken and presented in Figure 5. Additionally, Figure 6 shows the theoretical sum of the spectra of the lyophilized extract and background nanofibers produced from PVP and HPβCD. Comparing the nanofiber no. 5 and the theoretical spectra, no additional bands or shifting of existing bands were observed, which may indicate no intermolecular interactions between the extract and the excipients.

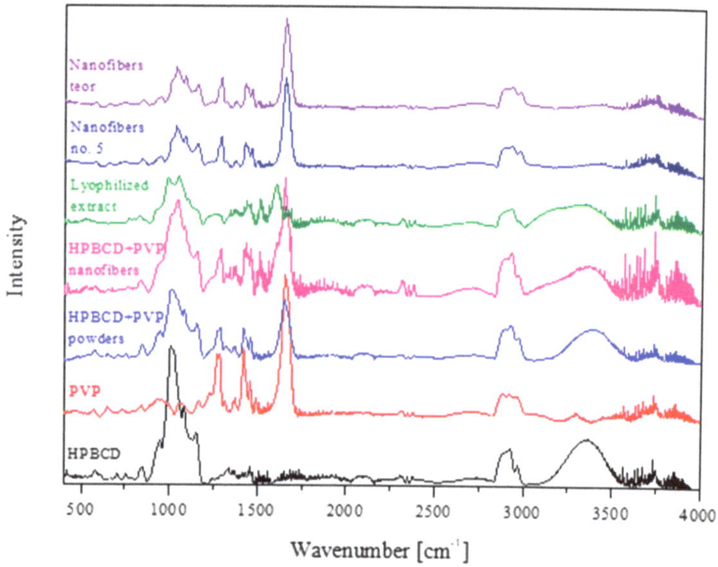

Figure 5. FTIR-ATR spectra for powder samples and nanofiber no. 5.

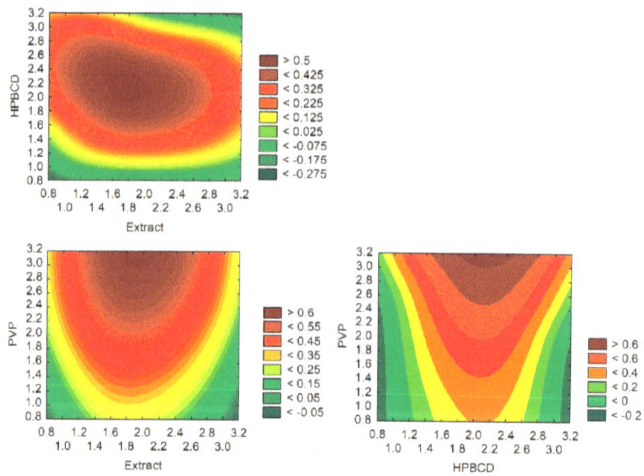

Figure 6. Model utility contour profiles for all effects for electrospun nanofibers optimalization.

After confirming that all the analyzed systems have a nanofiber structure, the analysis of the influence of the input parameters (Table 2) on the physicochemical properties of the obtained systems was started.

The first of the analyzed output parameters were the content of active components, which were marked in the prepared nanofibers and are collected in Table 6. While none of the effects was statistically significant (Figure S5, Supplementary Materials), one can notice the trend of the increase in active compounds content when HPβCD content increased in the prepared system. This is related to the possibility of increasing the solubility of the active compound by the greatest water solubility, high amorphization, wetting, solubilizing and complexing of this cyclodextrin [25,26].

Table 6. Content of Active Components in Nanofibers.

	Nanofibers 2	Nanofibers 3	Nanofibers 4	Nanofibers 5	Nanofibers 6	Nanofibers 9
	Content (µg/100 mg Nanofibers)					
	Solvent: methanol					
Polydatin	5.16 ± 0.11	8.35 ± 0.11	13.66 ± 0.87	33.32 ± 1.60	15.04 ± 3.48	9.45 ± 0.33
Resveratrol	3.70 ± 0.01	1.28 ± 0.01	5.93 ± 0.59	8.60 ± 0.39	1.68 ± 0.33	1.05 ± 0.04
	Solvent: artificial saliva solution at pH 6.8					
Polydatin	8.20 ± 0.01	6.96 ± 0.06	15.46 ± 0.01	27.46 ± 1.08	18.62 ± 0.27	19.24 ± 0.08
Resveratrol	0.81 ± 0.03	0.60 ± 0.03	1.02 ± 0.01	11.00 ± 0.52	1.79 ± 0.02	2.26 ± 0.05

Another critical research was dissolution studies of polydatin and resveratrol from nanofiber (Figure 7, Table 7). There is no guideline for the release method of nanofibers, so in this work, a modified basket method was used. While it is hard to map release medium volume and other physiological conditions located in the oral cavity, the temperature and artificial saliva solution at pH 6.8 closely meet the physiological requirements. It was possible to achieve a 3-fold improvement in the dissolution rate of polydatin, and in the case of resveratrol, there was as much as a 6-fold improvement compared to pure powder compounds. This phenomenon can be explained by the benefits of nanofiber formation, including high load capacity, encapsulation efficiency and a high surface area to volume ratio, all of which can lead to an increase in dissolution rate [13]. PVP-based nanofibers are known to improve the dissolution rate with a burst release of poorly water-soluble compounds [26,27]. This phenomenon can be explained by the following factors: (1) PVP has hygroscopic and hydrophilic properties, and the interactions between polymers and their solvents are more potent than their interactions with one another, so as a result, the polymer chain can quickly absorb solvent molecules, increasing the volume of the polymer matrix and loosening the polymer chains from their rolled shape; (2) a large surface area for the PVP to absorb water molecules increased the porosity for the water molecules to diffuse into the interior of the membrane, and void space for the polymer to swell and disentangle and for the dissolved substance molecules to disperse into the bulk dissolution medium can all be provided by the membrane's three-dimensional continuous web structure; (3) composites were created between the drug and the matrix polymer at the molecular level, allowing for the dissolution of both PVP and substance molecules [28].

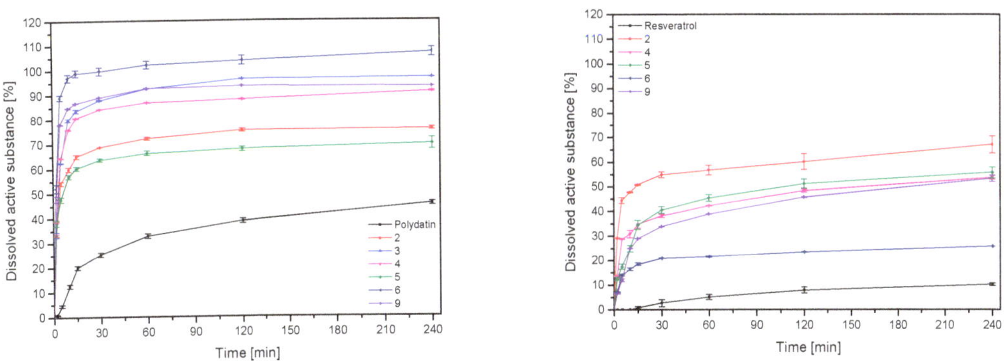

(a) (b)

Figure 7. Dissolution profiles of polydatin (**a**) and resveratrol (**b**) from the nanofibers at artificial saliva solution at pH 6.8.

Table 7. Total Amount of Released Polydatin and Resveratrol from Nanofibers at 15 Minutes.

	Nanofibers 2	Nanofibers 3	Nanofibers 4	Nanofibers 5	Nanofibers 6	Nanofibers 9
	Total Amount of Released Drug from 100 mg of Nanofibers (μg) at 15 min					
Polydatin	5.75 ± 0.07	5.83 ± 0.04	12.50 ± 0.03	16.56 ± 0.22	18.80 ± 0.29	17.47 ± 0.02
Resveratrol	0.31 ± 0.01	0	0.31 ± 0.02	3.80 ± 0.20	0.26 ± 0.01	0.56 ± 0.01

The total amount of released polydatin and resveratrol from nanofibers at 15 min was assessed to determine the significance of the input factors. Again, while none of the effects was statistically significant (Figure S6, Supplementary Materials), it can be seen that the increase in extract and HPβCD content in the prepared system was influenced by the rise in the content of released substances. This is again related to the solubilizing properties of the cyclodextrin used in systems, so we can see how crucial the presence of HPβCD is in the prepared nanofibers.

An important factor determining the application properties of a buccal product is the ability to bind the polymer to the mucin located in the mucosa. For this purpose, rheological studies of nanofiber mixtures were used, and the results are presented in the Figure 8.

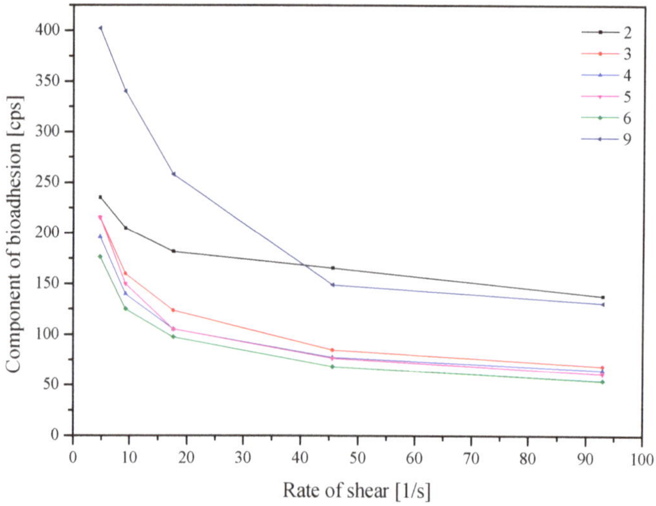

Figure 8. Component of bioadhesion of the nanofibers.

When analyzing the significance of the input factors (Figure S7, Supplementary Materials), it can be noticed that the increase in bioadhesion is noticeable with the increase in the amount of PVP and HPβCD in the samples. PVP is a polymer with good biodegradability, biocompatibility and low toxicity, commonly used as a matrix for controlled drug delivery. Its mucoadhesiveness is explained by the interaction of carbonyl groups with mucin through the presence of hydrogen bonds and Van der Walls forces [29,30]. While cyclodextrins are not typical mucoadhesive compounds, their mucoadhesive properties are notable, perhaps due to the molecules' hydrophilic outer part, which can form hydrogen bonds with hydroxyl groups on sugars and other O- and *N*-containing groups on the mucosal protein backbone [31].

As previously mentioned, the oral cavity becomes an exciting site for administering polyphenols, especially those undergoing rapid first-pass metabolism, which results in low bioavailability, as in the case of resveratrol [32]. For this purpose, apparent permeability coefficients for standards and active compounds from extract W10 and nanofibers were assessed (Table 8).

Table 8. Apparent Permeability Coefficients for Standards, as well as Active Compounds from Extract W10 and Nanofibers.

	Standards	W10	Nanofibers 2	Nanofibers 3	Nanofibers 4	Nanofibers 5	Nanofibers 6	Nanofibers 9
	Apparent Permeability Coefficient $P_{app} \times 10^{-6}$ (cm/s)							
Polydatin	0.0036 ± 0.0001	0.0096 ± 0.0007	3.6213 ± 0.4921	0.7235 ± 0.0460	0.1740 ± 0.0043	0.2060 ± 0.0155	0.0455 ± 0.0039	0.0088 ± 0.0003
Resveratrol	1.0924 ± 0.0778	0.0281 ± 0.0017	47.5107 ± 6.1468	11.7056 ± 0.6146	11.3306 ± 0.2093	13.4603 ± 0.9008	0.1387 ± 0.0070	0.1102 ± 0.0042

While none of the effects was statistically significant (Figure S8, Supplementary Materials), it can be seen that the content of the extract and HPβCD was the factor most influencing the permeation coefficients. Again, HPβCD improves the solubility of active compounds, and those in the dissolved form are more available for transport across the membrane by passive diffusion. Therefore, the use of cyclodextrins is one of the approaches to increase the penetration of resveratrol through biological membranes by the solubility improvement and permeability enhancement through the unstirred water layer [25,33]. There are also reports of a significant influence of the nanofibers structure on the resveratrol penetration through the skin, specifically through the stratum corneum and into the epidermal and dermal layers [13]. This is worth noting, as the increased content of PVP in the system limited permeation through the membrane, which is associated with increased system viscosity. Permeability is inversely proportional to the viscosity of the fluid, which confirms the obtained results.

Dissolution of a drug in an aqueous environment is almost always a prerequisite for oral absorption, and thus, insufficient water solubility often results in limited oral bioavailability. Intestinal permeability is, in addition to water solubility, a key parameter that governs oral absorption [34]. Thus, solubility and permeability are key factors affecting oral bioavailability [35]. Therefore, it is possible to extrapolate the obtained results and suggest the possibility of increasing the bioavailability with the use of buccal electrospun nanofibers together with the increased solubility and permeability of active compounds.

Finally, to confirm the maintenance of the antioxidant properties of the extract, such studies have also been carried out for nanofibers by using the DPPH method. The activity of W10 was assessed and amounted to IC_{50} = 0.13 ± 0.01 mg/mL. The IC_{50} for nanofibers in each case remained within the error limits obtained for the W10 extract. Therefore, it confirmed that the electrospinning process does not affect the biological properties of the processed extract.

To summarize the production of electrospun nanofibers step, a model of utility contours of profiles was made for all the tested effects (Figure 6). Based on the analysis of the Figure 5, the best composition of electrospun nanofibers was selected, and it is the ratio of extract/PVP/HPβCD 1:1:1 *w/w/w*, i.e., the design of nanofiber number 5.

4. Conclusions

The use of the Design of Experiments (DoE) in the extraction of plant material will be essential in advancing and modernizing the creation of standardized phytotherapy. In our study, the influence of the parameters (composition of the extraction mixture, its temperature and number of extraction cycles) on biological properties of the obtained extracts was assessed. It was shown that 70% of methanol in the extraction mixture, temperature 70 °C and 4 cycles are the optimal parameters for extraction of active compounds, such as resveratrol and polydatin from the *P. cuspidati* rhizome and root.

The DoE approach can also be successfully used to select the optimal electrospinning parameters. In the present study, PVP/HPβCD-loaded blend and core shell nanofibers with smooth and bead-less morphology were successfully fabricated for application as new and controlled drug delivery extract systems. The release rate of active compounds in electrospun nanofibers increased with the increase of HPβCD ratio due to enhanced hydrophilicity of the nanofibers. The best composition of electrospun nanofibers was selected, and it is the ratio of extract/PVP/HPβCD 1:1:1 *w/w/w*.

Thus, the PVP/HPβCD-based electrospun nanofibers might be strong enough to be easily inserted within the oral cavity and immediate release the incorporated bioactives, while ensuring patient compliance with the smooth structure of nanofibers and low stiffness during treatment. These characteristics, along with the proven antioxidant and anti-inflammatory properties of *P. cuspidati* extract, suggest the particular benefits of the buccal delivery system as a promising strategy to improve the bioavailability of bioactives.

Supplementary Materials: The following supporting information can be downloaded at: https://www.mdpi.com/article/10.3390/nu14193897/s1, Table S1: Validation parameters; Figure S1: Pareto plot of standardized effects for the sum of active compounds; Figure S2: Pareto plot of standardized effects for the TPC; Figure S3: Pareto plot of standardized effects for the antioxidant activity using DPPH assay; Figure S4: Pareto plot of standardized effects for the anti-hyaluronidase activity; Figure S5: Pareto plot of standardized effects for content of polydatin (a) and resveratrol (b) dissolved in methanol, and polydatin (c) and resveratrol (d) dissolved in artificial saliva solution at pH 6.8; Figure S6: Pareto plot of standardized effects for the total amount of released polydatin (a) and resveratrol (b) from nanofibers at 15 min; Figure S7: Pareto plot of standardized effects for component of bioadhesion; Figure S8: Pareto plot of standardized effects for the apparent permeability coefficients of polydatin (a) and resveratrol (b).

Author Contributions: Conceptualization, M.P.-W. and J.C.-P.; methodology, M.P.-W. and A.M.; writing—original draft preparation, M.P.-W., A.M. and J.C.-P.; writing—review and editing, M.P.-W. and J.C.-P.; supervision, J.C.-P.; project administration, M.P.-W.; funding acquisition, M.P.-W. and J.C.-P. All authors have read and agreed to the published version of the manuscript.

Funding: This research was funded by National Science Center (Poland), under Sonata grant (number 2020/39/D/NZ7/01824).

Institutional Review Board Statement: Not applicable.

Informed Consent Statement: Not applicable.

Data Availability Statement: Not applicable.

Conflicts of Interest: The authors declare no conflict of interest.

References

1. Garg, S.K.; Maurer, H.; Reed, K.; Selagamsetty, R. Diabetes and Cancer: Two Diseases with Obesity as a Common Risk Factor. *Diabetes Obes. Metab.* **2014**, *16*, 97–110. [CrossRef] [PubMed]
2. Pizzino, G.; Irrera, N.; Cucinotta, M.; Pallio, G.; Mannino, F.; Arcoraci, V.; Squadrito, F.; Altavilla, D.; Bitto, A. Oxidative Stress: Harms and Benefits for Human Health. *Oxid. Med. Cell Longev.* **2017**, *2017*, 8416763. [CrossRef] [PubMed]
3. Bhattacharyya, A.; Chattopadhyay, R.; Mitra, S.; Crowe, S.E. Oxidative Stress: An Essential Factor in the Pathogenesis of Gastrointestinal Mucosal Diseases. *Physiol. Rev.* **2014**, *94*, 329–354. [CrossRef] [PubMed]
4. Swallah, M.S.; Sun, H.; Affoh, R.; Fu, H.; Yu, H. Antioxidant Potential Overviews of Secondary Metabolites (Polyphenols) in Fruits. *Int. J. Food Sci.* **2020**, *2020*, 9081686. [CrossRef]
5. Park, M.-Y.; Kwon, H.-J.; Sung, M.-K. Evaluation of Aloin and Aloe-Emodin as Anti-Inflammatory Agents in Aloe by Using Murine Macrophages. *Biosci. Biotechnol. Biochem.* **2009**, *73*, 828–832. [CrossRef] [PubMed]
6. Zhang, H.; Li, C.; Kwok, S.-T.; Zhang, Q.-W.; Chan, S.-W. A Review of the Pharmacological Effects of the Dried Root of Polygonum Cuspidatum (Hu Zhang) and Its Constituents. *Evid.-Based Complement. Altern. Med.* **2013**, *2013*, e208349. [CrossRef]
7. Lin, Y.W.; Yang, F.J.; Chen, C.L.; Lee, W.T.; Chen, R.S. Free radical scavenging activity and antiproliferative potential of Polygonum cuspidatum root extracts. *J. Nat. Med.* **2010**, *64*, 146–152. [CrossRef]
8. Paczkowska-Walendowska, M.; Szymańska, E.; Winnicka, K.; Szwajgier, D.; Baranowska-Wójcik, E.; Ruchała, M.A.; Simon, M.; Cielecka-Piontek, J. Cyclodextrin as Functional Carrier in Development of Mucoadhesive Tablets Containing Polygoni Cuspidati Extract with Potential for Dental Applications. *Pharmaceutics* **2021**, *13*, 1916. [CrossRef]
9. Smoliga, J.M.; Blanchard, O. Enhancing the Delivery of Resveratrol in Humans: If Low Bioavailability Is the Problem, What Is the Solution? *Molecules* **2014**, *19*, 17154–17172. [CrossRef]
10. Sinha, S.; Ali, M.; Baboota, S.; Ahuja, A.; Kumar, A.; Ali, J. Solid Dispersion as an Approach for Bioavailability Enhancement of Poorly Water-Soluble Drug Ritonavir. *AAPS PharmSciTech* **2010**, *11*, 518–527. [CrossRef]
11. Zupančič, Š.; Baumgartner, S.; Lavrič, Z.; Petelin, M.; Kristl, J. Local Delivery of Resveratrol Using Polycaprolactone Nanofibers for Treatment of Periodontal Disease. *J. Drug Deliv. Sci. Technol.* **2015**, *30*, 408–416. [CrossRef]
12. Xiang, S.; Tang, H.W.; Zhou, J.; Li1, X.Z. Electrospinning of Hydroxypropyl-Beta-Cyclodextrin/Polyvinylpyrrolidone Resveratrol-Loaded Nanofibers: Preparation and Characterization. *Indian J. Pharm. Sci.* **2019**, *81*, 618–625. [CrossRef]

13. Lin, Y.-C.; Hu, S.C.-S.; Huang, P.-H.; Lin, T.-C.; Yen, F.-L. Electrospun Resveratrol-Loaded Polyvinylpyrrolidone/Cyclodextrin Nanofibers and Their Biomedical Applications. *Pharmaceutics* **2020**, *12*, 552. [CrossRef] [PubMed]
14. Karakucuk, A.; Tort, S. Preparation, Characterization and Antimicrobial Activity Evaluation of Electrospun PCL Nanofiber Composites of Resveratrol Nanocrystals. *Pharm. Dev. Technol.* **2020**, *25*, 1216–1225. [CrossRef]
15. Maria Leena, M.; Yoha, K.S.; Moses, J.A.; Anandharamakrishnan, C. Edible Coating with Resveratrol Loaded Electrospun Zein Nanofibers with Enhanced Bioaccessibility. *Food Biosci.* **2020**, *36*, 100669. [CrossRef]
16. Wang, C.; Ma, C.; Wu, Z.; Liang, H.; Yan, P.; Song, J.; Ma, N.; Zhao, Q. Enhanced Bioavailability and Anticancer Effect of Curcumin-Loaded Electrospun Nanofiber: In Vitro and In Vivo Study. *Nanoscale Res. Lett.* **2015**, *10*, 439. [CrossRef]
17. Torres-Martínez, E.J.; Bravo, J.M.C.; Medina, A.S.; González, G.L.P.; Gómez, L.J.V. A Summary of Electrospun Nanofibers as Drug Delivery System: Drugs Loaded and Biopolymers Used as Matrices. *Curr. Drug Deliv.* **2018**, *15*, 1360–1374. [CrossRef]
18. Paczkowska-Walendowska, M.; Gościniak, A.; Szymanowska, D.; Szwajgier, D.; Baranowska-Wójcik, E.; Szulc, P.; Dreczka, D.; Simon, M.; Cielecka-Piontek, J. Blackberry Leaves as New Functional Food? Screening Antioxidant, Anti-Inflammatory and Microbiological Activities in Correlation with Phytochemical Analysis. *Antioxidants* **2021**, *10*, 1945. [CrossRef]
19. Noriega, P.; Mafud, D. de F.; Souza, B. de; Soares-Scott, M.; Rivelli, D.P.; Barros, S.B. de M.; Bacchi, E.M. Applying Design of Experiments (DOE) to Flavonoid Extraction from Passiflora Alata and P. Edulis. *Rev. Bras. De Farmacogn.* **2012**, *22*, 1119–1129. [CrossRef]
20. Ćućuz, V.; Cvejić, J.; Torović, L.; Gojković-Bukarica, L.; Acevska, J.; Dimitrovska, A.; Aldawoud, T.M.S.; Galanakis, C.M. Design of Experiments (DoE) to Model Phenolic Compounds Recovery from Grape Pomace Using Ultrasounds. *J. Food Sci. Technol.* **2022**, *59*, 2913–2924. [CrossRef]
21. Li, Z.; Chen, X.; Liu, G.; Li, J.; Zhang, J.; Cao, Y.; Miao, J. Antioxidant Activity and Mechanism of Resveratrol and Polydatin Isolated from Mulberry (*Morus alba* L.). *Molecules* **2021**, *26*, 7574. [CrossRef]
22. Zykova, T.A.; Zhu, F.; Zhai, X.; Ma, W.-Y.; Ermakova, S.P.; Lee, K.W.; Bode, A.M.; Dong, Z. Resveratrol Directly Targets COX-2 to Inhibit Carcinogenesis. *Mol. Carcinog.* **2008**, *47*, 797–805. [CrossRef] [PubMed]
23. Gębalski, J.; Graczyk, F.; Załuski, D. Paving the Way towards Effective Plant-Based Inhibitors of Hyaluronidase and Tyrosinase: A Critical Review on a Structure–Activity Relationship. *J. Enzyme Inhib. Med. Chem.* **2022**, *37*, 1120–1195. [CrossRef]
24. Coles, S.R.; Jacobs, D.K.; Meredith, J.O.; Barker, G.; Clark, A.J.; Kirwan, K.; Stanger, J.; Tucker, N. A Design of Experiments (DoE) Approach to Material Properties Optimization of Electrospun Nanofibers. *J. Appl. Polym. Sci.* **2010**, *117*, 2251–2257. [CrossRef]
25. Paczkowska-Walendowska, M.; Dvořák, J.; Rosiak, N.; Tykarska, E.; Szymańska, E.; Winnicka, K.; Ruchała, M.A.; Cielecka-Piontek, J. Buccal Resveratrol Delivery System as a Potential New Concept for the Periodontitis Treatment. *Pharmaceutics* **2021**, *13*, 417. [CrossRef] [PubMed]
26. Bai, Y.; Wang, D.; Zhang, Z.; Pan, J.; Cui, Z.; Yu, D.-G.; Annie Bligh, S.-W. Testing of Fast Dissolution of Ibuprofen from Its Electrospun Hydrophilic Polymer Nanocomposites. *Polym. Test.* **2021**, *93*, 106872. [CrossRef]
27. Torres-Martínez, E.J.; Vera-Graziano, R.; Cervantes-Uc, J.M.; Bogdanchikova, N.; Olivas-Sarabia, A.; Valdez-Castro, R.; Serrano-Medina, A.; Iglesias, A.L.; Pérez-González, G.L.; Cornejo-Bravo, J.M.; et al. Preparation and Characterization of Electrospun Fibrous Scaffolds of Either PVA or PVP for Fast Release of Sildenafil Citrate. *e-Polymers* **2020**, *20*, 746–758. [CrossRef]
28. Yu, D.-G.; Branford-White, C.; White, K.; Li, X.-L.; Zhu, L.-M. Dissolution Improvement of Electrospun Nanofiber-Based Solid Dispersions for Acetaminophen. *AAPS PharmSciTech* **2010**, *11*, 809–817. [CrossRef]
29. Sizílio, R.H.; Galvão, J.G.; Trindade, G.G.G.; Pina, L.T.S.; Andrade, L.N.; Gonsalves, J.K.M.C.; Lira, A.A.M.; Chaud, M.V.; Alves, T.F.R.; Arguelho, M.L.P.M.; et al. Chitosan/Pvp-Based Mucoadhesive Membranes as a Promising Delivery System of Betamethasone-17-Valerate for Aphthous Stomatitis. *Carbohydr. Polym.* **2018**, *190*, 339–345. [CrossRef]
30. Vecchi, C.F.; Cesar, G.B.; de Souza, P.R.; Caetano, W.; Bruschi, M.L. Mucoadhesive Polymeric Films Comprising Polyvinyl Alcohol, Polyvinylpyrrolidone, and Poloxamer 407 for Pharmaceutical Applications. *Pharm. Dev. Technol.* **2021**, *26*, 138–149. [CrossRef]
31. Gavini, E.; Rassu, G.; Haukvik, T.; Lanni, C.; Racchi, M.; Giunchedi, P. Mucoadhesive Microspheres for Nasal Administration of Cyclodextrins. *J. Drug Target* **2009**, *17*, 168–179. [CrossRef] [PubMed]
32. Teng, Z.; Yuan, C.; Zhang, F.; Huan, M.; Cao, W.; Li, K.; Yang, J.; Cao, D.; Zhou, S.; Mei, Q. Intestinal Absorption and First-Pass Metabolism of Polyphenol Compounds in Rat and Their Transport Dynamics in Caco-2 Cells. *PLoS ONE* **2012**, *7*, e29647. [CrossRef]
33. Fenyvesi, F.; Nguyen, T.L.P.; Haimhoffer, Á.; Rusznyák, Á.; Vasvári, G.; Bácskay, I.; Vecsernyés, M.; Ignat, S.-R.; Dinescu, S.; Costache, M.; et al. Cyclodextrin Complexation Improves the Solubility and Caco-2 Permeability of Chrysin. *Materials* **2020**, *13*, 3618. [CrossRef] [PubMed]
34. Dahan, A.; Miller, J.M. The Solubility–Permeability Interplay and Its Implications in Formulation Design and Development for Poorly Soluble Drugs. *AAPS J.* **2012**, *14*, 244–251. [CrossRef] [PubMed]
35. Ndayishimiye, J.; Kumeria, T.; Popat, A.; Blaskovich, M.A.T.; Falconer, J.R. Understanding the Relationship between Solubility and Permeability of γ-Cyclodextrin-Based Systems Embedded with Poorly Aqueous Soluble Benznidazole. *Int. J. Pharm.* **2022**, *616*, 121487. [CrossRef] [PubMed]

nutrients

Article

Ethanol Extracts of Rice Bran and Whole Grain Adlay Seeds Mitigate Colonic Inflammation and Damage in Mice with Colitis

Hui-Chen Lo [1], Yu-Hsin Chen [2] and Wen-Tzu Wu [3,*]

1 Department of Nutritional Science, Fu Jen Catholic University, New Taipei City 242062, Taiwan
2 Taichung District Agricultural Research and Extension Station, Tatsuen Hsiang, Changhua County 515008, Taiwan
3 Department of Food Nutrition and Health Biotechnology, Asia University, Taichung 41354, Taiwan
* Correspondence: irenewu@asia.edu.tw; Tel.: +886-4-23323456 (ext. 48025)

Abstract: Ulcerative colitis (UC) is a chronic inflammatory bowel disease with frequent relapsing inflammation in the colon. Whole grains have been promoted as healthy and sustainable foods; however, the use of whole gains in UC is inconclusive. The aim of this study was to investigate the effects of ethanol extracts of rice bran (RBE) and whole-grain adlay seeds (ADE) on inflammation, oxidative stress, and colonic damage in UC. Male C57BL/6JNarl mice were intra-rectal injected twice with 2,4-dinitrobenzene sulfonic acid to induce (day 0) and reactivate (day 21) UC. Control mice were fed AIN-93M diet (R group) and injected with a vehicle. UC mice were fed AIN-93M diet (UC group) supplemented with RBE (RBE group) or ADE (ADE group) for 21 days. The results showed that the UC group had an increased disease activity index, plasma interleukin (IL)-6 and glutathione levels, microscopic injury scores, and inflammatory cytokine and chemokine levels in the colon and decreased colonic claudin-4 compared to the R group. RBE and ADE supplementation significantly reduced UC-elevated plasma IL-6 and colonic glutathione and pro-inflammatory cytokines and a chemokine. In addition, RBE and ADE supplementation significantly decreased T-helper-cell-associated cytokines in the plasma and colon. Moreover, RBE supplementation increased colonic IL-10 and tight junction protein claudin-4 levels, and ADE supplementation alleviated diarrhea in UC mice. In conclusion, these results suggest that RBE and ADE may mitigate colonic inflammation, oxidative stress, and damage in UC relapse.

Keywords: ulcerative colitis; rice bran; whole-grain adlay seeds; cytokines; tight junction; oxidative stress

Citation: Lo, H.-C.; Chen, Y.-H.; Wu, W.-T. Ethanol Extracts of Rice Bran and Whole Grain Adlay Seeds Mitigate Colonic Inflammation and Damage in Mice with Colitis. *Nutrients* **2022**, *14*, 3877. https://doi.org/10.3390/nu14183877

Academic Editors: Maria Digiacomo, Doretta Cuffaro and Marica Bakovic

Received: 4 August 2022
Accepted: 16 September 2022
Published: 19 September 2022

1. Introduction

Ulcerative colitis (UC), a subtype of inflammatory bowel disease (IBD), is a chronic, relapsing inflammatory disease that occurs primarily in the rectum and colon in a continuous pattern. The clinical symptoms of UC, including episodes of diarrhea, fever, fatigue, abdominal pain, cramping, bloody stool, reduced appetite, and weight loss, result in a significant decrease in quality of life and social functioning of patients. In addition, the presence of discontinuous and recurring inflammatory lesions in the rectal and colonic mucosa may increase the risk of colorectal cancer in UC patients [1]. The incidence and prevalence of UC are high in Europe and North America and are increasing in Asian countries [2]. In Taiwan, the prevalence rates of UC have increased more than six-fold in the past 20 years [2], and this increase has occurred predominantly in males [3].

UC is an autoimmune disease; however, its etiology remains unknown. Genetic influences, dysregulation of mucosal immunological barrier function, and overproduction of pro-inflammatory cytokines and reactive oxygen species (ROS) have been found to contribute to colonic tissue damage and ulceration [4]. Western diets that contain high amounts of fats, saccharides, and refined sugar with low dietary fiber content are considered one of the environmental factors that induce the onset and relapse of UC [5]. Diets that are rich in fruits, vegetables, and whole grains may be used to decrease the recurrence of IBD and the risk of colon cancer [6]. However, a diet composed of highly fermentable oligo-, di-, monosaccharides and polyols (FODMAP) may not be appropriate for UC [7]. Indeed, a cross-sectional study indicated that a modified diet containing approximately two-thirds of the daily recommendation in fiber, whole grains, fruits, and vegetables may reduce IBD symptoms [8].

Recent studies have shown that nutraceuticals with antioxidant and anti-inflammatory activities may have beneficial effects on IBD [9]. Whole grains contain the bran, germ, and endosperm, which provide various health-promoting nutrients, such as vitamins, minerals, fibers, and phytochemicals [8,10–12]. Rice bran, a nutrient-rich byproduct of the rice-milling process, has been considered a sustainable source of functional ingredients [13]. Its bioactive components—mainly phenolic and flavonoid compounds and vitamin E—have been found to have antioxidant, anti-inflammatory, anti-diabetic, and anticancer activities [14,15]. In rats, pretreatment of rice bran ethanol extracts significantly alleviated lipopolysaccharide-induced oxidative stress [13]. In mice with dextran sodium sulfate-induced UC, dietary supplementation of fermented rice bran alleviated mucosal inflammation; crypt loss; inflammatory cell infiltration; myeloperoxidase activity; thiobarbituric acid reactive substance (TBARS) levels; and T-helper-cell-associated cytokines, such as interleukin (IL)-1β, IL-6, and IL-17 [9,16]. The attenuated intestinal inflammation and enhanced tight junction barrier integrity suggest that rice bran may effectively protect against inflammation-induced intestinal fibrosis [16].

The adlay seed (*Coix lacryma-jobi* L.)—mainly the polished seed without bran—is an annual crop that is traditionally used as a medicinal food. Evidence has shown that the fraction with high phenolic and flavonoid contents from the ethanol extracts of adlay bran has an anti-inflammatory effect [12]. Polysaccharides and phenolic compounds extracted from adlay, especially the adlay bran, by ethanol, ethyl acetate, or acidified methanol have anti-inflammatory, immunological, and anti-hyperuricemia activities [11,17–19] and may improve epithelial barrier dysfunction in Caco-2 cells [11]. The anti-inflammatory and anti-apoptotic effects of ethanol extracts of the adlay hull have been demonstrated in rat brain PC-12 cells [20]. These results reveal that whole-grain adlay seeds may have beneficial effects for UC patients.

In clinical practice, no primary nutrition therapy is available for UC patients in the acute phase [5]. In this study, extracts of rice bran and whole-grain adlay seeds were created by using ethanol instead of water for extraction to exclude the water-soluble dietary fiber and to obtain higher yields of phenolic compounds. Here, mice with dinitrobenzene sulfonic acid (DNBS)-induced UC were used to evaluate the efficacy of these extracts on alleviating the disease activity index (DAI) and UC-induced alterations in plasma and colonic cytokine levels, macroscopic and microscopic injury, collagen levels, antioxidant status, and tight junction proteins in the colon. We hypothesized that ethanol extracts of rice bran (RBE) and whole-grain adlay seeds (ADE) may have antioxidant and anti-inflammatory activities that can alleviate the severity of UC.

2. Materials and Methods

2.1. Animals

The protocols and animal procedures were reviewed and approved by the Institutional Animal Care and Use Committee of Fu-Jen Catholic University (#A10803). Male C57BL/6JNarl mice at 6 weeks old (20 to 22 g) were purchased from the National Laboratory Animal Center (Taipei, Taiwan) and were housed in a room maintained at 22 °C on a 12:12 h light–dark cycle with free access to water and a normal chow diet (5001 Laboratory Rodent diet, Labdiet®, Richmond, IN, USA). After 2 weeks of acclimation, mice were divided into normal, healthy (R group, n = 10), and UC (3 groups, n = 12/group) groups and adapted with semi-purified AIN-93M diets by gradually introducing the diet over a 3-day period.

2.2. Ethanol Extract Preparations and Experimental Diets

Rice bran (stabilized, Herbfood Co., Taichung, Taiwan) and whole-grain adlay seeds (Taichung No. 5′ job's tears, *Coix lacryma-jobi* L.) were extracted with 70% ethanol, as described in the study of Chen et al. [21]. In brief, the powders of rice bran and whole-grain adlay seeds were defatted twice with hexane. Subsequently, the defatted materials were extracted with 10-fold volume of 70% ethanol overnight, at 4 °C, with shaking. After centrifugation at 10,000×g at 4 °C for 15 min, the supernatant was collected, filtered, ethanol-evaporated, and lyophilized. The dry extract was stored at −20 °C for further use.

The doses of ethanol extracts were calculated from the recommended daily doses of whole-grain brown rice (100 g per day, containing 16% of weight as rice bran) and whole-grain adlay seeds (50 g per day) for human adults. In the rice-milling process, approximately 16 g of rice bran was obtained from 100 g of whole-grain brown rice, and approximately 1.2 g of ethanol extract was obtained from the rice bran. In addition, approximately 0.64 g of ethanol extract was obtained from 50 g of whole-grain adlay seeds.

To prepare the experimental diets with ethanol extracts, the mouse metabolic rate was estimated based on the human metabolic rate, assuming a 12.3-fold difference. The average daily food intake was approximately 4 g for a mouse weighing 25 g. Therefore, 1.5 and 0.8 g of ethanol extracts from rice bran and whole-grain adlay seeds, respectively, were added into 1 kg of semi-purified AIN-93M diet, as shown in Table 1. All diet ingredients were purchased from Dyets Inc. (Bethlehem, PA, USA). The three different diets provided identical amounts of calories, protein, fat, fiber, minerals, and vitamins, with the exception of vitamin E.

Table 1. Composition of AIN-93M-modified diets.

Ingredient (g)	R/UC Groups	RBE Group	ADE Group
Corn starch	465.7	464.2	464.9
Rice bran extract	0	1.5	0
Whole-grain adlay seeds' extract	0	0	0.8
Dextrin	155	155	155
Casein, vitamin-free	140	140	140
Sucrose, granular	100	100	100
Cellulose	50	50	50
Soybean oil	40	40	40
AIN-93M Mineral Mix	35	35	35
AIN-93M Vitamin Mix	10	10	10
Choline bitartrate	2.5	2.5	2.5
L-cystine	1.8	1.8	1.8
Total	1000	1000	1000

R: control; UC: ulcerative colitis; RBE: ethanol extract of rice bran; ADE: ethanol extract of whole-grain adlay seeds.

2.3. Induction of UC

UC induction methods were based on the description by Martin et al. [22]. Following an overnight fast, mice in the UC, RBE, and ADE groups were anesthetized with 3% isoflurane (day 0) and intra-rectally injected with 200 mg/kg of 2,4-dinitrobenzene sulfonic acid (DNBS in 50% ethanol, Sigma, St. Louis, MO, USA) via PE-50 tubes. Mice in the R group were injected with 100 μL of 50% ethanol to control for the stress from the intra-rectal injection. After the injection, 2% agar gel with 6% sucrose was provided for 3 days to prevent the occurrence of hypoglycemia and dehydration. On day 21, UC was reactivated with 100 mg/kg of DNBS. All mice were sacrificed on day 24, and blood samples were collected by cardiac puncture for the collection of red blood cells and plasma. The colon was dissected, weighed, and collected for further analysis.

2.4. Experimental Design

From day 3 to day 24, the UC mice were fed semi-purified AIN-93M diets with or without different ethanol extracts. UC mice were divided into three groups and fed either the AIN-93M diet (UC group) or the AIN-93M diet supplemented with 70% alcohol extract of rice bran (RBE group) or whole-grain adlay seeds (ADE group).

2.5. Measurements

2.5.1. Body Weight, Food Intake, and Disease Activity Index (DAI)

During the experimental period, body weight was recorded twice per week, and food intake, stool consistency, and signs of hematochezia were observed daily. The DAI was calculated by using scores correlating with body weight loss, diarrhea, and hematochezia [23]. Stool consistency was graded by using the following scale: 0 for normal stool, 2 for loosely stool, and 4 for diarrhea. The hematochezia score was graded as 0 for no blood being seen, 1 for possible bleeding undetectable by the naked eye, 2 for visible bleeding, and 4 for hematochezia around the anus. The body weight loss score was graded as 0 for no weight loss, 1 for 1–5% of total body weight loss, 2 for 5–10%, 3 for 10–20%, and 4 for over 20% of total body weight loss. The DAI was then calculated by using the following equation. DAI = (body weight lost score + diarrhea score + hematochezia sore)/3.

2.5.2. Inflammatory Mediators in the Plasma and Colon

To evaluate the effects of the ethanol extracts on systemic and local inflammation, plasma and colonic cytokines, including the T-helper-cell-associated tumor necrosis factor (TNF)-α, IL-6, interferon (IFN)-γ, IL-10, and IL-12p70, were measured. DNBS-induced colitis is particularly useful in studying T-cell-dependent immune mechanisms [24]. Levels of inflammatory mediators and the chemokine monocyte chemoattractant protein-1 (MCP-1) were determined by using commercially available cytometric bead assays (BD Biosciences, Minneapolis) and analyzed on a flow cytometer (BD Accuri® C6, Becton, Dickinson and Company, Franklin Lakes, NJ, USA).

2.5.3. Macroscopic Injury Score of the Colon

After the mice were euthanized, the total colon length was recorded, and any injury was photographed with a digital camera (Coolplex 4500, Nikon, Tokyo, Japan). The macroscopic injury score was based on the study of Impellizzeri et al. [25]. The scoring system was graded as the following: 0 for no visible damage; 1 for small hyperemia without ulcers; 2 for linear ulcers and no major inflammation; 3 for linear ulcers with one site of inflammation; 4 for two or more sites of inflammation with ulceration covering >1 cm along the length of the colon; and 5 for two or more major sites of inflammation and ulceration or one major site of inflammation and ulceration extending >2 cm along the length of the colon and an additional point for each centimeter of ulceration beyond the initial 2 cm [25].

2.5.4. Collagenous Connective Tissue of the Colon: Masson's Trichrome Stain

After colon collection, 1 cm of colon from the region approximately 3 cm proximal to the distal segment of the colon was collected and fixed in 10% buffered formalin for routine paraffin embedding [24]. The 4 μm paraffin-embedded sections were stained with hematoxylin and eosin (H&E). Histological changes of colonic sections, including the depth of ulceration, area of ulceration, edema, and immune cell infiltration, were scored by using a graded scale of 0 to 4, as described by Iwazawa et al. [26]. In addition, the severity of inflammation, hyperplasia, and the loss of goblet cells in the mucosa were graded according to the modified method described by Impellizzeri et al. [25] and Shackelford et al. [27]. The interpretation of these scores was performed by an animal pathologist (National Taiwan University College of Medicine Laboratory Animal Center, Taipei, Taiwan) with an official pathology report.

2.5.5. Histopathological (Microscopic) Assessment of the Colon

To detect collagen fiber deposition, 4 μm colonic sections were stained with a Masson's trichrome staining kit (TRM-1-IFU, Logan, UT, USA), according to the manufacturer's instruction. In brief, the deparaffinized slides were incubated in preheated Bouin's fluid for 60 min, then cooled, rinsed, and stained with Weigert's iron hematoxylin. Next, the slides were incubated with Biebrich scarlet-acid fuchsin solution and differentiated in phosphomolybdic-phosphotungstic acid solution. Subsequently, 1% acetic acid solution was applied to the slides, which were then dehydrated with 95% alcohol and absolute alcohol, cleared in xylene, and mounted in synthetic resin. The immunoreaction was observed by using a light microscope (Eclipes TE2000-U; Nikon, Tokyo, Japan). The area of blue color, representing collagen, was semi-quantified in five fields per slide at 200× magnification, using Image-Pro Plus software version 6.0 (Media Cybernetics; Rockville, MD, USA). The ratio of collagen in each field was calculated as the blue area (collagen) divided by the total red (muscle fiber), blue (collagen), and dark-red-to-black (nuclei) areas combined.

2.5.6. Tight Junction Proteins in the Colon

To further investigate gut integrity at the molecular level, two transmembrane tight junction proteins, claudin-4 and occludin, were measured by using immunohistochemistry (IHC) analysis. After deparaffinization and dehydration, 4 μm colonic sections were treated with 10 mM sodium citrate at 100 °C for 15 min for antigen retrieval, 3% H_2O_2 for 10 min for quenching endogenous peroxidase, and 5% skim milk for 60 min for nonspecific blocking, followed by hybridization with the primary antibodies against claudin-4 (1:500, Thermo Scientific, Waltham, MA, USA) and occludin (1:50, Santa Cruz, Dallas, TX, USA). Horseradish peroxidase and rabbit anti-mouse biotinylated IgG (1:200), diaminobenzidine substrate stain, and hematoxylin were used as per the manufacturer's protocol (Thermo Scientific, Waltham, MA, USA). Immunoreactions were observed by using a light microscope (Nikon, Tokyo, Japan) with a cooling-charge-coupled-device camera. Semi-quantitative analysis of the images was performed in five fields per slide (Image-Pro Plus software version 6.0). The ratio of claudin-4 and occludin to total colonic area in each field was calculated.

2.5.7. Indicators of Oxidative Stress in the Colon

The levels of lipid peroxidation products, TBARS, were determined in the colon, using the methods presented by Ohkawa et al. [28]. Non-enzymatic antioxidants, i.e., reduced glutathione (GSH) and oxidized glutathione (GSSG), were measured fluorometrically, according to the methods of Hissin and Hilf [29]. The ratio of GSH to GSSG was calculated to evaluate oxidative status.

2.6. Statistical Analysis

Values are reported as the mean ± the standard error of the mean (SEM) for continuous measures and as the median and interquartile range (IQR) for nonparametric measures. For continuous measures, all groups were compared by one-way analysis of variance (ANOVA), using the SAS general linear model program (SAS Institute, Cary, NC, USA). Group means were considered significantly different at $p < 0.05$. The Duncan's multiple range test was used as a post hoc analysis to compare the differences among the groups when the ANOVA indicated an overall significant group effect, $p < 0.05$. The Kruskal–Wallis test was used for the analyses of macroscopic and microscopic injury scores, stool consistency, and signs of hematochezia, followed by the Mann–Whitney U test to determine significant differences ($p < 0.05$) between groups

3. Results

3.1. Food Intake, Body Weight, DAI, and Colon Length

The final sample size for each group was 10 mice in the R, UC, and RBE groups and 11 mice in the ADE group. After the intra-rectal injection of DNBS, five mice died within

2 days. The autopsy confirmed that the cause of death was large bowel perforation. During the experimental period, the body weight (Figure 1A) was not significantly different among groups, but the body-weight change (Figure 1B) from day 0 to day 24 was significantly lower in the UC group compared to the R group. The average daily food intake during the experimental period was not significantly different between the R and UC groups (Figure 1C). However, the ADE group had significantly greater food intake than the R, UC, and RBE groups.

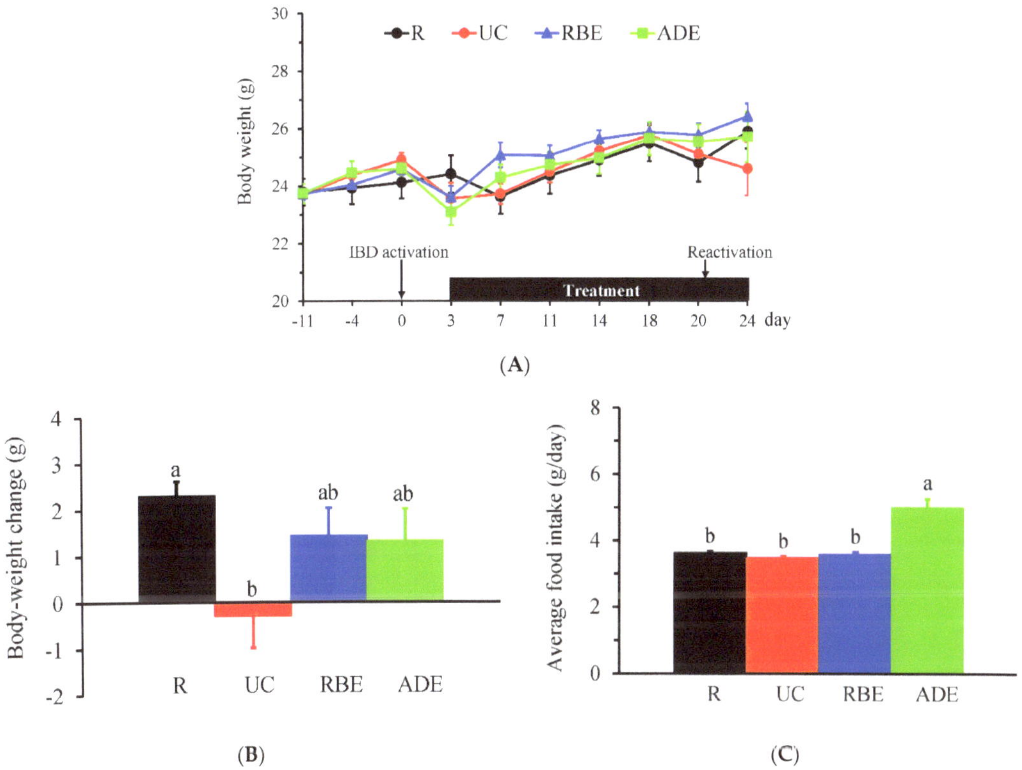

Figure 1. Body weight and food intake of control mice (R group) fed AIN-93M diet and UC mice fed either the AIN-93M diet (UC group) or the AIN-93M diet supplemented with alcohol extract of rice bran (RBE group) or whole-grain adlay seeds (ADE group). (**A**) Body weight before and on the day of UC induction (day 0) until the end of the experiment (day 24). (**B**) Body-weight change and (**C**) average daily food intake from day 0 to day 24. Values are mean ± SEM, n = 10−11 per group. Different superscript letters indicate significant differences (one-way ANOVA and Duncan's multiple range test, $p < 0.05$).

Diarrhea (stool consistency), bloody stool (hematochezia) and DAI scores and the length of the colon on day 24 are shown in Table 2. The UC group had significantly higher diarrhea, bloody stool, and DAI scores than the R group. The ADE group had a significantly decreased diarrhea score compared to the UC group. A shortened colon length is one of the signs of IBD; however, the colon length was not significantly different among groups.

Table 2. Scores of diarrhea, bloody stool, disease activity index, and colon length.

Group	Diarrhea	Bloody Stool	DAI	Colon Length (cm)
R	0 (0, 0) ^{b}	0 (0, 0) ^{b}	0 (0, 0) ^{b}	6.93 ± 0.26
UC	2 (0, 2) ^{a}	0.5 (0, 2) ^{a}	1 (0.33, 1) ^{a}	6.90 ± 0.38
RBE	0 (0, 2) ^{ab}	0 (0, 1) ^{ab}	0.67 (0.33, 1.33) ^{a}	6.53 ± 0.16
ADE	0 (0, 0) ^{b}	0 (0, 2) ^{ab}	0 (0, 1.33) ^{ab}	6.44 ± 0.19

Values are median (interquartile range, IQR) or mean ± SEM, n = 10−11 per group. DAI, disease activity index. Different superscript letters represent significant differences (Kruskal–Wallis test and Mann–Whitney U test, $p < 0.05$).

3.2. *Inflammatory Mediators in the Plasma*

There were no significant differences in the levels of TNF-α, IFN-γ, IL-10, IL-12p70, or MCP-1 between the R and UC groups (Table 3). The UC group had significantly increased IL-6 levels compared to the R group, and the RBE and ADE groups had significantly decreased IL-6 levels compared to the UC group. In addition, the RBE and ADE groups had significantly decreased the levels of TNF-α, IFN-γ, IL-10, IL-12p70, and MCP-1 compared to the R and UC groups (Table 3).

Table 3. Plasma levels of cytokines and chemokine MCP-1.

Group	TNF-α	IL-6	IFN-γ	IL-10	IL-12p70	MCP-1
	(pg/mL)					
R	38.4 ± 0.9 ^{a}	53.4 ± 0.5 ^{b}	12.00 ± 0.27 ^{a}	409 ± 5 ^{a}	106.8 ± 3.3 ^{a}	10.05 ± 0.13 ^{a}
UC	37.7 ± 3.2 ^{a}	67.1 ± 8.8 ^{a}	11.38 ± 0.29 ^{a}	393 ± 4 ^{a}	97.0 ± 3.9 ^{a}	10.81 ± 1.29 ^{a}
RBE	29.2 ± 1.3 ^{b}	53.3 ± 2.5 ^{b}	9.79 ± 0.17 ^{b}	364 ± 5 ^{b}	71.8 ± 4.2 ^{b}	8.30 ± 0.21 ^{b}
ADE	30.3 ± 1.8 ^{b}	50.2 ± 0.9 ^{b}	9.82 ± 0.45 ^{b}	364 ± 4 ^{b}	81.6 ± 4.7 ^{b}	8.36 ± 0.32 ^{b}

Values (pg/mL) are mean ± SEM, n = 10−11 per group. TNF, tumor necrosis factor; IL, interleukin; IFN, interferon; MCP, monocyte chemoattractant protein. Different superscript letters represent significant differences (one-way ANOVA and Duncan's multiple range test, $p < 0.05$).

3.3. *Macroscopic and Microscopic Injury Scores of the Colon*

The macroscopic injury scores, including hyperemia, ulcers, and inflammation, were used to evaluate colon damage. There were no significant differences in the colonic macroscopic injury scores among groups. To evaluate microscopic injury, colonic sections were stained with H&E. The scores of edema, immune-cell infiltration, loss of goblet cells, inflammation, and total injury in the colon were significantly increased in the UC, RBE, and ADE groups compared to the R group (Table 4); however, there were no significant differences among the UC, RBE, and ADE groups. In addition, there were no significant differences in the depth of ulceration, area of ulceration, and hyperplasia scores in the colon among the groups.

Table 4. Microscopic injury scores of the colon.

Group	Area of Ulceration	Edema	Immune-Cell Infiltration	Goblet-Cell Loss	Inflammation	Total Injury Score
R	0 (0, 0)	0 (0, 0) ^{b}	0 (0, 0) ^{b}	0 (0, 0) ^{b}	0 (0, 0) ^{b}	0 (0, 0) ^{b}
UC	0 (0, 1)	1 (0, 1) ^{a}	1 (0, 3) ^{a}	0.5 (0, 1) ^{a}	2 (0, 3) ^{a}	4.5 (0, 10) ^{a}
RBE	0 (0, 1)	1 (0, 1) ^{a}	1 (0, 3) ^{a}	1 (0, 1) ^{a}	2 (0, 3) ^{a}	6 (0, 9) ^{a}
ADE	0 (0, 1)	1 (1, 1) ^{a}	2 (1, 3) ^{a}	1 (1, 2) ^{a}	2 (1, 3) ^{a}	8 (4, 11) ^{a}

Values are median (IQR), n = 10−11 per group. Different superscript letters represent significant differences (Kruskal–Wallis test and Mann–Whitney U test, $p < 0.05$).

3.4. Collagen, Claudin-4, and Occludin Contents in the Colon

To evaluate the degree of colonic damage, Masson's trichrome staining was used to detect collagen deposition, i.e., an index of fibrosis. As shown in Figure 2A, the blue color represents collagen, the red represents muscle fibers, and the dark-red-to-black areas represent nuclei. The protein expression of tight junction proteins, i.e., claudin-4 (Figure 2B) and occludin (Figure 2C), was measured by IHC staining with brown color in the colonic sections. When calculating the area of collagen in total colonic area of the microscopic field, the UC and ADE groups had a significantly greater collagen-positive area than the R group (Figure 2D). Moreover, the UC and ADE groups had a significantly decreased proportion of claudin-4 compared to the R and RBE groups (Figure 2D). However, the occludin-positive area was not significantly different among groups.

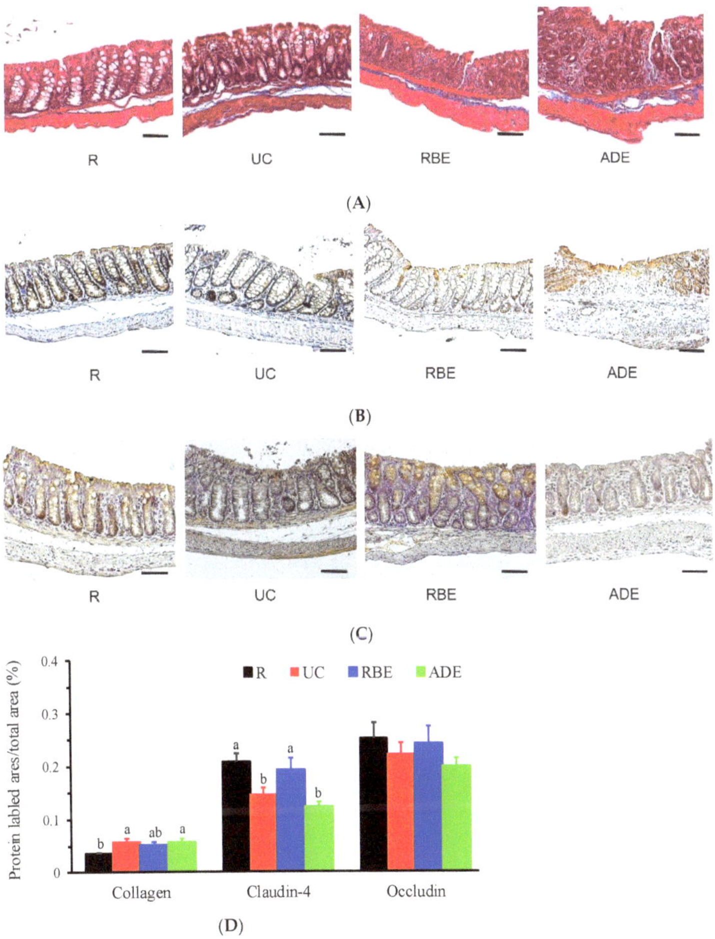

Figure 2. Protein expression of collagen, claudin-4 and occludin in the colon of control mice (R group) fed AIN-93M diet and UC mice fed either the AIN-93M diet (UC group) or the AIN-93M diet supplemented with alcohol extract of rice bran (RBE group) or whole-grain adlay seeds (ADE group). Photos (100×, bars = 50 μm) of Masson's trichrome staining for (**A**) collagen, and immunohistochemical staining of (**B**) claudin-4 and (**C**) occludin in the colon and (**D**) the ratio of indicated protein area to total colonic area. Values are mean ± SEM, n = 10−11 per group. Different superscript letters represent significant differences (one-way ANOVA and Duncan's multiple range test, $p < 0.05$).

3.5. GSH, GSSG and TBARS Levels in the Colon

Levels of antioxidant indicators, GSH, GSSG, the ratio of GSH to GSSG, and lipid peroxidation indicators (TBARS) in the colon are shown in Table 5. The UC group had significantly increased GSH levels and a higher GSH:GSSG compared to the R, RBE, and ADE groups. There were no significant differences in GSSG and TBARS levels in the colon among groups.

Table 5. Levels of reduced and oxidized glutathione and TBARS in the colon.

Group	GSH (μmol/g Colon)	GSSG (μmol/g Colon)	GSH/GSSG	TBARS (nmol/g Colon)
R	33.8 ± 2.0^{b}	6.23 ± 0.44	5.60 ± 0.31^{b}	132.5 ± 13.7
UC	43.9 ± 3.4^{a}	5.49 ± 0.37	8.60 ± 0.50^{a}	143.3 ± 29.6
RBE	30.0 ± 1.3^{b}	5.60 ± 0.31	5.10 ± 0.39^{b}	138.1 ± 22.6
ADE	35.3 ± 1.8^{b}	6.63 ± 0.19	5.30 ± 0.30^{b}	158.2 ± 25.0

Values are mean ± SEM, n = 10–11 per group. GSH, reduced glutathione; GSSG, oxidized glutathione, GSH/GSSG, ratio of the reduced to oxidized glutathione; TBARS, thiobarbituric acid reactive substance. Different superscript letters represent significant differences (one-way ANOVA and Duncan's multiple range test, $p < 0.05$).

3.6. Inflammatory Mediators in the Colon

Cytokine and chemokine levels in the colon are shown in Table 6. The UC group had approximately 0.8- to 1.5-fold higher TNF-α, IL-6, IFN-γ, IL-12p70, and MCP-1 levels compared to the R group. In contrast, IL-10 in the colon of the UC group was 50% lower than that in the R group. The UC-induced increases in colonic levels of TNF-α and IL-12p70 were significantly decreased in the RBE and ADE groups. Furthermore, the UC-induced increases in IL-6 and MCP-1 and the UC-induced decrease in IL-10 were significantly ameliorated in the RBE group.

Table 6. Levels of cytokines and chemokine MCP-1 in the colon.

Group	TNF-α	IL-6	IFN-γ	IL-10	IL-12p70	MCP-1
	(pg/g Colon)					
R	31.2 ± 1.5^{b}	8.92 ± 0.85^{b}	2.08 ± 0.16^{b}	3.08 ± 0.15^{a}	56.8 ± 4.3^{c}	0.25 ± 0.08^{b}
UC	55.4 ± 4.1^{a}	18.11 ± 2.84^{a}	5.49 ± 0.37^{a}	1.52 ± 0.14^{b}	112.0 ± 8.9^{a}	0.51 ± 0.10^{a}
RBE	40.3 ± 3.3^{b}	10.85 ± 2.57^{b}	4.81 ± 0.28^{a}	3.42 ± 0.20^{a}	72.4 ± 9.2^{b}	0.32 ± 0.08^{b}
ADE	44.8 ± 2.6^{b}	13.02 ± 1.99^{ab}	4.92 ± 0.34^{a}	2.15 ± 0.23^{ab}	83.9 ± 6.7^{b}	0.40 ± 0.17^{ab}

Values are mean ± SEM, n = 10–11 per group. TNF, tumor necrosis factor; IL, interleukin; IFN, interferon; MCP, monocyte chemoattractant protein. Different superscript letters represent significant differences (one-way ANOVA and Duncan's multiple range test, $p < 0.05$).

4. Discussion

The prevalence of UC is gradually increasing in Taiwan. Its symptoms and relapses threaten the quality of life and increase the risk of colorectal cancer in UC patients [30]. In clinical practice, UC treatment includes corticosteroids, antibiotics, and immunosuppressants; however, none of these drugs are curative, and their long-term use may result in adverse side effects and complications [31]. Foods with antioxidant, anti-inflammatory, and immunomodulatory activities, such as whole grains, are considered to have beneficial effects for UC patients [8,32]. In the present study using DNBS-induced UC mice, oral administration of ethanol extracts from rice bran and whole-grain adlay seeds significantly alleviated the DNBS-induced alterations in T-helper-cell-associated cytokines. In addition, the ethanol extracts of rice bran significantly improved DNBS-induced decreases in the tight junction protein claudin-4's levels in the colon.

Increasing evidence suggests that inappropriate food choice, for instance, a Western diet, may induce the overproduction of pro-inflammatory cytokines and dysregulate immune function, resulting in recurrence, delayed recovery, or worsening disease severity in UC patients [9]. In addition, a low FODMAP diet is recommended for UC patients during the active period to control gastrointestinal symptoms [32,33]. Whole grains, for example, rice bran and adlay, are traditional Chinese foods that have been demonstrated to have antioxidant and anti-inflammatory activities [9,12,16,18,34]. However, the high dietary fiber content in whole grains may limit their use in UC patients. After defatting and excluding polysaccharides, starch, and fiber, ethanol extracts of rice bran and whole-grain adlay seeds contain copious amounts of phenolic and flavonoid compounds and vitamin E. These materials may have potential biological functions in UC.

DNBS is a hapten that can trigger the host innate and adaptive immune responses and has been used in mice to mimic the relapsing pattern observed in UC patients [22,24]. In our previous study, DNBS-induced UC mice were administered various water extracts derived from grains and plants, including *Echinacea purpurea*, *Salvia miltiorrhiza*, and adlay polysaccharides; however, clinical symptoms and colon damage in UC mice were not improved [35]. We speculated that the high content of dietary fiber from these water extracts might not be appropriate for UC mice during the relapsed state. Thus, the present study focused on rice bran and whole-grain adlay seeds that were extracted with 70% ethanol to reduce FODMAP and further evaluate their effects on colonic injury and inflammation.

In the present study, body weight was significantly reduced in UC mice during the experimental period, and DAI and microscopic injury scores, including edema, immune cell infiltration, goblet cell loss, inflammation, and total injury in the colon, were increased. In addition, UC mice had significantly increased plasma IL-6 levels, elevated colonic pro-inflammatory and T-helper-cell-associated cytokines and chemokine MCP-1 levels, and a decreased anti-inflammatory cytokine IL-10 level. Tight junctions act as a semipermeable barrier and function to divide the apical and basolateral domains of the plasma membrane [36]. Collagen, an index of fibrosis, has been shown to deposit in the colon from the mucosal to the muscular layers in UC patients [37]. In the present study, UC mice had a significant decrease in the levels of the tight junction protein claudin-4 and a significant increase in collagen in the colon. The changes in clinical symptoms, microscopic colonic injury, cytokine profiles in the plasma and colon, tight junction protein, and collagen reveal that the DNBS-induced UC mice had similar clinical and histological features to UC patients.

GSH, a major antioxidant in the body, contains a free thiol group that can be oxidized to GSSG. In the present study, UC mice had unaltered colonic TBARS and increased colonic GSH levels, suggesting an adaptive response to alleviate oxidative stress, such as lipid peroxidation, in the colon during UC reactivation. The low DAI scores and unchanged colonic length, colonic TBARS, and plasma proinflammatory cytokines suggest that the DNBS-induced UC mice may have mild-to-moderate disease severity and colon damage, which were not as severe as the findings in the study of Martin et al. [22]

Whole grains are a good source of dietary fiber, minerals, vitamins, and phytochemicals. Rice bran, a by-product of rice milling, has drawn attention due to its surplus value. In human umbilical vein endothelial cells, rice-bran-derived phenolic extracts showed antioxidant and anti-inflammatory effects that protected against endothelial dysfunction [10]. In a randomized double-blind controlled trial, γ-oryzanol, a triterpenic alcohol mixture with ferulic acid esters and phytosterols extracted from rice bran, improved antioxidant status in hyperlipidemic subjects [14,38]. The antioxidant and anti-inflammatory effects of rice bran components, including oligosaccharides, γ-oryzanol, ferulic acid, geranylgeraniol, farnesol, isoprenoids, and lectins, have been reviewed [15,39]. In the present study, an ethanol extract of rice bran significantly decreased plasma and colonic levels of IL-6; T-helper-cell-associated cytokines, such as TNF-α and IL-12p70; and the chemokine MCP-1, a marker secreted by monocytes, memory T cells, and dendritic cells in UC mice [9,16,40]. In addition, the ethanol extract of rice bran significantly increased claudin-4 protein expression and alleviated UC-elevated GSH content in the colon. The decreased GSH and unaltered TBARS reveal that an ethanol extract of rice bran may reduce the need for high levels of antioxidants, e.g., GSH, to maintain the colonic redox status in UC. The results of plasma and colonic cytokines and the chemokine MCP-1 and colonic claudin-4 and GSH content suggest that the ethanol extract of rice bran may alleviate UC-induced impairment in colonic integrity, inflammation, and oxidative stress.

Adlay seeds, also called Job's tear seeds or Coix seeds, have been used as nourishment foods and in traditional Chinese medicine and possess antioxidant and anti-inflammatory activities [18,34]. In Caco-2 cells, polysaccharides of adlay bran significantly decreased the secretion of pro-inflammatory cytokines and IL-10 and alleviated TNF-α-induced epithelial barrier dysfunction [11]. In a rat model of rheumatoid arthritis, an ethanol extract of adlay seed significantly decreased serum pro-inflammatory cytokines and increased the activities of antioxidant enzymes [34]. In the present study, UC mice administered the ethanol extract from whole-grain adlay seeds had an approximately 25% increase in food intake without subsequent weight gain, suggesting a potential role for adlay seed extract in weight control. In addition, diarrhea; T-helper-cell-associated cytokines in the plasma; and DNBS-induced increases in TNF-α, IL-12p70, and GSH in the colon were decreased by dietary supplementation with an ethanol extract of whole-grain adlay seeds in UC mice. These results suggest that an ethanol extract of whole-grain adlay seeds may be used to alleviate clinical symptoms and colonic inflammation, at least partially, in UC patients.

There are several limitations in the present study. First, components of the ethanol extracts were not determined, thus limiting the ability to distinguish the main bioactive compound(s) in rice bran and whole-grain adlay seeds. Second, the severity of the UC induced in the study was mild to moderate, so that the UC mice would present an adaptive redox response, such as unaltered TBARS and increased GSH levels in the colon. This adaptive response limited the degree of improvement observed in oxidative-stress markers in UC mice administered the ethanol extracts. Third, the mucosal and muscular layers of the colon were not separated when determining cytokine and chemokine levels. The mucosal layer is the major site with immunological processes in the colon. Therefore, the cytokines, chemokine MCP-1, GSH, and TBARS results in the colon may be underestimated. Even so, cytokine and chemokine levels were simultaneously determined in the plasma and colon, allowing us to reveal that the systemic and local inflammatory and immune responses were different in UC.

5. Conclusions

Our findings suggest that oral administration of ethanol extracts from rice bran and whole-grain adlay seeds may partially alleviate the clinical symptoms, morphological changes of the colon, and colonic damage in DNBS-induced UC mice. In addition, these ethanol extracts may eliminate DNBS-induced alterations in T-helper-cell-associated cytokines and GSH in the colon. Furthermore, the ethanol extract of rice bran has beneficial effects in regard to preserving colon integrity via increasing the tight junction protein claudin-4 in UC mice. The bioactive components and mechanisms of ethanol extracts from rice bran and whole-grain adlay seeds in modulating the homeostasis of inflammatory mediators and colonic oxidant–antioxidant profiles in UC, as well as the safety and tolerability of these extracts, are worth further investigation.

Author Contributions: Methodology, investigation, writing—original draft, conceptualization, software, visualization, funding acquisition, data curation, and formal analysis, H.-C.L.; conceptualization, methodology, and writing—review and editing, Y.-H.C.; methodology, investigation, validation, supervision, and writing—review and editing, W.-T.W. All authors have read and agreed to the published version of the manuscript.

Funding: This work was funded by Taichung District Agricultural Research and Extension Station, Council of Agriculture, Executive Yuan, Taiwan (108AS-12.3.1-CS-D1).

Institutional Review Board Statement: The animal study protocol was approved by the Institutional Review Board of Fu-Jen Catholic University (#A10803).

Informed Consent Statement: Not applicable.

Acknowledgments: The authors would like to thank Kuan-Lin Wang, Chi-Chin Yu, Ruo-wei Wang, Yu-Hsuan Su, Yu-Shuo Liu, Jing-Yao Zhang, Bing-Xiang Lin, and Bing-Yi Chen for their technical assistance in performing experiments.

Conflicts of Interest: The authors declare no conflict of interest.

References

1. Ramos, G.P.; Papadakis, K.A. Mechanisms of disease: Inflammatory bowel diseases. *Mayo Clin. Proc.* **2019**, *94*, 155–165. [CrossRef]
2. Yen, H.H.; Weng, M.T.; Tung, C.C.; Wang, Y.T.; Chang, Y.T.; Chang, C.H.; Shieh, M.J.; Wong, J.M.; Wei, S.C. Epidemiological trend in inflammatory bowel disease in Taiwan from 2001 to 2015: A nationwide populationbased study. *Intest. Res.* **2019**, *17*, 54–62. [CrossRef] [PubMed]
3. Yen, H.H.; Hsu, T.C.; Chen, M.W.; Su, P.Y.; Chen, Y.Y. Clinical features and treatment of inflammatory bowel disease in a low-incidence area: A hospital-based retrospective cohort study in Taiwan. *Medicine* **2021**, *100*, e25090. [CrossRef] [PubMed]
4. Shizuma, T. Anti-colitis effects of brown rice reported by experimental studies. *J. Rice Res.* **2014**, *2*, 2. [CrossRef]
5. Keshteli, A.H.; Madsen, K.L.; Dieleman, L.A. Diet in the pathogenesis and management of ulcerative colitis; a review of randomized controlled dietary interventions. *Nutrients* **2019**, *11*, 1498. [CrossRef] [PubMed]
6. Hnatyszyn, A.; Hryhorowicz, S.; Kaczmarek-Rys, M.; Lis, E.; Slomski, R.; Scott, R.J.; Plawski, A. Colorectal carcinoma in the course of inflammatory bowel diseases. *Hered. Cancer Clin. Pract.* **2019**, *17*, 18. [CrossRef]
7. Vandeputte, D.; Joossens, M. Effects of low and high FODMAP diets on human gastrointestinal microbiota composition in adults with intestinal diseases: A systematic review. *Microorganisms* **2020**, *8*, 1638. [CrossRef]
8. Kamp, K.J.; Pennings, B.; Javelli, D.; Wyatt, G.; Given, B. Dietary patterns, beliefs and behaviours among individuals with inflammatory bowel disease: A cross-sectional study. *J. Hum. Nutr. Diet.* **2021**, *34*, 257–264. [CrossRef]
9. Islam, J.; Koseki, T.; Watanabe, K.; Ardiansyah; Budijanto, S.; Oikawa, A.; Alauddin, M.; Goto, T.; Aso, H.; Komai, M.; et al. Dietary supplementation of fermented rice bran effectively alleviates dextran sodium sulfate-induced colitis in mice. *Nutrients* **2017**, *9*, 747. [CrossRef]
10. Saji, N.; Francis, N.; Blanchard, C.L.; Schwarz, L.J.; Santhakumar, A.B. Rice bran phenolic compounds regulate genes associated with antioxidant and anti-Inflammatory activity in human umbilical vein endothelial cells with induced oxidative stress. *Int. J. Mol. Sci.* **2019**, *20*, 4715. [CrossRef]
11. Li, Y.; Tian, X.; Li, S.; Chang, L.; Sun, P.; Lu, Y.; Yu, X.; Chen, S.; Wu, Z.; Xu, Z.; et al. Total polysaccharides of adlay bran (*Coix lachryma-jobi* L.) improve TNF-alpha induced epithelial barrier dysfunction in Caco-2 cells via inhibition of the inflammatory response. *Food Funct.* **2019**, *10*, 2906–2913. [CrossRef] [PubMed]
12. Chen, H.J.; Chung, C.P.; Chiang, W.; Lin, Y.L. Anti-inflammatory effects and chemical study of a flavonoid-enriched fraction from adlay bran. *Food Chem.* **2011**, *126*, 1741–1748. [CrossRef]
13. Yadav, S.; Appukuttan, J.P. Inhibition of LPS induced neurochemical imbalance and oxidative stress by pigmented and non-pigmented rice bran extracts. *J. Food Biochem.* **2019**, *43*, e12735. [CrossRef] [PubMed]
14. Sohail, M.; Rakha, A.; Butt, M.S.; Iqbal, M.J.; Rashid, S. Rice bran nutraceutics: A comprehensive review. *Crit. Rev. Food Sci. Nutr.* **2017**, *57*, 3771–3780. [CrossRef] [PubMed]
15. Sapwarobol, S.; Saphyakhajorn, W.; Astina, J. Biological functions and activities of rice bran as a functional ingredient: A review. *Nutr. Metab. Insights* **2021**, *14*, 11786388211058559. [CrossRef]
16. Agista, A.Z.; Rusbana, T.B.; Islam, J.; Ohsaki, Y.; Sultana, H.; Hirakawa, R.; Watanabe, K.; Nochi, T.; Ardiansyah; Budijanto, S.; et al. Fermented rice bran supplementation prevents the development of intestinal fibrosis due to DSS-induced inflammation in mice. *Nutrients* **2021**, *13*, 1869. [CrossRef]
17. Yao, Y.; Zhu, Y.; Gao, Y.; Ren, G. Effect of ultrasonic treatment on immunological activities of polysaccharides from adlay. *Int. J. Biol. Macromol.* **2015**, *80*, 246–252. [CrossRef]
18. Choi, G.; Han, A.R.; Lee, J.H.; Park, J.Y.; Kang, U.; Hong, J.; Kim, Y.S.; Seo, E.K. A comparative study on hulled adlay and unhulled adlay through evaluation of their LPS-induced anti-inflammatory effects, and isolation of pure compounds. *Chem. Biodivers.* **2015**, *12*, 380–387. [CrossRef]
19. Zhao, M.; Zhu, D.; Sun-Waterhouse, D.; Su, G.; Lin, L.; Wang, X.; Dong, Y. In vitro and in vivo studies on adlay-derived seed extracts: Phenolic profiles, antioxidant activities, serum uric acid suppression, and xanthine oxidase inhibitory effects. *J. Agric. Food Chem.* **2014**, *62*, 7771–7778. [CrossRef]
20. Tsay, G.J.; Lin, Y.T.; Hsu, C.H.; Tang, F.Y.; Kuo, Y.H.; Chao, C.Y. Adlay hull extracts attenuate beta-amyloid-induced neurotoxicity and oxidative stress in PC12 cells through antioxidative, anti-inflammatory, and antiapoptotic activities. *Biochem. Biophys. Rep.* **2021**, *26*, 101020. [CrossRef]
21. Chen, M.H.; Choi, S.H.; Kozukue, N.; Kim, H.J.; Friedman, M. Growth-inhibitory effects of pigmented rice bran extracts and three red bran fractions against human cancer cells: Relationships with composition and antioxidative activities. *J. Agric. Food Chem.* **2012**, *60*, 9151–9161. [CrossRef] [PubMed]
22. Martin, R.; Chain, F.; Miquel, S.; Lu, J.; Gratadoux, J.J.; Sokol, H.; Verdu, E.F.; Bercik, P.; Bermudez-Humaran, L.G.; Langella, P. The commensal bacterium Faecalibacterium prausnitzii is protective in DNBS-induced chronic moderate and severe colitis models. *Inflamm. Bowel Dis.* **2014**, *20*, 417–430. [CrossRef] [PubMed]

23. Kim, J.J.; Shajib, M.S.; Manocha, M.M.; Khan, W.I. Investigating intestinal inflammation in DSS-induced model of IBD. *J. Vis. Exp.* **2012**, *60*, 3678. [CrossRef] [PubMed]
24. Morampudi, V.; Bhinder, G.; Wu, X.; Dai, C.; Sham, H.P.; Vallance, B.A.; Jacobson, K. DNBS/TNBS colitis models: Providing insights into inflammatory bowel disease and effects of dietary fat. *J. Vis. Exp.* **2014**, e51297. [CrossRef] [PubMed]
25. Impellizzeri, D.; Siracusa, R.; Cordaro, M.; Peritore, A.F.; Gugliandolo, E.; Mancuso, G.; Midiri, A.; Di Paola, R.; Cuzzocrea, S. Therapeutic potential of dinitrobenzene sulfonic acid (DNBS)-induced colitis in mice by targeting IL-1beta and IL-18. *Biochem. Pharmacol.* **2018**, *155*, 150–161. [CrossRef]
26. Iwazawa, R.; Kozakai, S.; Kitahashi, T.; Nakamura, K.; Hata, K.I. The therapeutic effects of adipose-derived stem cells and recombinant peptide pieces on mouse model of DSS colitis. *Cell Transplant.* **2018**, *27*, 1390–1400. [CrossRef]
27. Shackelford, C.; Long, G.; Wolf, J.; Okerberg, C.; Herbert, R. Qualitative and quantitative analysis of nonneoplastic lesions in toxicology studies. *Toxicol. Pathol.* **2002**, *30*, 93–96. [CrossRef]
28. Ohkawa, H.; Ohishi, N.; Yagi, K. Assay for lipid peroxides in animal tissues by thiobarbituric acid reaction. *Anal. Biochem.* **1979**, *95*, 351–358. [CrossRef]
29. Hissin, P.J.; Hilf, R. A fluorometric method for determination of oxidized and reduced glutathione in tissues. *Anal. Biochem.* **1976**, *74*, 214–226. [CrossRef]
30. Kim, D.H.; Cheon, J.H. Pathogenesis of inflammatory bowel disease and recent advances in biologic therapies. *Immune Netw.* **2017**, *17*, 25–40. [CrossRef]
31. Ribeiro, D.; Proenca, C.; Rocha, S.; Lima, J.; Carvalho, F.; Fernandes, E.; Freitas, M. Immunomodulatory effects of flavonoids in the prophylaxis and treatment of inflammatory bowel diseases: A comprehensive review. *Curr. Med. Chem.* **2018**, *25*, 3374–3412. [CrossRef]
32. Damas, O.M.; Garces, L.; Abreu, M.T. Diet as adjunctive treatment for inflammatory bowel disease: Review and update of the latest literature. *Curr. Treat. Options Gastroenterol.* **2019**, *17*, 313–325. [CrossRef] [PubMed]
33. Whelan, K.; Staudacher, H. Low FODMAP diet in irritable bowel syndrome: A review of recent clinical trials and meta-analyses. *Curr. Opin. Clin. Nutr. Metab. Care* **2022**, *25*, 341–347. [CrossRef] [PubMed]
34. Zhang, C.; Zhang, W.; Shi, R.; Tang, B.; Xie, S. *Coix lachryma-jobi* extract ameliorates inflammation and oxidative stress in a complete Freund's adjuvant-induced rheumatoid arthritis model. *Pharm. Biol.* **2019**, *57*, 792–798. [CrossRef] [PubMed]
35. Lo, H.-C.; Wang, K.-L.; Chen, Y.-C. An Animal Platform for Inflammatory Bowel Disease. In Proceedings of the 2019 R&D Results Presentation of Human Ecology College in Fu Jen Catholic University, New Taipei City, Taiwan, 22 November 2019.
36. Lean, Q.Y.; Eri, R.D.; Randall-Demllo, S.; Sohal, S.S.; Stewart, N.; Peterson, G.M.; Gueven, N.; Patel, R.P. Orally administered enoxaparin ameliorates acute colitis by reducing macrophage-associated inflammatory responses. *PLoS ONE* **2015**, *10*, e0134259. [CrossRef] [PubMed]
37. Manetti, M.; Rosa, I.; Messerini, L.; Ibba-Manneschi, L. Telocytes are reduced during fibrotic remodelling of the colonic wall in ulcerative colitis. *J. Cell Mol. Med.* **2015**, *19*, 62–73. [CrossRef]
38. Tanaka, T.; Oyama, T.; Sugie, S. Dietary tricin suppresses inflammation-related colon carcinogenesis in mice. *J. Nutr. Sci. Vitaminol.* **2019**, *65*, S100–S103. [CrossRef]
39. Kurtys, E.; Eisel, U.L.M.; Hageman, R.J.J.; Verkuyl, J.M.; Broersen, L.M.; Dierckx, R.; de Vries, E.F.J. Anti-inflammatory effects of rice bran components. *Nutr. Rev.* **2018**, *76*, 372–379. [CrossRef]
40. Xing, L.; Fu, L.; Cao, S.; Yin, Y.; Wei, L.; Zhang, W. The anti-inflammatory effect of bovine bone-gelatin-derived peptides in LPS-induced RAW264.7 macrophages cells and dextran sulfate sodium-induced C57BL/6 mice. *Nutrients* **2022**, *14*, 1479. [CrossRef]

nutrients

Article

Flammulina velutipes Mycorrhizae Attenuate High Fat Diet-Induced Lipid Disorder, Oxidative Stress and Inflammation in the Liver and Perirenal Adipose Tissue of Mice

Zhen Luo [1], Qingying Gao [1], Yuanfei Li [2], Yifei Bai [3], Jing Zhang [1], Weina Xu [1] and Jianxiong Xu [1,*]

1 Shanghai Key Laboratory of Veterinary Biotechnology, School of Agriculture and Biology, Shanghai Jiao Tong University, Shanghai 200240, China
2 Institute of Biological Technology, Nanchang Normal University, Nanchang 330032, China
3 College of Food Science and Technology, Shanghai Ocean University, Shanghai 201306, China
* Correspondence: jxxu1962@sjtu.edu.cn; Tel.: +86-021-34205767

Abstract: *Flammulina velutipes* (FV) is edible mushroom that has nutritional and medicinal values. FV mycorrhizae, the by-products of FV, are an abundant source and receive less attention. The objective of this study was to investigate the composition of FV mycorrhizae, and its effects on high fat diet (HFD)-induced lipid disorder, oxidative stress, and inflammatory cytokines, both in the liver and perirenal adipose tissue (PAT) of mice. The results showed that FV mycorrhizae contain abundant trace elements, polysaccharide, amino acids and derivatives, and organic compounds. It was found that 4% FV mycorrhizae (HFDFV) supplementation decreased HFD-induced liver weight and triglyceride (TG) in the plasma, liver and PAT, altered plasma and hepatic fatty acids profiles, promoted gene expression involved in lipid hydrolysis, fatty acid transportation and β-oxidation in the liver and reduced lipid synthesis in the liver and PAT. HFDFV attenuated HFD-induced oxidative stress and pro-inflammatory cytokine by increasing GSH/GSSG, and decreasing levels of MDA and IL6 both in the liver and PAT, while it differentially regulated gene expression of IL1β, IL6, and CCL_2 in liver and PAT. The results indicated that FV mycorrhizae are effective to attenuate HFD-induced lipid disorder, oxidative stress and inflammation in the liver and PAT, indicating their promising constituents for functional foods and herbal medicine.

Keywords: *Flammulina velutipes* mycorrhizae; oxidative stress; fatty acids metabolism; liver; perirenal adipose tissue

Citation: Luo, Z.; Gao, Q.; Li, Y.; Bai, Y.; Zhang, J.; Xu, W.; Xu, J. *Flammulina velutipes* Mycorrhizae Attenuate High Fat Diet-Induced Lipid Disorder, Oxidative Stress and Inflammation in the Liver and Perirenal Adipose Tissue of Mice. *Nutrients* **2022**, *14*, 3830. https://doi.org/10.3390/nu14183830

Academic Editors: Maria Digiacomo and Doretta Cuffaro

Received: 13 August 2022
Accepted: 13 September 2022
Published: 16 September 2022

1. Introduction

Obesity is a major worldwide health challenge. In the past few decades, its prevalence has increased, posing a significant economic burden. Obesity increases the risk of metabolic diseases such as type 2 diabetes mellitus, fatty liver disease and hypertension [1]. The adipose tissues and liver play central roles in lipid metabolism and whole-body energy homeostasis, which is responsible for the fundamental pathogenesis of obesity [2]. Thus, the attenuation of lipid disorders in these organs is an effective way to relieve obesity-related metabolic syndromes [3]. Non-steroidal anti-inflammatory drugs (NSAIDs) are widely used in metabolic disorders and inflammation because of their analgesic, anti-inflammatory and antipyretic effects, but adverse effects also exist, such as drug–drug interactions and organ damage, highlighting the need for and importance of complementary and alternative medicine [4–6]. Recently, plant extracts and phytochemicals have attracted extensive attention in the management of weight loss and prevention of metabolic diseases [7].

Flammulina velutipes (FV) is an excellent source of bioactive compounds such as vitamins, phenol, fibers, polysaccharides and minerals, and exhibits various biological activities such as anti-oxidative, immunomodulatory, and lowering-cholesterol, which are considered as an alternative strategy for the prevention and treatment of lipid disorders [8,9]. Application of plant-based wastes as functional foods has also received growing interest, with food and agricultural industries generating large quantities of by-products. Furthermore, the stem or mycorrhizae of mushrooms have shown different chemical components and structural characteristics such as polysaccharides conjugation, solubility and antioxidant activities compared with the caps of the mushroom, resulting in differences in their absorption and metabolism [10,11]. Although FV mycorrhizae were able to effectively alleviate HFD-induced lipid metabolism disorders, regulate the gut microbiota and activate the immune function of T lymphocyte [12,13], the chemical composition of FV mycorrhizae, its bioavailability and metabolism, and its effects on the liver and adipose tissue are still lacking. In addition, the perirenal adipose tissues (PAT) play an important role in chronic inflammatory, renal disease and cardiovascular events due to its unique anatomy, physiology and location [14,15]. The effect of FV mycorrhizae on lipid metabolism and inflammation in the PAT is unknown. Thus, this study aimed to explore the composition of FV mycorrhizae, and its effect on HFD-induced lipid disorders, oxidative stress and inflammation, both in the liver and PAT of mice.

2. Materials and Methods

2.1. Composition Analysis

The major compositions, including crude protein (CP), ethanol extract (EE), crude fiber (CF), Ash, Ca and P, harmful substances such as aflatoxin B1 (AFB1), zearalenone (ZEN), deoxynivalenol (DON) and Salmonella, were determined as described previously [16]. The heavy metals (Pb, Cd and Hg) and trace elements (Fe, Mn, Cu, Zn, Se, Na, Mg and K) in FV mycorrhizae were determined using inductively coupled plasma mass spectrometry in Centre Testing International (Qingdao) Co., Ltd. An untargeted and targeted metabolomics approach based on ultrahigh-performance liquid chromatography and equipped with quadrupole time-of-flight mass spectrometry (UHPLC-QTOF-MS) and multiple reaction monitoring (MRM) was used to analyze the composition and quantify the contents of amino acids and derivatives in FV mycorrhizae, as we described previously [17].

2.2. Preparation of Polysaccharide

Polysaccharides from dried FV mycorrhizae were extracted according to the water-extraction and alcohol-precipitation method. Briefly, the dried sample were crushed and dissolved in ethanol solution overnight, then extracted by hot water in sample:water = 1:10 for 4 h at 60 °C. The precipitate was re-extracted twice and the supernatants were collected and pooled. Then, the solution was concentrated and precipitated with four volumes of ethanol at 4 °C to obtain crude water-extracts. The extracts were re-dissolved in water, deproteinized, dialyzed, concentrated and freeze dried to obtain crude water polysaccharide. The purity of crude polysaccharide (total soluble sugar) was determined by the sulfuric acid-anthrone method. Total sugar content mg/mg = C*V/M, where C is the calculated result according to the standard curve; M is the actual weighing mass; V is the sample dissolved volume.

2.3. Analysis of Monosaccharide Compositions

The monosaccharide compositions were determined according to previous study [18]. Briefly, approximately 5 mg of dried polysaccharide was hydrolyzed with 2 M trifluoroacetic acid at 121 °C for 2 h in a sealed tube. The residue was re-dissolved in deionized water and filtered through 0.22 μm microporous filtering film for measurement. Then, the samples were analyzed by high-performance anion-exchange chromatography (HPAEC) on a CarboPac PA-20 anion-exchange column (3 by 150 mm; Dionex) using a pulsed amperometric detector (PAD; Dionex ICS 5000 system, ThermoFisher Scientific, Waltham, MA, USA). The monosaccharide content was calculated according to the calibration curve of monosaccharide standard.

2.4. SEC-MALLS-RI Measurement

The samples were dissolved in 0.1M $NaNO_3$ aqueous solution at 1 mg/mL concentration and filtered through a filter of 0.45 μm pore size. The samples were dissolved in DMSO solution containing lithium bromide (0.5% *w*/*w*) at 1 mg/mL and filtered through 0.45 μm filter. The homogeneity and molecular weight were measured using SEC-MALLS-RI. The weight and number-average molecular weight (Mw and Mn) and polydispersity index (Mw/Mn) in 0.1 M $NaNO_3$ aqueous solution were measured on a DAWN HELEOS-II laser photometer equipped with three tandem columns (300 × 8 mm, Shodex OH-pak SB-805, 804 and 803; Showa Denko K.K., Tokyo, Japan), which was held at 45 °C using a model column heater. Data were acquired and processed using ASTRA6.1.

2.5. Fourier-Transform Infrared Spectra (FT-IR)

The dried polysaccharides were mixed with potassium bromide powder and then pressed into 1 mm pellets for FT-IR measurement (Nicolet iZ-10, ThermoFisher Scientific, Waltham, MA, USA) in the range of 400 to 4000 cm^{-1}.

2.6. Animal Experiments

Twenty-seven 8-week-old C57BL/6J mice were obtained from Huafukang Bioscience (Beijing, China). The mice were housed in box cages at 22–25 °C under 12 h light/night cycles with ad libitum access to food and water. After 1-week adaptation, they were randomly divided into three groups: control (CON), high fat diet (HFD) and HFD supplemented with +4% FV mycorrhizae powder (HFDFV). Mice that received basic diet served as CON. The CON group (12% fat, #D12450B) and HFD group (60% fat, #D12492) were provided by Huafukang Bioscience (Beijing, China). The HFDFV was evenly mixed HFD with 4% FV mycorrhizae powder, and the blend were extruded to cylinder manually. The experiment lasted for 16 weeks. Body weight and food intake were recorded every 3 days. At the end of the experimental period, the mice were sacrificed followed by exsanguination. The blood samples were collected. Then, livers, PAT and epididymal adipose tissues (EAT) were weighted, washed with PBS, sectioned and stored at −80 °C. The colon length was measured. Fresh tissues were also immediately fixed in 4% paraformaldehyde for further analysis.

2.7. Histopathological Examination

The fresh livers were fixed in 4% paraformaldehyde and embedded in paraffin, sectioned and stained with hematoxylin and eosin (H&E) for histopathologic analysis. The liver samples were also embedded in the frozen liver section for oil red staining to visualize lipid droplets.

2.8. Triglyceride (TG), Cholesterol (TC) and Transaminase Determination

The plasma was collected after centrifuging at 3500× *g* for 20 min. The liver and PAT tissues were homogenized in PBS buffer and the supernatants were gathered after centrifuging 5000× *g* for 10 min. Protein concentration was measured through the bicinchoninic acid (BCA) protein assay kit. The content of TG, TC and activity of alanine aminotransferase (ALT) in these samples were determined according to the manufacturers' instructions (Nanjing Jiancheng Bioengineering Institute, Nanjing, China), as in our previous description [19].

2.9. Oxidative Stress Parameters Determination

The oxidative stress parameters were determined according to the manufacturers' instructions (Nanjing Jiancheng Bioengineering Institute, Nanjing, China), as in our previous description [20]. Protein concentration was measured through the bicinchoninic acid (BCA) protein assay kit. Superoxide dismutase (SOD), malondialdehyde (MDA), H_2O_2, GSH and GSSG were measured and the absorbance at 450, 530, 405, 405 and 405 nm was recorded, respectively.

2.10. Cytokines Determination

The cytokines such as IL1β, IL6, TNFα and CCL_2 in the liver and PAT tissues were determined using commercially available enzyme-linked immune sorbent assay (ELISA) kits (Nanjing Jiancheng Bioengineering Institute, Nanjing, China), as in our previous description [20]. Briefly, liver and PAT tissues were homogenized in PBS and then centrifuged at 12,000× *g* for 15 min. The microplates were coated with IL1β, IL6, TNFα and CCL_2, followed by detection with a horseradish peroxidase-labeled substrate after incubation for 10 min at 37 °C. Absorbance values were read in a spectrophotometer at 450 nm.

2.11. Untargeted Metabolomics

An untargeted metabolomics approach was used to study the changes in plasma metabolites among three dietary treatments. Analyses were performed using UHPLC (Agilent 1290 Infinity LC) coupled to a QTOF (AB Triple TOF 6600) in Shanghai Applied Protein Technology (APT, Shanghai). Plasma samples were mixed with pre-cooling methanol/acetonitrile/water solution (2:2:1, *v*/*v*). The mixtures were vortexed using an ultrasonator for 30 min and placed at −20 °C for 10 min, centrifuged at 14,000× *g* for 20 min. The vacuum dried supernatants were dissolved in 100 μL acetonitrile solution (acetonitrile:water= 1:1, *v*/*v*) for further analysis. The determination methods and process were carried out according to our previous description [17].

2.12. Targeted Metabolomics

A targeted metabolomics approach was used to study the compositions of medium long chain fatty acids in the liver between HFDFV and HFD. Briefly, 10% H_2SO_4-CH_3OH solution (600 μL) were added into the 50 mg sample, suspended for 1 min and bathed in water at 62 °C for 2 h. After cooling, anhydrous sodium sulfate and 600 μL n-hexane were added and suspended for 1 min. Then, the solution was centrifuged at 3500 r/min for 5 min. The supernatants were dried vacuum, dissolved in 200 μL n-hexane and analyzed by GC-MS. The samples were separated by gas chromatography on a DB-5MS capillary column (30 m × 0.25 mm ID × 0.25 μm). A standard sample mixture was used to identify the fatty acids profiles. The concentrations were calculated based on the chromatogram peak areas.

2.13. RNA Isolation, cDNA Synthesis and Real-Time PCR

The total RNA from liver and PAT tissues was extracted following the instructions of the RNA extraction kit. The concentration of RNA was quantified using a spectrophotometer (ThermoFisher Scientific, Waltham, MA, USA). Then, 1 μg of RNA was reverse transcribed into cDNA using the PrimeScrip RT reagent Kit. The real-time quantitative PCR reaction was applied to quantify the gene expression with the LightCycler96 system (Roche). The reaction was performed in a total volume of 20 μL, including 10 μL SYBR Green mix, 2 μL cDNA, 7.2 μL H_2O, and 0.4 μL each of forward and reverse primers. Amplification conditions were initially 95 °C for 30 s, 40 cycles of 95 °C for 5 s and 60 °C for 20 s, and melting curve data. The primers were designed and listed in Supplementary Table S1. β-actin was used as a housekeeping gene to normalize target gene transcript levels. The values were expressed using the formula $2^{-(\triangle\triangle Ct)}$, where $\triangle\triangle Ct = (Ct_{Target} - Ct_{\beta\text{-actin}})_{treatment} - (Ct_{Target} - Ct_{\beta\text{-actin}})_{HFD}$.

2.14. Statistical Analysis

Data were analyzed with one-way analysis of variance (ANOVA) followed by post hoc Duncan's test using the statistical software SPSS 17.0 (SPSS Inc., Chicago, IL, USA), while data of medium long chain fatty acids profiles was analyzed with independent sample *t*-test. Data were presented as mean ± SEM. $p < 0.05$ was considered statistically significant.

3. Results

3.1. Compositional Analysis of FV Mycorrhizae

The compositions such as CP, EE, CF and ash in FV mycorrhizae are listed in Supplementary Table S2. The levels of AFB1, ZEN and Salmonella were not detected in FV mycorrhizae, and the DON level was very low. The eight essential amino acids (EAAs) Lys, Phe, Met, Thr, Ile, Leu, Val and Trp, non-essential amino acids (NEAAs) such as Tyr, Asp, Asn, Pro, Ala, Ser, Gly, Arg, Glu, Gln, His and cystine were identified (Supplementary Table S3), but the Cys was not detected. The amino acid derivatives such as citrulline, ornithine, hydroxyproline, aminoadipic acid, creatine, creatinine, choline and taurine, biogenic amines such as putrescine were also identified. However, the spermidine was not detected. Untargeted metabolomics showed that 945 compounds (12 superclass and 57 class) were identified (Figure 1A,B), including organic acids and derivatives (25.98%), lipids and lipid-like molecules (21.53%), undefined (11.77%), organoheterocyclic compounds (11.13%), organic oxygen compounds (9.76%), and so on. The specific compositions are listed in the Supplementary Materials.

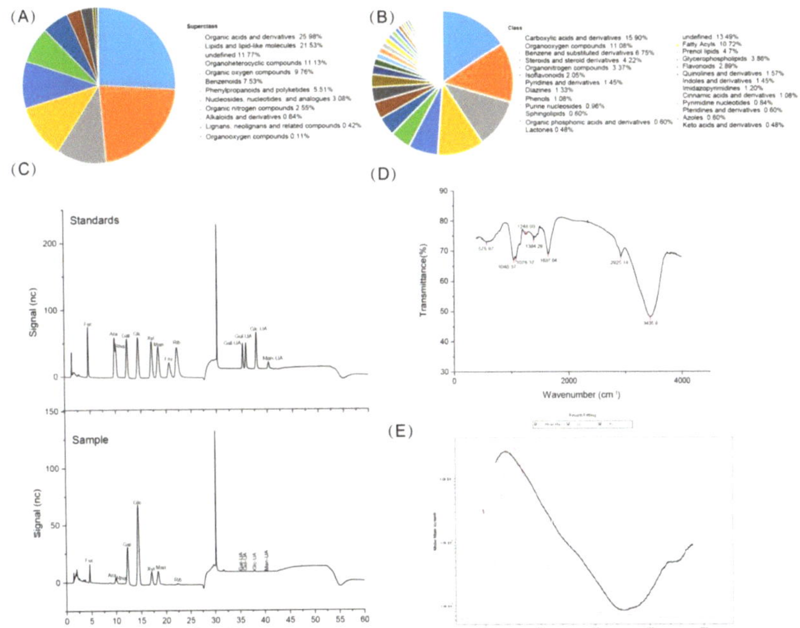

Figure 1. The composition of FV mycorrhizae and characteristics of polysaccharides from FV mycorrhizae were determined. (**A**) Superclass; (**B**) Class; (**C**) The ion chromatography of monosaccharide mixture standards and polysaccharides from FV mycorrhizae; (**D**) The FT-IR spectrum; (**E**) Chromatograms of the molar mass distribution.

In total, 1.354 g crude polysaccharide was obtained after extraction and impurity removal from 50 g FV mycorrhizae powder. Thus, the extraction rate was about 2.7%. The purity of polysaccharide was calculated as 52.9% according to the formula. The ion chromatography of monosaccharide mixture standards and polysaccharides from FV mycorrhizae are shown in Figure 1C. They were composed of glucose (Glc), galactose (Gal), mannose (Man), xylose (Xyl), fucose (Fuc), arabinose (Ara), ribose (Rib), rhamnose (Rha), guluronic acid (Gul-UA), mannuronic acid (Man-UA), galacturonic acid (Gal-UA) and glucuronic acid (Glu-UA) in a molar ratio of 46.02:20.67:11.3:7.38:6.95:2.83:1.1:1.09:1.11:1.02:0.32:0.22 (Table 1), indicating that the acidic polysaccharides were rich in glucose. Polysaccharides had three characteristic absorptions peaks, which represented by a strong absorption peak at 3435 cm^{-1} for the stretching vibrations of O-H, absorption peak at 2925.14 cm^{-1} for C-H stretching vibration and absorption peak at 1637.04 cm^{-1} for the stretching vibration of C=O in this study (Figure 1D). Galactose and glucose showed the strongest peak at 1078.37 cm^{-1} and 1040.37 cm^{-1}, which is consistent with the monosaccharide composition analysis. The peak at around 1384.28 cm^{-1} and 1248.09 cm^{-1} could be due to the C-O-H of carboxylic acid, S=O or acetyl group (CH3-CO) stretching vibrations, according to a previous study [21]. The molecular mass distribution and chain conformation are shown in Figure 1E. The number-average molar masses (Mw) and weight-average molar masses (Mn) were 1.809×10^6 and 1.632×10^5 g/mol. The Rw, Rn and Rz were 42.6, 45.3 and 37.2 nm (Table 1). The polydispersity value (Mw/Mn) was 11.085, indicating the extract had wide molecular weight distribution.

Table 1. Characteristics of polysaccharides extracted from FV mycorrhizae.

Item	Results
Sugar composition (%)	
Glc	46.02
Gal	20.67
Man	11.3
Xyl	7.38
Fuc	6.95
Ara	2.83
Rib	1.1
Rha	1.09
Fru	0
Gul-UA	1.11
Man-UA	1.02
Gal-UA	0.32
Glu-UA	0.22
Molecular characteristics	
Mn (g/mol)	1.632×10^5
Mw (g/mol)	1.809×10^6
Mz (g/mol)	1.427×10^7
Polydispersity (Mw/Mn)	11.085
Rn (nm)	45.3
Rw (nm)	42.6
Rz (nm)	37.2

3.2. *Effect of FV Mycorrhizae on Mice Body Weight, Food Intake, Organs Weight and Plasma Lipid Disorders*

As shown in Figure 2B, the mice body weight was not significantly different ($p > 0.05$) among the three groups from the beginning to the end of the experimental period. Compared with the CON, HFD had no effect on food intake ($p > 0.05$), while HFDFV significantly decreased food intake compared with the HFD ($p < 0.05$) (Figure 2C). Liver HE and oil red staining are shown in Figure 2D. HFDFV improved HFD-induced structural damage in liver by HE staining. The weight of PAT and EAT were not affected by dietary treatment ($p > 0.05$) and the weight of liver was significantly decreased by 16.5% in HFDFV compared with HFD ($p < 0.05$) (Figure 2E). HFD significantly increased the length of the colon compared with CON ($p < 0.05$), but HFDFV had no effect on the length of the colon compared with HFD ($p > 0.05$) (Figure 2F). HFDFV significantly decreased the HFD-induced content of plasma TG ($p < 0.05$) (Figure 2G) but had no effect on the content of plasma TC ($p > 0.05$) (Figure 2H). Additionally, HFDFV significantly decreased the HFD-induced activity of ALT in plasma ($p < 0.05$) (Figure 2I). These results indicated that FV mycorrhizae treatment reduced HFD-induced lipid disorder and liver dysfunction in mice.

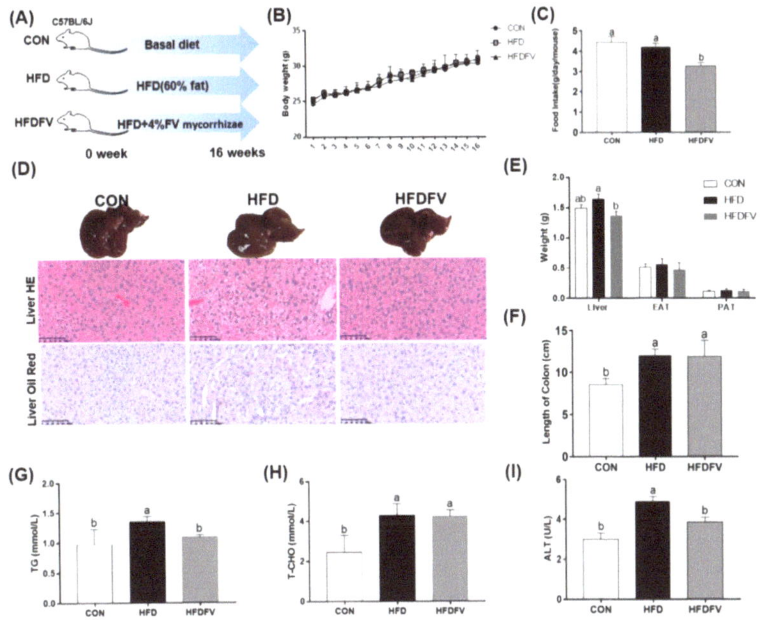

Figure 2. Effect of FV mycorrhizae on the HFD-induced body weight, food intake, liver histopathology and plasma lipids abundance in mice. (**A**) The experimental design. CON: control group, mice were received basic diet. HFD: high fat diet group, mice were received 60% high fat diet. HFDFV: high fat diet + 4% FV mycorrhizae group, mice were received 60% high fat diet supplemented with 4% FV mycorrhizae powder; (**B**) The body weight; (**C**) Food intake; (**D**) The liver morphology, HE and oil red staining; (**E**) The weight of liver, PAT and EAT; (**F**) The length of colon; (**G**) The content of plasma TG; (**H**) The content of plasma TC; (**I**) The activity of plasma ALT. C57BL/6J mice were divided into 3 groups and were fed either basal diet (CON, 10% fat), high-fat diet (HFD, 60% fat) or HFD + 4% FV (HFDFV) for 16 weeks. Values with different letters differ significantly ($p < 0.05$). Values were expressed as mean ± SEM (n = 6–8).

3.3. *Effect of FV Mycorrhizae on Lipid Metabolism and Oxidative Stress Parameters in the Liver and PAT*

The lipid metabolism and oxidative stress parameters in the liver and PAT were further studied. In the liver, HFDFV significantly decreased the HFD-induced content of TG and TC ($p < 0.05$) (Figure 3A,B). Oxidative stress parameters showed that HFDFV significantly decreased the HFD-induced content of H_2O_2, MDA and the activity of SOD, increased GSH/GSSG in the liver ($p < 0.05$) (Figure 3C–F). In the PAT, HFDFV significantly decreased the HFD-induced content of TG and TC ($p < 0.05$) (Figure 3G,H). Oxidative stress parameters showed that HFDFV significantly decreased the HFD-induced content of MDA and increased the activity of SOD and GSH/GSSG in the PAT ($p < 0.05$). The content of H_2O_2 in the PAT was not affected by dietary treatments ($p > 0.05$) (Figure 3I–L). These results indicated that FV mycorrhizae reduced HFD-induced fat accumulation and oxidative injury and improved anti-oxidative capacity in the liver and PAT of mice.

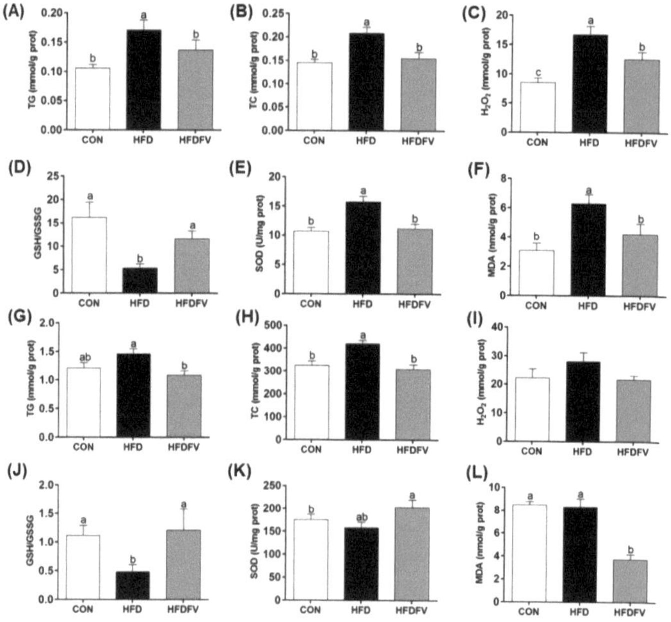

Figure 3. Effect of FV mycorrhizae on the lipids content and oxidative stress parameters in the liver and PAT of mice. (**A–F**) Lipid content and oxidative stress parameters in the liver; (**G–L**) Lipid content and oxidative stress parameters in the PAT. TG, triglyceride; TC, total cholesterol. Values with different letters differ significantly ($p < 0.05$). Values were expressed as mean ± SEM (n = 6–8).

3.4. Effect of FV Mycorrhizae on Gene Expression of Lipid Metabolism and Mitochondrial Biogenesis in the Liver and PAT

The gene expression of fatty acids transport (CD36, Slc27a1 and FABP1), hydrolysis (ATGL, HSL and MGLL), β-oxidation (CPT1α and MACD) and synthesis (SREBP1, FASN and Acaca) in the liver and PAT were further studied (Figure 4). In the liver, compared with the CON, HFD significantly decreased the gene expression of CD36, ATGL, CPT1α, MACD and Acaca, while HFDFV significantly increased the gene expression of CD36, ATGL, CPT1α and MACD, decreased the expression of FASN in the liver ($p < 0.05$) (Figure 4A–C). The gene expression of Slc27a1, FABP1, MGLL, HSL and SREBP1 in the liver were not affected by dietary treatment ($p > 0.05$). In the PAT, compared with the CON, HFD had no effect on the gene expression of Slc27a1, CD36, FABP1, ATGL, MGLL, HSL, CPT1α and MACD ($p > 0.05$). Compared with the HFD group, HFDFV significantly decreased the gene expression of SREBP1, FASN and Acaca in the PAT ($p < 0.05$) (Figure 4D–F). These results suggested that FV promoted fatty acids transport, hydrolysis and β-oxidation in the liver and inhibited fatty acids synthesis both in the liver and PAT.

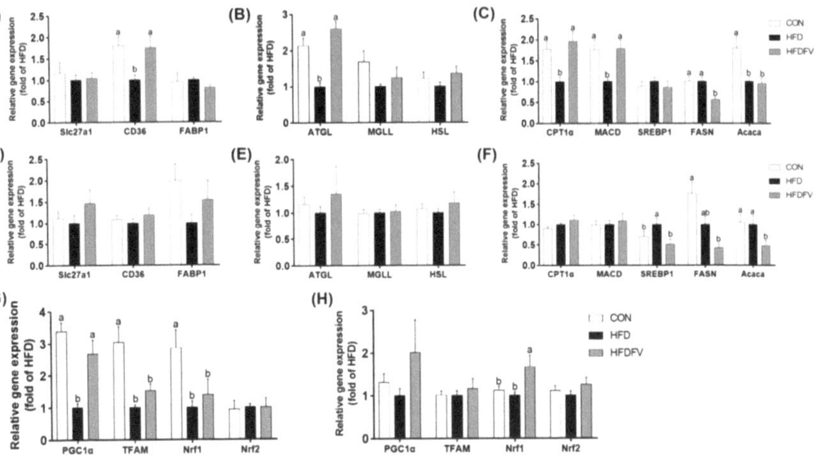

Figure 4. Effects of FV mycorrhizae on HFD-induced gene expression of lipid transport (slc27a1, CD36 and FABP1), lipases (ATGL, MGLL and HSL), β-oxidation (CPT1a and MACD), synthesis (FASN, Acaca and SREBP1) and mitochondrial biogenesis in the liver and PAT of mice. (**A**) Gene expression of lipid transport in the liver; (**B**) Gene expression of lipases in the liver; (**C**) Gene expression of β-oxidation and synthesis in the liver; (**D**) Gene expression of lipid transport in the PAT; (**E**) Gene expression of lipases in the PAT; (**F**) Gene expression of β-oxidation and synthesis in the PAT. (**G**) Gene expression of mitochondrial biogenesis in the liver; (**H**) Gene expression of mitochondrial biogenesis in the PAT. Values with different letters differ significantly ($p < 0.05$). Values were expressed as mean ± SEM (n = 6–8).

Compared with the CON, HFD significantly decreased the gene expression of mitochondrial biogenesis PGC1α, TFAM and Nrf1 ($p < 0.05$), but did not affect Nrf2 in the liver ($p > 0.05$). HFDFV significantly increased HFD-induced gene expression of PGC1α in the liver ($p < 0.05$) and had no effect on the gene expression of TFAM, Nrf1 and Nrf2 ($p > 0.05$) (Figure 4G). In the PAT, compared with the CON, HFD had no effect on the gene expression of PGC1α, TFAM, Nrf1 and Nrf2 ($p > 0.05$). HFDFV significantly increased the HFD-induced gene expression of Nrf1 in the PAT ($p < 0.05$) and had no effect on the gene expression of PGC1α, TFAM and Nrf2 in the PAT ($p > 0.05$) (Figure 4H).

3.5. Effect of FV Mycorrhizae on Inflammation in the Liver and PAT

The pro-inflammatory M1 markers (IL1β, IL6, TNFα and CCL_2) and anti-inflammatory M2 markers (TGFβ1, Fn1, CD206 and IL10) were further studied in the liver and PAT. Compared with the CON, HFD had no significant effect on the gene expression of IL1β, IL6, TNFα and CCL_2 in the liver ($p > 0.05$). Compared with HFD, HFDFV significantly decreased the gene expression of IL1β, IL6 and CCL_2 in the liver ($p < 0.05$) and had no effect on the gene expression of TNFα in the liver ($p > 0.05$) (Figure 5A). Compared with the CON, HFD had no effect on the gene expression of TGF-β1, Fn1, CD206 and IL10 in the liver ($p > 0.05$), whereas HFDFV significantly increased the gene expression of TGFβ1 and IL10 in the liver ($p < 0.05$) (Figure 5B). In the PAT, compared with the CON, HFD had no effect on the gene expression of IL1β, IL6, TNFα and CCL_2 ($p > 0.05$). Compared with HFD, HFDFV significantly increased the gene expression of IL6 and TNFα in the PAT ($p < 0.05$) and had no effect on the gene expression of IL1β and CCL_2 in the PAT ($p > 0.05$) (Figure 5C). HFDFV had no effect on the gene expression of TGFβ1, IL10 and CD206 in the PAT ($p > 0.05$), but significantly increased the gene expression of Fn1 in the PAT ($p < 0.05$) (Figure 5D).

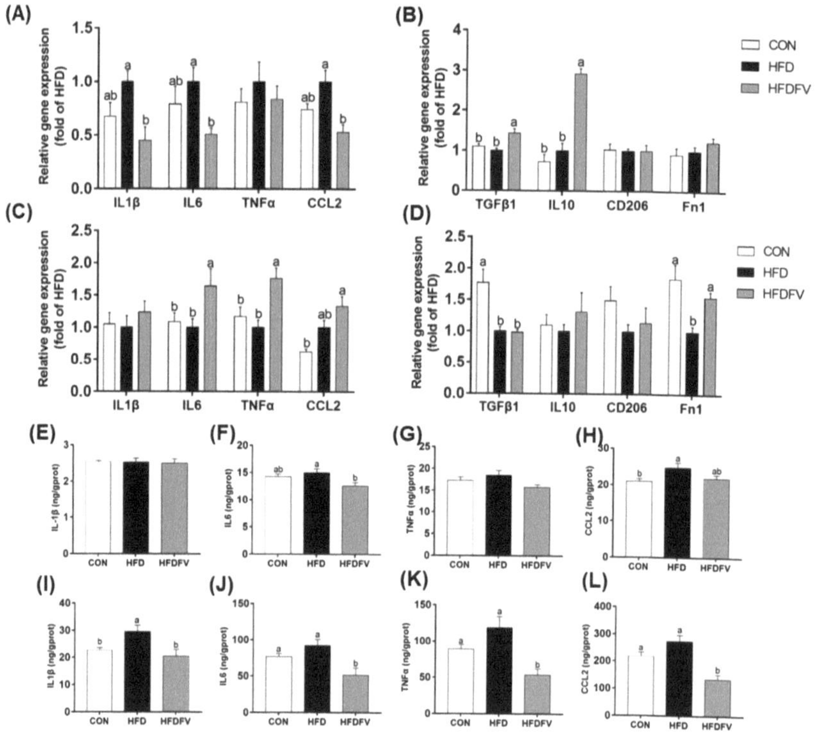

Figure 5. Effects of FV mycorrhizae on HFD-induced inflammatory cytokines in the liver and PAT of mice. (**A**) Gene expression of pro-inflammatory cytokines in the liver; (**B**) Gene expression of anti-inflammatory cytokines in the liver; (**C**) Gene expression of pro-inflammatory cytokines in the PAT; (**D**) Gene expression of anti-inflammatory cytokines in the PAT; (**E–H**) The contents of pro-inflammatory cytokines in the liver; (**I–L**) The contents of pro-inflammatory cytokines in the PAT. Values with different letters differ significantly ($p < 0.05$). Values were expressed as mean ± SEM (n = 6–8).

The contents of pro-inflammatory cytokines in the liver and PAT were further studied by ELISA. Compared with the CON, HFD increased CCL_2 content in the liver ($p < 0.05$), had no effect on the contents of IL1β, IL6 and TNFα ($p > 0.05$). HFDFV significantly decreased the content of IL6 in the liver compared with HFD ($p < 0.05$) and had no effect on the contents of IL1β, IL6 and CCL_2 in the liver ($p > 0.05$) (Figure 5E–H). In the PAT, HFDFV significantly decreased the content of IL1β, IL6, TNFα and CCL_2 in the PAT compared with HFD ($p < 0.05$) (Figure 5I–L).

3.6. *Effect of FV Mycorrhizae on Differential Metabolites in Plasma*

The untargeted metabolomics method was used to identify plasma differential metabolites among three groups. In total, 1000 metabolites were ultimately obtained including lipids and lipid-like molecules (26.2%), organic acids and derivatives (21.5%), organoheterocyclic compounds (14.8%), benzenoids (12%), undefined (8.8%), organic oxygen compounds (6%), phenylpropanoids and polyketides (4.1%), organic nitrogen compounds (3.7%), nucleosides, nucleotides, and analogues (1.6%), alkaloids and derivatives (0.3%) and so on (Figure 6A). PCA with 7-fold cross-validation showed the model interpretation rates were R2X = 0.582 and 0.54 (HFD vs. CON) and 0.545 and 0.536 (HFDFV vs. HFD) under negative and positive ions, respectively. OPLS-DA with 7-fold cross-validation showed Q2 = 0.733 and 0.66 (HFD and CON) and 0.389 and 0.384 (HFDFV vs. HFD) under negative and positive ions (Figure 6B–E). There was no overfitting in the permutation test under the negative and positive ions model. These results indicate that dietary treatment significantly changed the plasma metabolites. A VIP > 1 and $p < 0.05$ were used as criteria for differential metabolite screening.

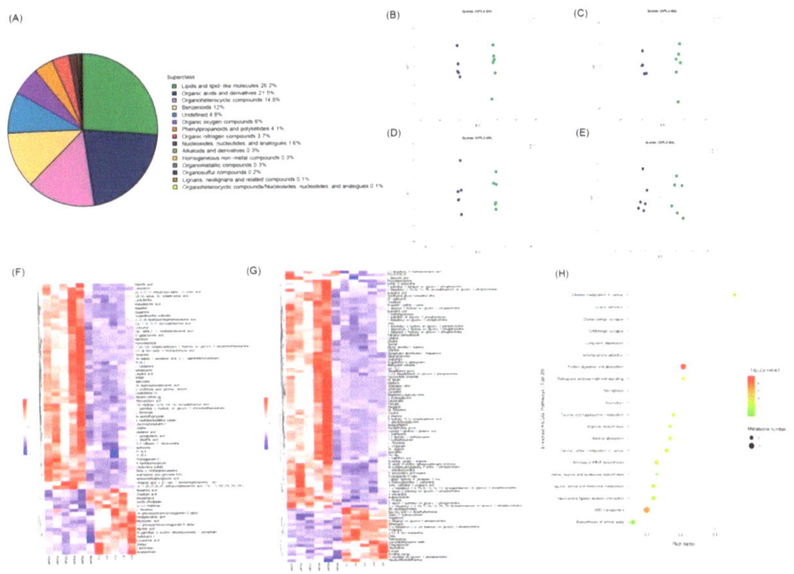

Figure 6. The differential metabolites were determined by untargeted metabolomics. (**A**) The classification of identified metabolites; OPLS-DA analysis HFD vs. CON (**B**,**C**) and HFDFV vs. HFD (**D**,**E**) under negative and positive ions model; (**F**,**G**) Hierarchical clustering of differential metabolites in HFD vs. CON under negative and positive ions, respectively; (**H**) KEGG enrichment analysis of HFD vs. CON.

Compared with the CON, 167 differential metabolites were identified in HFD under the negative and positive ions model (Figure 6F,G and Table 2). Compared with CON, the plasma fatty acids such as palmitic acid (C16:0) and 16-hydroxyhexadecanoic acid, unsaturated fatty acids (UFA) such as linoleic acid (C18:2), arachidonic acid (C20:4), 14,17,20,23,26,29-dotriacontahexaenoic acid (14z,17z,20z,23z,26z,29z)-, Cis-7,10,13,16-docosatetraenoic acid, Cis-.delta.2-11-methyldodecenoic acid, 9-hydroxy-10,12-octadecadienoic acid, 12s-hydroxy-5z,8z,10e,14z-eicosatetraenoic acid, amino acids such as Gln, Glu, Asn, Ile, taurine, Thr, Met, Arg and Pro were increased, while myristic acid (C14:0), myristoleic acid, pentadecanoic acid (C15:0) and eicosenoic acid (C20:1) decreased in the HFD group, indicating that HFD induced plasma fatty acid disorders and increased levels of amino acids in mice. KEGG pathways showed that HFD affected pathways involving protein digestion and absorption, ABC transporters, aminoacyl-tRNA biosynthesis, central carbon metabolism in cancer and so on (Figure 6H). Compared with the HFD group, 39 differential metabolites under the negative and positive ion model were identified in the HFDFV group (Table 2). HFDFV decreased HFD-induced plasma fatty acids such as linoleic acid (C18:2) and cis-.delta.2-11-methyldodecenoic acid, suggesting that FV mycorrhizae supplementation can effectively inhibit HFD-induced fatty acid disorders in mice. However, there were no KEGG enrichment pathways between HFDFV and HFD in the present study.

Table 2. Identification of differential metabolites between HFDFV and HFD under negative and positive ionization mode.

Adduct	Name	VIP	FC	*p*-Value	*m*/*z*
Negative					
[M-H]-	[6]-gingerol	5.634407	1.422392	0.000207	293.1763
[M-H-$C_6H_{12}O$]-	8,11-tridecadienoic acid, 13-(3-pentyl-2-oxiranyl)-, (8z,11z)-	1.649861	1.3444	0.00024	221.1545
[M-H]-	Tetradec-5-ynoic acid	1.299375	0.357147	0.00122	223.1702
[M-H-$C_{17}H_{27}SO$]-	Probucol	2.159909	1.65893	0.003397	236.1054
[M-H]-	Cis-.delta.2-11-methyldodecenoic acid	1.599766	0.186622	0.004097	211.1701
[M-H]-	12-hydroxydodecanoic acid	1.252276	1.765689	0.006782	215.1654
[M-H-CO_2]-	3,5-di-tert-butyl-4-hydroxybenzoic acid	1.876308	0.365134	0.021764	205.1596
[M-H-$C_{18}H_{34}O_2$]-	1,2-dioleoyl-sn-glycero-3-phosphate	1.975814	2.621357	0.022762	417.2414
[M-H]-	1h-indole-1-pentanoic acid, 3-(1-naphthalenylcarbonyl)-	3.057796	0.318523	0.023438	370.1697
[M+Cl]-	1-palmitoyl-2-arachidonoyl-sn-glycero-3-phosphocholine	1.347016	0.665036	0.031371	816.5336
[M-H]-	Linoleic acid	14.5538	0.618988	0.033811	279.2333
[M-H]-	2′,2′-difluoro-2′-deoxyuridine	2.68574	0.656394	0.03705	263.0445
[M-H]-	Deoxythymidine 5′-phosphate (dTMP)	2.468689	3.258614	0.040277	321.0441
[M-H]-	3-hydroxycapric acid	1.544758	1.462081	0.041071	187.1337
Positive					
[M+H]+	Lauryldimethylamine oxide	4.245431	0.489801	0.000229	230.2479
[M+H]+	N-acetyl-o-fluoro-dl-phenylalanine	3.723481	1.206781	0.003209	226.084
[M+H]+	Coumarin	1.934626	0.525499	0.00427	147.0554
[M+H-NH_3]+	Porphobilinogen	3.582239	1.384773	0.005902	210.0796
(M+H)+	L-Proline	1.783547	0.545155	0.005949	116.0709
[M+Na]+	Stavudine	1.570389	1.749773	0.009195	247.0607
[M+H-H_2O]+	Prostaglandin e3	1.374113	0.454302	0.012679	333.2059
[M+H]+	Pro-Trp	1.402203	0.572081	0.01372	302.3052
[M+H]+	Tetraethylene glycol monomethyl ether	5.241345	0.431367	0.013853	209.1385
[M+H]+	Fenpropimorph	1.4829	1.619531	0.014177	304.2845
[M+H]+	Pentapropylene glycol	2.740352	0.51482	0.016788	309.227
[M+H-H_2O]+	Fingolimod	1.43912	0.595578	0.016981	290.2688
[M+H]+	3,4-dimethylmethcathinone	1.876348	0.319043	0.020052	192.1596
[M+H]+	Triethylene glycol monobutyl ether	3.491736	1.698806	0.021376	207.1592
[M+H]+	Tributyl phosphate	1.08038	1.705361	0.022941	267.1719
[M+H]+	Decanoyl m-nitroaniline	1.016153	0.557516	0.023725	293.211
(M-H+2Na)+	1-Stearoyl-sn-glycerol 3-phosphocholine	13.17318	1.343167	0.024106	568.3402
[M+H]+	.gamma.-nonalactone	1.413858	1.068891	0.030097	157.1337
[M+H]+	Lpc 18:1	20.32652	1.257069	0.03399	522.3565
[M+NH4]+	Desferrioxamine d2	2.815889	4.933469	0.034191	604.3541
[M+NH4]+	1,2-dilinoleoylglycerol	1.123168	0.244778	0.036202	634.5408
[M+H]+	2-fluoroamphetamine	1.286169	1.631088	0.039465	154.0863
[M+H-CH_2O_2]+	4-hydroxynonenal alkyne	1.267965	3.974206	0.042399	107.0859
[M+Na]+	Anisomycin	1.322298	3.083059	0.04541	288.144
[M+H]+	Fenpropidin	9.809554	0.31689	0.049266	274.2742

3.7. Effect of FV Mycorrhizae on Medium Long Chain Fatty Acids in the Liver

Furthermore, the medium long chain fatty acids including saturated fatty acids (SFA), trans fatty acids (TFA), monounsaturated fatty acids (MUFA) and polyunsaturated fatty acids (PUFA) in the liver between HFD and HFDFV groups were determined by targeted metabolomics (Table 3). Compared with the HFD group, HFDFV significantly increased the contents of lauric acid (C12:0), TFA such as linoelaidic acid (C18:2TT), nervonic acid (C24:1) and docosapentaenoate (C22:5*n*-3) in the liver ($p < 0.05$). The contents of 11C,14C-eicosadienoic acid (C20:2), docosapentaenoate (C22:5*n*-6) and adrenic acid (C22:4) were decreased in HFDFV compared with HFD ($p < 0.05$). Other fatty acids were not significantly different between the two groups ($p > 0.05$).

Table 3. Effects of FV mycorrhizae on HFD-induced medium long chain fatty acids profiles (mg/g Tissue) in the liver.

Item		HFD	HFDFV	p
SFA		2973.287 ± 356.423	2865.914 ± 262.981	0.813
Octanoic acid	C8:0	0.259 ± 0.033	0.22 ± 0.019	0.335
Decanoic acid	C10:0	0.489 ± 0.095	0.468 ± 0.039	0.838
Lauric acid	C12:0	0.247 ± 0.018	0.337 ± 0.036	0.047
Tridecanoic acid	C13:0	0.368 ± 0.026	0.361 ± 0.018	0.827
Myristic acid	C14:0	3.577 ± 0.551	5.19 ± 0.922	0.164
Pentadecanoic Acid	C15:0	1.907 ± 0.225	2.522 ± 0.178	0.057
Palmitic acid	C16:0	299.472 ± 32.524	308.18 ± 18.933	0.822
Margaric Acid	C17:0	5.693 ± 0.699	6.715 ± 0.698	0.325
Stearic acid	C18:0	2641.364 ± 320.93	2524.592 ± 244.726	0.778
Arachidic Acid	C20:0	5.342 ± 0.661	3.793 ± 0.547	0.101
Docosanoic acid	C22:0	10.2 ± 1.475	8.68 ± 1.272	0.453
Tricosanoic Acid	C23:0	1.262 ± 0.162	1.581 ± 0.17	0.204
Lignoceric Acid	C24:0	3.107 ± 0.526	3.276 ± 0.584	0.834
MUFA		39.561 ± 3.541	40.531 ± 1.791	0.812
Myristelaidic Acid	C14:1	0.608 ± 0.029	0.571 ± 0.038	0.450
10-Pentadecenoic Acid	C15:1	4.307 ± 0.912	4.265 ± 0.632	0.970
Palmitoleic Acid	C16:1	5.177 ± 0.607	5.23 ± 0.36	0.941
Heptadecanoic acid (cis-10)	C17:1	3.704 ± 0.45	4.069 ± 0.321	0.524
Oleic acid	C18:1	0.261 ± 0.109	0.625 ± 0.162	0.092
cis-11-Eicosenoic acid	C20:1	22.144 ± 2.478	20.347 ± 1.561	0.553
Nervonic acid	C24:1	3.36 ± 0.503	5.425 ± 0.629	0.028
TFA		131.714 ± 19.462	208.164 ± 13.972	0.010
Hexadecanoic acid (trans-9)	C16:1T	10.689 ± 2.988	22.393 ± 5.15	0.078
Trans-10-HeptadecenoicAcid(C17:1T)	C17:1T	1.611 ± 0.214	1.944 ± 0.214	0.297
Trans-10-Nonadecenoic acid	C19:1T	2.193 ± 0.307	1.77 ± 0.215	0.286
Trans-11-Eicosenoic Acid	C20:1T	1.653 ± 0.217	1.982 ± 0.262	0.356
Linoelaidic Acid	C18:2TT	115.569 ± 16.333	180.075 ± 11.339	0.009
PUFA		1420.182 ± 231.802	1401.189 ± 153.088	0.947
γ-Linolenic acid	C18:3	14.133 ± 2.116	16.241 ± 1.172	0.404
α-Linolenic acid	C18:3	296.486 ± 38.875	234.847 ± 14.542	0.168
Eicosapentaenoate	C20:5	358.805 ± 40.702	377.603 ± 32.96	0.727
Arachidonic acid	C20:4	443.149 ± 148.283	415.357 ± 134.038	0.892
Cis-11,14,17-Eicosatrienoic Acid	C20:3	54.141 ± 6.682	48.17 ± 5.572	0.508
11C,14C-Eicosadienoic Acid	C20:2	4.505 ± 0.593	2.011 ± 0.379	0.005
Docosapentaenoate (C22:5n-6)	C22:5n-6	72.843 ± 7.426	39.97 ± 3.493	0.002
Docosapentaenoate (C22:5n-3)	C22:5n-3	158.709 ± 20.87	254.869 ± 16.861	0.005
Adrenic Acid	C22:4	15.307 ± 1.763	10.138 ± 1.475	0.048
13C,16C-Docosadienoic Acid	C22:2	2.103 ± 0.362	1.985 ± 0.51	0.854

SFA, MUFA, TFA and PUFA were the sum of corresponding fatty acids in this table. Values were expressed as mean ± SEM (n = 6).

4. Discussion

Lipid metabolic disorder is one of the important factors that cause obesity. Liver and adipose tissue are the major sites responsible for lipid metabolism, transportation and storing. In this study, although the body weight of mice was not significantly different among treatments, the results showed that HFDFV decreased HFD-induced liver weight, plasma TG and ALT, TG and TC in liver and PAT. Furthermore, metabolomics results showed that HFD increased the plasma palmitic acid (C16:0) and 16-hydroxyhexadecanoic acid, linoleic acid (C18:2) and arachidonic acid (C20:4), decreased myristic acid (C14:0), pentadecanoic acid (C15:0) and eicosenoic acid (C20:1), while HFDFV reversed HFD-induced plasma linoleic acid (C18:2) and cis-.delta.2-11-methyldodecenoic acid. SFA such as palmitic acid is abundant in human serum and western-style diet, and appeared more powerful than UFA at inducing metabolic disorders [22,23]. A high level of 16-hydroxyhexadecanoic acid was related to chronic kidney disease [24]. High intake of essential fatty acids linoleic acid

(C18:2) and its derivative arachidonic acid (C20:4) induced inflammation, adipogenesis and lipid accumulation in male mice, although the role of linoleic acid in obesity development is still contentious [25,26]. Myristic acid (C14:0) and odd chain fatty acid pentadecanoic acid (C15:0) are negatively associated with liver fat, injury and non-alcoholic steatohepatitis [27,28]. These studies suggested that FV mycorrhizae mainly inhibited HFD-induced lipid disorders through decreasing plasma UFA in mice. Meanwhile, HFDFV increased the contents of lauric acid (C12:0), linoelaidic acid (C18:2TT), nervonic acid (C24:1) and docosapentaenoate (C22:5*n*-3) in the liver, decreased 11C,14C-eicosadienoic acid (C20:2), docosapentaenoate (C22:5*n*-6) and adrenic acid (C22:4). Lauric acid is rich in coconut oil and has a anti-inflammation, antibacterial activity and anti-obesity role [29,30]. Nervonic acid was reported to negatively correlate with obesity; its supplementation could effectively treat obesity through increasing fatty acids β-oxidation [31,32]. Docosapentaenoate (C22:5*n*-3) is an intermediate product between eicosapentaenoic acid (EPA) and docosapentaenoic acid (DHA) that exhibit good efficacy in lowering lipids as well as their isomer docosapentaenoate (C22:5*n*-6) [33]. Adrenic acid is an important factor to promote the disease progression of NAFLD through the induction of inflammation [34]. Indeed, present results showed that HFDFV inhibited pro-inflammatory gene expression (IL1β, IL6 and CCL_2) and increased anti-inflammatory gene expression (TGFβ and IL10) in liver. Thus, these studies suggested that FV mycorrhizae was effective to improve metabolic disorders and decrease plasma UFA, possibly by promoting fatty acids β-oxidation and inhibiting inflammation in liver.

Slc27a1, CD36 and FABP1 are required for fatty acid uptake and transport [35]. The decrease in gene expression of CD36 in liver in this study indicated a restriction in fat mass gain when fed a high fat diet. Metabolic lipases such as ATGL, HSL and MGLL are responsible for lipid catabolism by hydrolyzing TG into diacylglycerols (DAG), monoacylglycerols (MAGs) and free FAs and glycerol, respectively [36]. CPT1α is the rate-limiting enzyme of fatty acid β-oxidation involved in mitochondrial fatty acid uptake [37]. MACD regulates the first step of fatty acid β-oxidation by converting acyl-coenzyme A into trans-enoyl-CoA [38]. Acaca is the rate-limiting enzyme of de novo fatty acid biosynthesis that catalyzes the carboxylation of acetyl-CoA to malonyl-CoA [39]. FASN is a cytosolic metabolic enzyme that catalyzes de novo fatty acid synthesis, which is transcriptionally regulated by SREBP1 [40]. In this study, FV increased HFD-induced gene expression of CD36, ATGL, CPT1α and MACD, decreased the gene expression of FASN in the liver and inhibited the gene expression of SREBP1, FASN and Acaca in the PAT, suggesting that FV mycorrhizae effectively inhibited lipid accumulation by promoting fatty acid transportation, lipid hydrolysis and β-oxidation in the liver and reducing lipid synthesis in the liver and PAT. These were similar to previous studies, which showed that FV polysaccharide, powder and mixture attenuated HFD-induced obesity and hyperlipidemia through increasing oxidation and decreasing synthesis in the liver and PAT [13,41–43].

Excessive FFA intake also induces lipotoxicity, resulting in increased oxidative stress and inflammation [44]. SOD is responsible for preventing O_2^--induced oxidative damage. MDA is the final products of PUFA peroxidation indicating lipid peroxidation [45]. The decreased ratio of GSH/GSSG indicated an impairment of redox status in the cell [46]. In this study, HFDFV reversed the HFD-induced activities of SOD and the content of H_2O_2, MDA and GSH/GSSG in the liver and MDA and GSH/GSSG in the PAT, suggesting that FV mycorrhizae attenuated HFD-induced oxidative stress, and improved anti-oxidative ability in the liver and PAT. Obesity was also a kind of low-grade chronic inflammatory response, with increasing circulating levels of inflammatory cytokines, such as TNFα and IL6. PAT was active to secrete adipokines and cytokines compared with other visceral adipose tissue, playing an important role in renal metabolism and cardiovascular system by paracrine or endocrine pathways [47]. In this study, HFDFV decreased the contents of IL1β, IL6, TNFα and CCL_2 in the PAT, and IL6 content in the liver, suggesting FV mycorrhizae inhibited HFD-induced inflammation, mainly through inhibiting the secretion of pro-inflammatory cytokines from PAT. Of note, HFDFV increased the gene expression of IL6, TNFα in the

PAT, which is contrary to the liver. In fact, the mushroom extracts were reported to differentially regulate immune cells such as T-lymphocytes, monocytes, NK and T cells and cytokines secretion dependent on mushroom compound compositions, extract methods, absorption and bioavailability, synergistic or antagonistic effects [48]. On the other hand, cytokines such as TNFα were also reported to drive lipolysis. The adipose tissues-liver crosstalk is responsible for lipid metabolic balance, but it is also sensitive to dietary types, composition and amount [49]. Thus, the specific ingredients and mechanism responsible for FV mycorrhizae regulating host immune response between adipose tissue and liver remain unknown and need further investigation. Of note, HFDFV increased plasma gingerols and probucol and decreased coumarin compared with the HFD group. 6-gingerols are the major pungent phenolic compounds present in the rhizomes of ginger, which showed antioxidant, anti-inflammatory, and antihyperglycemic activity [50]. Probucol is a well-known natural product antioxidant that can effectively treat hyperlipidaemia and NAFLD [51]. Coumarin is a plant-derived natural compound that showed antioxidant, antibacterial and anti-inflammation ability. These results suggest that the anti-oxidative and hypolipidemic ability of FV mycorrhizae possibly attributed to these antioxidants. However, these ingredients were not found in the composition of FV mycorrhizae in our results, indicating the metabolism and transform of mycorrhizae or their low contents under detection, which need further verification.

5. Conclusions

In this study, the results showed that FV mycorrhizae contained multifarious nutritive components, including abundant trace elements, polysaccharide, amino acids and derivatives, and organic small molecule compounds, which attenuated HFD-induced lipid disorders in the plasma and liver, inhibited oxidative stress and differentially regulated inflammation in the liver and PAT of mice, indicating the application of promising constituents of FV mycorrhizae for functional foods and herbal medicine.

Supplementary Materials: The following supporting information can be downloaded at: https://www.mdpi.com/article/10.3390/nu14183830/s1, Table S1: Primers used in this study; Table S2: The nutritional composition, harmful substances and heavy metals in FV mycorrhizae; Table S3: Amino acids contents and derivatives in FV mycorrhizae (μmol/g). Reference [52] are cited in the supplementary materials.

Author Contributions: Z.L., J.Z., W.X. and J.X. conceived and designed the experiments; Z.L, Q.G. and Y.B. performed the experiments; Z.L. analyzed the data and wrote the manuscript. Z.L., Y.L. and J.X revised the manuscript. All authors have read and agreed to the published version of the manuscript.

Funding: This work was supported by the Shanghai Agriculture Applied Technology Development Program, China (X2019-02-08-00-08-F01155) and China Postdoctoral Science Foundation (2020M671133).

Institutional Review Board Statement: The animal study protocol was approved by the guidelines of Shanghai Jiao Tong University Institutional Animal Care and Use Committee. The ethical approval code in the manuscript is 202201188.

Informed Consent Statement: Not applicable.

Data Availability Statement: Not applicable.

Conflicts of Interest: The authors declare no conflict of interest.

References

1. Blüher, M. Obesity: Global epidemiology and pathogenesis. *Nat. Rev. Endocrinol.* **2019**, *15*, 288–298. [CrossRef] [PubMed]
2. Azzu, V.; Vacca, M.; Virtue, S.; Allison, M.; Vidal-Puig, A. Adipose tissue-liver cross talk in the control of whole-body metabolism: Implications in nonalcoholic fatty liver disease. *Gastroenterology* **2020**, *158*, 1899–1912. [CrossRef] [PubMed]
3. Cusi, K. Role of obesity and lipotoxicity in the development of nonalcoholic steatohepatitis: Pathophysiology and clinical implications. *Gastroenterology* **2012**, *142*, 711–725.e6. [CrossRef] [PubMed]

4. Seebacher, F.; Zeigerer, A.; Kory, N.; Krahmer, N. Hepatic lipid droplet homeostasis and fatty liver disease. *Semin. Cell Dev. Biol.* **2020**, *108*, 72–81. [CrossRef] [PubMed]
5. Bindua, S.; Mazumderb, S.; Bandyopadhyay, U. Non-steroidal anti-inflammatory drugs (NSAIDs) and organ damage: A current perspective. *Biochem. Pharmacol.* **2020**, *180*, 114147. [CrossRef]
6. Datta, A.; Flynn, N.R.; Barnette, D.A.; Woeltje, K.F.; Miller, G.P.; Swamidass, S.J. Machine learning liver-injuring drug interactions with non-steroidal anti-inflammatory drugs (NSAIDs) from a retrospective electronic health record (EHR) cohort. *PLoS Comput. Biol.* **2021**, *17*, e1009053. [CrossRef]
7. Wang, H.-N.; Xiang, J.-Z.; Qi, Z.; Du, M. Plant extracts in prevention of obesity. *Crit. Rev. Food Sci. Nutr.* **2022**, *62*, 2221–2234. [CrossRef] [PubMed]
8. Yadav, D.; Negi, P.S. Bioactive components of mushrooms: Processing effects and health benefits. *Food Res. Int.* **2021**, *148*, 110599. [CrossRef]
9. Tang, C.; Hoo, P.C.-X.; Tan, L.T.-H.; Pusparajah, P.; Mehmood Khan, T.; Lee, L.-H.; Goh, B.-H.; Chan, K.-G. Golden needle mushroom: A culinary medicine with evidenced-based biological activities and health promoting properties. *Front. Pharmacol.* **2016**, *7*, 474. [CrossRef]
10. Zhang, L.; Hu, Y.; Duan, X.; Tang, T.; Shen, Y.; Hu, B.; Liu, A.; Chen, H.; Li, C.; Liu, Y. Characterization and antioxidant activities of polysaccharides from thirteen boletus mushrooms. *Int. J. Biol. Macromol.* **2018**, *113*, 1–7. [CrossRef]
11. Sande, D.; Oliveira, G.P.D.; Moura, M.A.F.; Martins, B.D.A.; Matheus, L.T.N.S.; Takahashi, J.A. Edible mushrooms as a ubiquitous source of essential fatty acids. *Food Res. Int.* **2019**, *125*, 108524. [CrossRef] [PubMed]
12. Yan, Z.-F.; Liu, N.-X.; Mao, X.-X.; Li, Y.; Li, C.-T. Activation effects of polysaccharides of *Flammulina velutipes* mycorrhizae on the T lymphocyte immune function. *J. Immunol. Res.* **2014**, *2014*, 285421. [CrossRef] [PubMed]
13. Wang, W.; Yang, S.; Song, S.; Zhang, J.; Jia, F. *Flammulina velutipes* mycorrhizae dietary fiber improves lipid metabolism disorders in obese mice through activating AMPK signaling pathway mediated by gut microbiota. *Food Biosci.* **2021**, *43*, 101246. [CrossRef]
14. Liu, B.-X.; Sun, W.; Kong, X.-Q. Perirenal fat: A unique fat pad and potential target for cardiovascular disease. *Angiology* **2019**, *70*, 584–593. [CrossRef]
15. Grigoraș, A.; Balan, R.A.; Căruntu, I.-D.; Giușcă, S.E.; Lozneanu, L.; Avadanei, R.E.; Rusu, A.; Riscanu, L.A.; Amalinei, C. Perirenal adipose tissue—current knowledge and future opportunities. *J. Clin. Med.* **2021**, *10*, 1291. [CrossRef] [PubMed]
16. Yang, S. *Feed Analysis and Feed Quality Testing Technology*; Beijing Agricultural University Press: Beijing, China, 1993.
17. Luo, Z.; Gao, Q.; Zhang, H.; Zhang, Y.; Zhou, S.; Zhang, J.; Xu, W.; Xu, J. Microbe-derived antioxidants attenuate cobalt chloride-induced mitochondrial function, autophagy and BNIP3-dependent mitophagy pathways in BRL3A cells. *Ecotox. Environ. Safe* **2022**, *232*, 113219. [CrossRef] [PubMed]
18. Wang, L.; Zhang, B.; Xiao, J.; Huang, Q.; Li, C.; Fu, X. Physicochemical, functional, and biological properties of water-soluble polysaccharides from Rosa roxburghii Tratt fruit. *Food Chem.* **2018**, *249*, 127–135. [CrossRef]
19. Luo, Z.; Xu, X.; Zhao, S.; Sho, T.; Luo, W.; Zhang, J.; Xu, W.; Hon, K.; Xu, J. Inclusion of microbe-derived antioxidant during pregnancy and lactation attenuates high-fat diet-induced hepatic oxidative stress, lipid disorders, and NLRP3 inflammasome in mother rats and offspring. *Food Nutr. Res.* **2019**, *63*, 1–11. [CrossRef]
20. Luo, Z.; Zhu, W.; Guo, Q.; Luo, W.; Zhang, J.; Xu, W.; Xu, J. Weaning induced hepatic oxidative stress, apoptosis, and aminotransferases through MAPK signaling pathways in piglets. *Oxidative Med. Cell. Longev.* **2016**, *2016*, 4768541. [CrossRef]
21. Ma, F.; Zhang, Y.; Liu, N.; Zhang, J.; Tan, G.; Kannan, B.; Liu, X.; Bell, A.E. Rheological properties of polysaccharides from Dioscorea opposita Thunb. *Food Chem.* **2017**, *227*, 64–72. [CrossRef]
22. Luukkonen, P.K.; Sädevirta, S.; Zhou, Y.; Kayser, B.; Ali, A.; Ahonen, L.; Lallukka, S.; Pelloux, V.; Gaggini, M.; Jian, C.; et al. Saturated fat is more metabolically harmful for the human liver than unsaturated fat or simple sugars. *Diabetes Care* **2018**, *41*, 1732–1739. [CrossRef] [PubMed]
23. De Almeida, I.T.; Cortez-Pinto, H.; Fidalgo, G.; Rodrigues, D.; Camilo, M.E. Plasma total and free fatty acids composition in human non-alcoholic steatohepatitis. *Clin. Nutr.* **2002**, *21*, 219–223. [CrossRef] [PubMed]
24. Grams, M.E.; Tin, A.; Rebholz, C.M.; Shafi, T.; Köttgen, A.; Perrone, R.D.; Sarnak, M.J.; Inker, L.A.; Levey, A.S.; Coresh, J. Metabolomic alterations associated with cause of CKD. *Clin. J. Am. Soc. Nephrol.* **2017**, *12*, 1787. [CrossRef] [PubMed]
25. Naughton, S.S.; Mathai, M.L.; Hryciw, D.H.; McAinch, A.J. Linoleic acid and the pathogenesis of obesity. *Prostaglandins Other Lipid Mediat.* **2016**, *125*, 90–99. [CrossRef] [PubMed]
26. Zhuang, P.; Shou, Q.; Lu, Y.; Wang, G.; Qiu, J.; Wang, J.; He, L.; Chen, J.; Jiao, J.; Zhang, Y. Arachidonic acid sex-dependently affects obesity through linking gut microbiota-driven inflammation to hypothalamus-adipose-liver axis. *BBA-Mol. Basis Dis.* **2017**, *1863*, 2715–2726. [CrossRef]
27. Yoo, W.; Gjuka, D.; Stevenson, H.L.; Song, X.; Shen, H.; Yoo, S.Y.; Wang, J.; Fallon, M.; Ioannou, G.N.; Harrison, S.A.; et al. Fatty acids in non-alcoholic steatohepatitis: Focus on pentadecanoic acid. *PLoS ONE* **2017**, *12*, e0189965. [CrossRef]
28. Rioux, V.; Pédrono, F.; Legrand, P. Regulation of mammalian desaturases by myristic acid: N-terminal myristoylation and other modulations. *BBA Mol. Cell Biol. Lipids* **2011**, *1811*, 1–8. [CrossRef]
29. Zhang, F.; Song, M.; Chen, L.; Yang, X.; Li, F.; Yang, Q.; Duan, C.; Ling, M.; Lai, X.; Zhu, X.; et al. Dietary supplementation of lauric acid alleviates the irregular estrous cycle and the impaired metabolism and thermogenesis in female mice fed with high-fat diet (HFD). *J. Agr. Food Chem.* **2020**, *68*, 12631–12640. [CrossRef]

30. Nitbani, F.O.; Siswanta, D.; Solikhah, E.N. Isolation and antibacterial activity test of lauric acid from crude coconut oil (*Cocos nucifera* L.). *Procedia Chem.* **2016**, *18*, 132–140. [CrossRef]
31. Keppley, L.J.W.; Walker, S.J.; Gademsey, A.N.; Smith, J.P.; Keller, S.R.; Kester, M.; Fox, T.E. Nervonic acid limits weight gain in a mouse model of diet-induced obesity. *FASEB J.* **2020**, *34*, 15314–15326. [CrossRef]
32. Oda, E.; Hatada, K.; Kimura, J.; Aizawa, Y.; Thanikachalam, P.V.; Watanabe, K. Relationships between serum unsaturated fatty acids and coronary risk factors negative relations between nervonic acid and obesity-related risk factors. *Int. Heart J.* **2005**, *46*, 975–985. [CrossRef] [PubMed]
33. Chen, J.; Jiang, Y.; Liang, Y.; Tian, X.; Peng, C.; Ma, K.Y.; Liu, J.; Huang, Y.; Chen, Z.Y. DPA n-3, DPA n-6 and DHA improve lipoprotein profiles and aortic function in hamsters fed a high cholesterol diet. *Atherosclerosis* **2012**, *221*, 397–404. [CrossRef] [PubMed]
34. Horas, H.; Nababan, S.; Nishiumi, S.; Kobayashi, T.; Yoshida, M.; Azuma, T. Adrenic acid as an inflammation enhancer in non-alcoholic fatty liver disease. *Arch. Biochem. Biophys.* **2017**, *623–624*, 64–75. [CrossRef] [PubMed]
35. Pei, K.; Gui, T.; Li, C.; Zhang, Q.; Feng, H.; Li, Y.; Wu, J.; Gai, Z. Recent progress on lipid intake and chronic kidney disease. *BioMed Res. Int.* **2020**, *2020*, 3680397. [CrossRef]
36. Fuchs, C.D.; Claude, T.; Trauner, M. Role of metabolic lipases and lipolytic metabolites in the pathogenesis of NAFLD. *Trends Endocrin. Met.* **2014**, *25*, 576–585. [CrossRef] [PubMed]
37. Bonnefont, J.-P.; Djouadi, F.; Prip-Buus, C.; Gobin, S.; Munnich, A.; Bastin, J. Carnitine palmitoyltransferases 1 and 2: Biochemical, molecular and medical aspects. *Mol. Aspects Med.* **2004**, *25*, 495–520. [CrossRef] [PubMed]
38. Abo, A.O.; Lopaschuk, G.D. Role of CoA and acetyl-CoA in regulating cardiac fatty acid and glucose oxidation. *Biochem. Soc. Trans.* **2014**, *42*, 1043–1051.
39. Colbert, C.L.; Kim, C.-W.; Moon, Y.-A.; Henry, L.; Palnitkar, M.; McKean, W.B.; Fitzgerald, K.; Deisenhofer, J.; Horton, J.D.; Kwon, H.J. Crystal structure of Spot 14, a modulator of fatty acid synthesis. *Proc. Natl. Acad. Sci. USA* **2010**, *107*, 18820–18825. [CrossRef]
40. You, M.; Crabb, D.W. Recent advances in alcoholic liver disease II. minireview: Molecular mechanisms of alcoholic fatty liver. *Am. J. Physiol.-Gastrointest. Liver Physiol.* **2004**, *287*, G1–G6. [CrossRef]
41. Zhao, R.; Ji, Y.; Chen, X.; Hu, Q.; Zhao, L. Polysaccharide from *Flammulina velutipes* attenuates markers of metabolic syndrome by modulating the gut microbiota and lipid metabolism in high fat diet-fed mice. *Food Funct.* **2021**, *12*, 6964–6980. [CrossRef]
42. Yeh, M.-Y.; Ko, W.-C.; Lin, L.-Y. Hypolipidemic and antioxidant activity of enoki mushrooms (*Flammulina velutipes*). *BioMed Res. Int.* **2014**, *2014*, 352385. [CrossRef] [PubMed]
43. Shimizu, T.; Mori, K.; Ouchi, K.; Kushida, M.; Tsuduki, T. Effects of dietary intake of Japanese mushrooms on visceral fat accumulation and gut microbiota in mice. *Nutrients* **2018**, *10*, 610. [CrossRef] [PubMed]
44. Schönfeld, P.; Wojtczak, L. Fatty acids as modulators of the cellular production of reactive oxygen species. *Free Radical Bio. Med.* **2008**, *45*, 231–241. [CrossRef] [PubMed]
45. Gaweł, S.; Wardas, M.; Niedworok, E.; Wardas, P. Malondialdehyde (MDA) as a lipid peroxidation marker. *Wiad. Lek.* **2004**, *57*, 453–455. [PubMed]
46. Schafer, F.Q.; Buettner, G.R. Redox environment of the cell as viewed through the redox state of the glutathione disulfide/glutathione couple. *Free Radical Bio. Med.* **2001**, *30*, 1191–1212. [CrossRef]
47. Hammoud, S.H.; AlZaim, I.; Al-Dhaheri, Y.; Eid, A.H.; El-Yazbi, A.F. Perirenal adipose tissue inflammation: Novel insights linking metabolic dysfunction to renal diseases. *Front. Endocrinol.* **2021**, *12*, 942. [CrossRef]
48. Martel, J.; Ko, Y.-F.; Ojcius, D.M.; Lu, C.-C.; Chang, C.-J.; Lin, C.-S.; Laisin, H.C.; Young, J.D. Immunomodulatory properties of plants and mushrooms. *Trends Pharmacol. Sci.* **2017**, *38*, 967–981. [CrossRef]
49. Duwaerts, C.C.; Maher, J.J. Macronutrients and the adipose-liver axis in obesity and fatty liver. *Cell. Mol. Gastroenter.* **2019**, *7*, 749–761. [CrossRef]
50. Naidu, P.B.; Uddandrao, V.V.S.; Naik, R.R.; Suresh, P.; Meriga, B.; Begum, M.S.; Pandiyan, R.; Saravanan, G. Ameliorative potential of gingerol: Promising modulation of inflammatory factors and lipid marker enzymes expressions in HFD induced obesity in rats. *Mol. Cell. Endocrinol.* **2016**, *419*, 139–147. [CrossRef]
51. Zhang, S.; Li, L.; Chen, W.; Xu, S.; Feng, X.; Zhang, L. Natural products: The role and mechanism in low-density lipoprotein oxidation and atherosclerosis. *Phytother. Res.* **2021**, *35*, 2945–2967. [CrossRef]
52. Liu, Y.; Wang, L.; Luo, M.; Chen, N.; Deng, X.; He, J.; Zhang, L.; Luo, P.; Wu, J. Inhibition of PAI-1 attenuates perirenal fat inflammation and the associated nephropathy in high-fat diet-induced obese mice. *Am. J. Physiol.-Endocrinol. Metab.* **2019**, *316*, E260–E267. [CrossRef]

nutrients

MDPI

Article

In Vitro Digestibility and Bioaccessibility of Nutrients and Non-Nutrients Composing Extruded Brewers' Spent Grain

Maria Belen Gutierrez-Barrutia [1,2], Sonia Cozzano [1], Patricia Arcia [1,3] and Maria Dolores del Castillo [2,*]

1 Departamento de Ingeniería, Universidad Católica del Uruguay, Montevideo 11600, Uruguay
2 Instituto de Investigacion en Ciencias de la Alimentacion, CSIC-UAM, Nicolas Cabrera 9, 28049 Madrid, Spain
3 Latitud-LATU Foundation, Av. Italia 6201, Montevideo 11500, Uruguay
* Correspondence: mdolores.delcastillo@csic.es; Tel.: +34-910-017-900 (ext. 4394553)

Abstract: This study aimed to evaluate the effect of the extrusion process on the bioaccessibility of brewers' spent grain (BSG) nutrients (carbohydrates and proteins) and non-nutrients (bioactive compounds). BSG and extruded BSG (EBSG) were digested in vitro simulating human oral-gastro-intestinal digestion and colonic fermentation. The duodenal bioaccessibility of glucose, amino acids and phenolic compounds was analyzed. The fermentability of the dietary fiber was assessed by analysis of short-chain fatty acids. Additionally, assessment of the bioaccessibility of phenolic compounds after colonic fermentation was undertaken. The antioxidant, anti-inflammatory and antidiabetic properties of the bioaccessible compounds were studied. Extrusion caused no change in the digestibility of gluten and glucose bioaccessibility ($p > 0.05$). Moreover, the bioaccessibility of amino acids and phenolic compounds significantly increased ($p < 0.05$) due to extrusion. However, higher short-chain fatty acid content was formed in colonic fermentation of BSG ($p < 0.05$) compared to EBSG. The latter inhibited intracellular ROS formation in IEC-6 cells and showed anti-inflammatory properties in RAW264.7 cells. With respect to antidiabetic properties, glucose absorption was lower, and the inhibition of carbohydrases higher ($p < 0.05$), in the presence of EBSG compared to BSG. The effects of EBSG and BSG digests on glucose transporters were not significantly different ($p > 0.05$). In conclusion, extrusion positively affected the nutritional value and health-promoting properties of BSG.

Keywords: amino acids; anti-inflammatory; antioxidant; brewers' spent grain; bioaccessibility; bioactive compounds; diabetes; extrusion; glucose transporters; short-chain fatty acids

Citation: Gutierrez-Barrutia, M.B.; Cozzano, S.; Arcia, P.; del Castillo, M.D. In Vitro Digestibility and Bioaccessibility of Nutrients and Non-Nutrients Composing Extruded Brewers' Spent Grain. *Nutrients* **2022**, *14*, 3480. https://doi.org/10.3390/nu14173480

Academic Editors: Maria Digiacomo and Doretta Cuffaro

Received: 22 July 2022
Accepted: 19 August 2022
Published: 24 August 2022

1. Introduction

In line with the seventeen Sustainable Development Goals of the United Nations [1], food system transformation is required involving a transition towards more sustainable and healthy diets to ensure food and nutrition security for all. Food systems are being challenged by malnutrition, environmental changes, and resource scarcity, and their resilience to emergencies, such as pandemics, climate change and geopolitical forces, is essential [2]. Additionally, noncommunicable diseases (NCDs) account for 41 million deaths each year with 77% of all NCD deaths occurring in low- and middle-income countries [3]. There is a clear relationship between food insecurity and NCDs related to nutrition. Therefore, it is essential for everyone concerned to move towards sustainable and healthy diets worldwide.

Brewers' spent grain (BSG) is the most abundant byproduct of the brewing industry. Its major components are dietary fiber (50%) and proteins (30%) [4], two macronutrients in very high demand by consumers and which are essential for achieving nutrition security. In addition, BSG is a valuable source of non-nutrients, including bioactive compounds, such as hydroxycinnamic acids [4]. Therefore, BSG has gained attention and is starting to be commercialized for human consumption in different forms [5–7].

A previous study has raised the possibility of the use of extruded BSG as a safe, healthy and sustainable ingredient in the human diet [8]. Extrusion is a thermomechanical process which applies high temperature and pressure over a short period of time [9]. It was found that, after extrusion, the soluble dietary fiber and extractable phenolic content of BSG was increased, while safe microbial standards were achieved, and no processing contaminants (acrylamide) were formed. Moreover, extruded BSG showed low free glucose content and protein quality parameters similar to those recorded for eggs, milk and soy [8].

Further studies which evaluate the effect of the processing of brewers' spent grain and its incorporation into the food matrix might have on the digestibility and bioaccessibility of nutrients and bioactive compounds are needed. It has been reported that extrusion may improve protein digestibility and cause the release of bound phenolics [10–12], favoring their bioaccessibility in the gastrointestinal tract. Nevertheless, the structure and properties of the cell wall may be unaltered by mastication and digestion, affecting the bioaccessibility of macronutrients from plants [13]. Furthermore, the bioaccessibility of polyphenols may be reduced due to dietary fiber binding [14]. When indigested dietary fiber from the small intestine reaches the colon it can be fermented by microbiota, producing short-chain fatty acids [13]. Moreover, microbial enzymes help to release native polyphenols from the food matrix, which is an essential mechanism for them to pass through the intestinal barrier [15].

It is important to establish whether BSG's bioaccessible compounds can exert health-promoting properties. In previous studies, the intestinal bioaccessibility of phenolic compounds from other cereal fractions with anti-inflammatory and antioxidant properties was reported [16,17]. These properties are important in preventing chronic NCDs involving energetic metabolism, such as diabetes [18]. Short-chain fatty acids can also regulate the characteristics of metabolic syndrome and modulate immune response to prevent colorectal cancer [19].

This study aimed to evaluate the duodenal and colonic bioaccessibility of nutrients (glucose and amino acids) and non-nutrients (phenolic compounds) derived from BSG and extruded BSG, and their health-promoting potential in key physiological processes associated with the development of NCDs, such as antioxidant, anti-inflammatory, and antidiabetic effects. The findings provide useful information for understanding the feasibility of using extruded BSG (EBSG) as an ingredient for achieving global nutrition security.

2. Materials and Methods

2.1. Materials

The chemicals used were of reagent grade. Digestive enzymes (α-amylase from human saliva-A0521, porcine gastric pepsin-P6887, porcine pancreatin-P1625, bile porcine extract-B8631), Folin reagent, 2,20-azinobis-(3-ethylbenzothiazoline-6-sulfonic acid) diammonium salt (ABTS), fluorescein (FL) disodium salt, 2,20-azobis (2-methylpropionamidine) dihydrochloride (AAPH), O-phtalaldehyde, acarbose, p-nitrophenyl-α-d-glucopyranoside and rat intestine acetone powder were purchased from Sigma-Aldrich (St. Louis, MO, USA). The glucose oxidase-peroxidase kit used was obtained from Spinreact (Girona, Spain). Quinine sulphate dihydrated, p-nitrophenyl and a Pierce®® Microplate BCA Protein Assay Kit—Reducing Agent Compatible were bought from Thermo Scientific™ (Waltham, MA, USA). Ferulic acid was obtained from Fluka Honeywell, Buchs, Switzerland.

For cellular studies, Dulbecco's Modified Eagle's Medium (DMEM), L-glutamine, antibiotics (penicillin and streptomycin) and trypsin were obtained from Gibco Laboratory (Invitrogen Co., Grand Island, NY, USA) and fetal bovine serum (FBS) was obtained from Hyclone (GE Healthcare, Chicago, IL, USA). 3-(4,5-dimethylthiazol-2-yl)-2,5-diphenyltetrazolium bromine (MTT), 2,7-dichlorofluorescin diacetate (DCFH-DA), sulfanilamide, *N*-1-(naphthyl)ethylenediamine-dihydrochloride, phosphoric acid, sodium nitrite and lipopolysaccharide from E. coli O55:B5 (LPS) were purchased from Sigma-Aldrich (St. Louis, MO, USA). Glucose transport inhibition standards (phloretin $\geq$ 99% and phloridzin dihydrate 99%) were purchased from Sigma-Aldrich (St. Louis, MO, USA).

2.2. Food Ingredient

Brewers' spent grain (BSG) was obtained from Fábricas Nacionales de Cerveza (Minas, Uruguay) from a pool of one day's production of lager beer. It was stabilized by drying in a convection oven (Atmos Promat 1, Alfa-Laval Gruppe, Glinde, Germany) at 45 °C ± 2 °C. Then, it was ground in a laboratory mill (Retsch ZM 200, Thermo Fisher Scientific™, Waltham, MA, USA) until its particle size was 0.5 mm.

Extruded BSG (EBSG) was obtained using a single screw extruder (Brabender Co Cordero E330, Brabender, Duisburg, Germany) with a die diameter of 4 mm, screw diameter of 1.8 cm and screw length equal to 40 cm. The thermomechanical processing conditions were 15.8% of sample moisture content, 164.3 revolutions per minute (rpm) screw speed and 122.5 °C barrel temperature, while the feeding rate was kept constant (100 rpm) [8]. Three series of extrusion processes were performed to obtain a pool of EBSG from different days, which were ground to 0.5 mm and kept at −20 °C for further analysis.

2.3. Duodenal Bioaccessibility of Nutrients and Non-Nutrients

2.3.1. In Vitro Oral-Gastro-Intestinal Digestion

BSG and extruded brewers' spent grain (EBSG) were subjected to in vitro oral-gastro-intestinal digestion according to Hollebeeck et al. [20] as modified by Martinez-Saez et al. [21]. The digests were centrifuged at 10,000 rpm, at 4 °C for 20 min. The undigestible fractions obtained at this stage were kept at −80 °C for further analysis. Supernatants were cleaned by taking out bile salts. Finally, a soluble fraction was obtained which corresponded to the intestinal digests of BSG (DBSG) and EBSG (DEBSG), composing the duodenal bioaccessible nutrients and bioactive compounds. Experiments were performed in triplicate. The effect of pH changes was evaluated employing samples without enzyme addition.

2.3.2. Glucose

The DBSG and DEBSG free glucose content was estimated by glucose oxidase/peroxidase reaction using a commercial kit (Spinreact, Girona, Spain). In a microplate, 5 µL of sample was mixed with 300 µL of glucose oxidase/peroxidase reagent. The mixture was incubated at 37 °C for 10 min. Finally, absorbance was measured at 505 nm using an Epoch 2 Microplate spectrophotometer (BioTek Winooski, VT, USA). A glucose calibration curve was used for quantification (0–11 mM). Analyses were performed in triplicate and the results expressed in mM.

2.3.3. Amino Acids and Protein

Soluble Protein Content

The DBSG and DEBSG soluble protein content was determined using a bicinchoninic acid (BCA) microplate Pierce assay kit compatible with reducing agents based on Smith et al. [22]. Nine microliters of sample or standard were added to the center of the microplate well with 4 µL of compatible reagent solution. The plate was shaken for 1 min and was incubated for 15 min at 37 °C. Afterwards, 260 µL of BCA working reagent solution was added to each well and the microplate was incubated for 30 min at 37 °C. Finally, the plate was cooled at room temperature and absorbance was measured at 562 nm. A BSA calibration curve was undertaken for quantification (0–2 mg BSA/mL). Experiments were performed in triplicate. The protein digestibility percentage was calculated as follows [23]:

$$Protein\ digestibility\ (\%) = \frac{Soluble\ protein\left(\frac{g}{100\ g\ of\ digested\ sample}\right)}{Protein\ content\left(\frac{g}{100\ g\ of\ sample}\right)} \times 100 \quad (1)$$

Gluten Content

Gluten was determined for DBSG and DEBSG according to the enzyme-linked competitive immunoassay R5 method as described in Mena et al. [24].

Free Amino Acid Content

The total amino acids released during BSG and EBSG in vitro oral-gastro-intestinal digestion was determined by o-phthalaldehyde (OPA) assay as described by Shene et al. [25], adapted to a micro-method. OPA reagent was prepared by dissolving 5 mg of OPA in 100 µL of ethanol (96%), 5 µL of β-2- mercaptoethanol and 10 mL of 50 mM carbonate buffer (pH 10.5). In a quartz microplate, 50 µL of sample were mixed with 200 µL of OPA reagent. Absorbance was measured during 1 min at 340 nm using a microplate reader (BioTek Epoch 2 Microplate spectrophotometer, Winooski, VT, USA). Analyses were performed in triplicate. An *N*-acetyl-lysine (0–1.20 mM) curve was done for quantification and the results were expressed in mM of *N*-acetyl-lysine equivalents.

The amino acid (AA) profile of DBSG and DEBSG was determined. Half µL samples of DBSG and DEBSG were analyzed using a Biochrom30 series amino acid analyzer (Biochrom Ltd., Cambridge Science Park, Cambridge, UK), based on Spackman et al. [26]. The determinations were performed in triplicate and the results were expressed in mM of AA.

Advanced Glycation End Products (AGEs) Content

Fluorescent AGEs were measured as described by Martinez-Saez et al. [27]. One hundred and fifty µL of DBSG and DEBSG were added on a black microplate and fluorescence was measured using a microplate reader (BioTek Cytation5 Cell Imaging Multi-Mode Reader, Winooski, VT, USA) at 360 ± 40 nm and 460 ± 40 nm as the excitation and emission wavelengths, respectively. Analyses were performed in triplicate. A calibration curve was used for quantification with dihydrated quinine sulphate (0–640 µM).

2.3.4. Phenolic Compounds

2.3.4.1. Total Polyphenolic Content

Total phenolic content (TPC) was determined for duodenal digests (DBSG and DEBSG) by the Folin–Ciocalteu method based on Singleton et al. [28] adapted to micromethod [29]. A ferulic acid (FA) calibration curve was used for quantification (0–3.6 mM). Experiments were performed in triplicate and the results expressed as mM FA equivalent (FAeq) and mg of FAeq per gram of digested sample.

2.3.4.2. Analysis of Phenolic Compounds by HPLC-QTOF Assay

DBSG and DEBSG phenolic compound identification by HPLC-QTOF was performed using HPLC equipment (Agilent 1200, Waldronn, Germany) equipped with a quaternary pump (G1311A), coupled degasser (G1322A), thermostated automatic injector (G1367B), thermostated column module (G1316A) and diode array detector (G1315B). The equipment was coupled to a mass spectrometer (Agilent G6530A Accurate Mass QTOF LC/MS) with an atmospheric pressure electrospray ionization source with JetStream technology. Control software used were the Masshunter Data Acquisition (B.05.00) and Masshunter Qualitative Analysis (B.07.00) programs. Samples and standard solution were injected (20 µL) in a ZORBAX Eclipse XDB-C18 column (150 mm × 4.6 mm × 5 µm) at 40 °C. The solvent systems were 0.1% formic acid (solvent A) and 0.1% formic acid diluted in acetonitrile (solvent B). The elution gradient (time, % of solvent A) was: 0 min, 95%; 20 min, 85%; 30 min, 70%; 35 min, 50%, 37 min, 95%, 45 min, 90%. Identification was performed by comparison of molecular formulae, retention times and previous references. A calibration curve for ferulic acid (1–16 µg/mL) was done. Semi quantitative results were obtained by comparing chromatogram areas for each compound and sample.

2.4. *Colonic Bioaccessibility of Nutrients and Non-Nutrients*

2.4.1. In Vitro Simulation of Colonic Fermentation

Undigestible fractions obtained by applying the procedure in Section 2.3.1 were submitted to in vitro simulation of colonic fermentation based on several studies [30–32]. Fecal material was obtained from seven healthy volunteers who had not been treated with antibiotics for the last three months. Volunteers were provided with sterile containers, BD

GasPak™ (NJ, USA) EZ Anaerobe Container System Sachets and two zip bags. Fecal samples were collected a minimum of 16 h prior to analysis and kept at 4 °C under anaerobic conditions until analysis. Experimentation took place in an anaerobic cabin (BACTRON Anaerobic Environmental Chamber, SHELLAB, OR, USA). The fecal inoculum was composed of equal parts of fecal material from each volunteer and phosphate buffer (0.1 M, pH 7) at 30% (*w/v*). It was homogenized on a Stomacher homogenizer (Stomacher 400 Circulator, SEWARD, Worthing, UK) for 10 min. Fermentation medium (pH 7) was made of peptone (15 g/L), cysteine (0.312 mg/L) and sodium sulfide (0.312 mg/L). Five hundred milligrams of undigested fractions were mixed with 7.5 mL of fermentation medium and 2 mL of fecal inoculum in a screw tap tube. The tubes were left under constant agitation for 24 h at 37 °C. Microbial activity was stopped by immersion in ice. Tubes were centrifuged at 4500 rpm for 10 min and supernatants were filtered (0.2 µm). The filtrate corresponded to the colonic fermented digests of BSG (FBSG) and EBSG (FEBSG). Aliquots were stored at −20 °C for further analysis. Experiments were performed in triplicate.

Microbiota Analysis

Microorganisms present in the fecal inoculum were determined immediately after collection, following Tamargo et al. [33]. Samples were diluted to different concentrations in physiological solution (0.9%). These dilutions were plated on different types of media: trypticase soy agar (TSA) (Difco™ BD, NJ, USA) for total aerobes; Wilkins Chalgren agar (Difco™ BD) for total anaerobes; MacConkey agar (Difco™ BD) for *Enterobacteriaceae*; Enterococcus agar (Difco™ BD) for *Enterococcus* spp.; MRS agar (pH = 5,4) (Pronadisa, CONDA, Madrid, Spain) for lactic acid bacteria; tryptose sulfite cycloserine agar (TSC) (Pronadisa, CONDA) for *Clostridium* spp.; BBL CHROMAgar (Difco™,BD) for *Staphylococcus* spp.; Bifidobacterium agar modified by Beerens (Difco™ BD) for *Bifidobacterium* spp. and LAMVAB for specific fecal *Lactobacillus* spp. [34]. All plates were incubated at 37 °C for 24 h to 72 h with an anaerobic gas pack system (BD, NJ, USA), except for BBL CHROMAgar and TSA, which were incubated in aerobic conditions. Plate culture was performed in triplicate and the results were expressed as the logarithm of colony forming units per mL [log(CFU)/mL].

2.4.2. Sugars

Identification and quantification of sugars (glucose, fructose, arabinose, xylose) in FBSG and FEBSG was performed by ion exchange liquid chromatography (LC-IC). The equipment used consisted of a Metrohm Advanced Compact ion chromatographic instrument with Bioscan module (817 IC. Metrohm, Herisau, Switzerland), equipped with a pulse amperometric detector (PAD) (945 Professional Detector Vario, Metrohm), a pump (IC Pump 812), 889 IC Sample Center injector (Metrohm), and coupled with a degasser (IC-837, Metrohm). Between two and 100 µL of samples and D(+) glucose (1.08337.0250, Merck), D(−) fructose (1.04007.0250, Merck) and D(−) arabinose standards (375,763.1206, Sigma-Aldrich) were injected in a Metrosep Carb 2 column (250 × 4 mm). The mobile phase used was 300 mM sodium hydroxide and 1 mM sodium acetate, applied at a flow rate of 0.5 mL/min for 35 min. Identification was performed by comparison of the retention time with the patterns used. Data were analyzed using Metrodata IC Net 2.3 and MagIIC Net 2.3 (Metrohm, Herisau, Switzerland) software. Analyses were performed in triplicate.

2.4.3. Organic Acids

A Metrohm Advanced Compact ion chromatography instrument (867 IC. Metrohm, Herisau, Switzerland) equipped with a conductivity detector (IC-819), 889 IC sample center injector, coupled with a degasser (IC-837, Metrohm, Herisau, Switzerland), was used for analysis of organic acids. FBSG, FEBSG, lactate standard (07096-100 ML, Sigma-Aldrich) and succinate standard (43057-100 ML, Sigma-Aldrich) were injected at a volume of 20 µL in a Metrosep organic acids column (250 × 4 mm, 5 µm). The mobile phase used was 0.5 mM sulfuric acid and 15% acetone, at 0.5 mL/min for 20 min. Identification was performed by comparison of the retention time with patterns used. The control software used were

Metrodata IC Net 2.3 and MagIIC Net 2.3 (Metrohm, Herisau, Switzerland). Analyses were performed in triplicate.

2.4.4. Short-Chain Fatty Acids

Short-chain fatty acids (SCFAs) were determined in colonic-fermented digests of BSG (FBSG) and EBSG (FEBSG). Fecal inoculum SCFAs were also determined to establish the baseline. Four hundred μL of samples or standards were acidified with 200 μL of phosphoric acid (0.5%). One hundred μL of methyl valeric (8092 μM) was added as an internal standard and was extracted with 1000 μL of n-butanol. Identification and quantification of SCFAs was performed using a gas chromatograph (Agilent 6890A, Santa Clara, CA, USA) equipped with a flame ionization detector (260 °C), an automatic injector (G2613A, Santa Clara, CA, USA) and a DB-WAXtr column (60 m × 0.325 mm × 0.25 μm) (Agilent Technologies, Santa Clara, CA, USA). Two μL of samples or standards were injected using a splitless method at 250 °C. The carrier gas used was helium at a constant flow rate (1.5 mL/min). The initial column temperature was 50 °C held for 2 min, increased to 150 °C at a rate of 15 °C/min, to 200 °C at 5 °C/min and then increased to 240 °C at a rate of 15 °C/min and kept for 20 min. The control software used was MSD Chemstation E.02.00.493 Agilent (Santa Clara, CA, USA). Analyses were performed in triplicate.

2.4.5. Phenolic Compounds

Phenolic compounds of fermented colonic digests (FBSG and FEBSG) were analyzed by the Folin–Ciocalteu method as described in Section 2.3.4.1. The phenolic compounds were identified as reported in Section 2.3.4.2.

2.5. Bioactivity of Bioaccessible Compounds

2.5.1. Antioxidant Capacity

ABTS Method

The BSG and EBSG intestinal and colonic fermented digests' antioxidant capacity was measured using the ABTS method proposed by Re et al. [35] following the procedure described by Martínez-Saez et al. [21]. A ferulic acid calibration curve was used for quantification (0–80 μM). Experiments were performed in triplicate and the results expressed in mM FAeq and μmol FAeq/g of the digested sample

ORAC Method

Assessment of total antioxidant capacity by oxygen radical absorbance capacity (ORAC) was performed according to Ou et al. [36] for the BSG and EBSG intestinal- and colonic-fermented digests. A ferulic acid calibration curve was used for quantification (0–30 μM). All measurements were performed in triplicate and the results expressed as mM FAeq.

Intracellular Reactive Oxygen Species (ROS) Formation

Physiological intracellular reactive oxygen species (ROS) were measured on normal rat small intestine epithelial cells (IEC-6 cells) using a fluorescent probe DCFH-DA, as described by Iriondo-DeHond et al. [36], with slight modifications. Briefly, IEC-6 cells were seeded at a density of 2×10^4 cells/well on a 96-well plate and cultured in complete medium (DMEM with 4.5 g/L of glucose, 10% *v/v* of FBS, 1% *v/v* of L-glutamine and 1% *v/v* of antibiotics) for 24 h (37 °C, 5% CO_2). Afterwards, the medium was aspirated, and cells were loaded with 100 μL solution containing DBSG and DEBSG at a concentration of 15% *v/v* and medium without FBS, at a ratio equal to 1:10. Digests concentration (15% *v/v*) was not cytotoxic, as shown in (Supplementary Material Figure S1), determined by MTT assay [37]. Following a 24 h incubation period, the well contents were aspirated, and cells were pre-loaded with 100 μL of medium without FBS containing 2 μL of DCFH-DA (0.3 mg/mL in DMSO). Cells were incubated for 45 min. Afterwards, culture medium was removed, cells were washed with PBS and were treated with 100 μL containing DBSG

(15% v/v) and DEBSG (15% v/v) and medium without FBS at a ratio equal to 1:10 for 30 min. Tert-butylhydroperoxide (tBOOH) 1 mM was used as an oxidation control and ferulic acid (10–100 µM) was used as an antioxidant control. Ferulic acid concentrations were shown to be non-cytotoxic; the results are shown in (Supplementary Material Figure S2). A control was set up by adding only medium without FBS to establish ROS formation under assay conditions. Afterwards, fluorescence was measured at 485 nm/528 nm (BioTek Synergy HT Multi-Mode Microplate Reader). Finally, 20 µL of MTT reagent were added per well and the plates were incubated for 90 min. Supernatants were removed, 100 µL of DMSO were added to each well and the absorbance measured at 570 nm using a BioTek Epoch 2 Microplate spectrophotometer (Winooski, VT, USA) [37]. All measurements were performed in triplicate and for three different cell passages. Intracellular ROS formation was calculated as follows:

$$\%\ ROS\ formation = \frac{Fluorescence_{Sample}/MTT\ Absorbance_{Sample}}{Fluorescence_{Control}/MTT\ Absorbance_{Control}} \times 100 \quad (2)$$

2.5.2. Anti-Inflammatory Properties

The anti-inflammatory properties of DBSG and DEBSG were determined by quantifying the nitrogen oxide (NO) production in macrophages (RAW264.7) as described by Benayad et al. [38]. Briefly, RAW264.7 cells were seeded on a 96-well plate (8×10^4 cell/well) and cultured in complete medium (DMEM with 4.5 g/L of glucose, 10% v/v of FBS, 1% v/v of L-glutamine and 1% v/v of antibiotics) for 24 h (37 °C, 5% CO_2). Afterwards, cells were treated with 150 µL of medium without FBS containing 1µg/mL lipopolysaccharide (LPS) from *Escherichia coli* O55:B5 and DBSG (15% v/v), DEBSG (15% v/v) or ferulic acid (10–100 µM). The intestinal digests and ferulic acid concentrations were shown not to be cytotoxic (Supplementary Material, Figure S3 and S4). Then, cells were incubated for 24 h (37 °C, 5% CO_2). Negative and positive controls were tested consisting of medium without FBS and 1µg/mL of LPS in medium without FBS, respectively. After the incubation period, 100 µL of supernatants from the wells were removed and combined with 100 µL of Griess reagent (1% (w/v) sulfanilamide and 0.1% w/v N-1-(naphthyl)ethylenediamine-dihydrochloride in 2.5% v/v H_3PO_4). The mixtures were incubated at room temperature in the dark for 15 min and absorbance was measured at 550 nm in a BioTek Epoch 2 Microplate spectrophotometer (Winooski, VT, USA). A NO-in-DMEM-without FBS calibration curve was used for quantification (0–10 µg/mL). Analyses were performed in triplicate and for three different cell passages.

2.5.3. Antidiabetic Properties

2.5.3.1. Carbohydrase Activity

An enzymatic extract from rat intestine powder was obtained prior to analysis as described by Martínez-Saez [21]. The α-amylase, α-glucosidase and sucrase activity of rat intestinal extract was (424.02 ± 42.52) µmol of maltose × mL^{-1} × min^{-1}, (0.181 ± 0.005) µmol of p-nitrophenyl × mL^{-1} × min^{-1} and (0.254 ± 0.028) µmol of glucose × mL^{-1} × min^{-1}, respectively. The methods used are available in (Supplementary Material Figures S5–S7).

Assessment of DBSG and DEBSG's α-glucosidase inhibitory activity was performed according to Lordan et al. [39], with slight modifications. Fifty µL of tested sample was mixed with 50 µL of 4-Nitrophenyl α-D-glucopyranoside (5 mM) and incubated at 37 °C for 5 min. An amount of 100 µL of rat intestinal extract (0.06 U/mL) was added to the previous mixtures. After incubating for 30 min at 37 °C, 80 µL of Na_2CO_3 1 M was added. Absorbance was measured at 405 nm using a BioTek Epoch 2 Microplate spectrophotometer (Winooski, VT, USA). An acarbose calibration curve was done for quantification (0.00–0.30 mM).

Assessment of the intestinal digest α-amylase inhibitory activity was performed following Lordan et al. [39]. Briefly, 100 µL of tested sample or acarbose (0.001–0.25 mM) was incubated (10 min, 25 °C) with 100 µL of 1% potato starch. Afterwards, 100 µL of rat intestinal extract (40 U/mL) was added and the mixture was incubated (10 min, 25 °C) using an Eppendorf ThermoMixer™ C (Eppendorf, Madrid, Spain). Two hundred µL of

DNS reagent was added and the mixture was incubated for 10 min at 100 °C. After cooling, 50 µL of sample was introduced in a microplate well, followed by 200 µL of distilled water. Absorbance was measured at 540 nm with a BioTek Epoch 2 Microplate spectrophotometer (Winooski, VT, USA).

Finally, a sucrase inhibitory activity assay was performed based on Li et al. [40]. Forty µL of sucrose (480 mM) was mixed with 10 µL of rat intestinal extract (0.254 U/mL) and 50 µL of tested sample. The mixture was incubated in an Eppendorf ThermoMixer™ C for 60 min at 37 °C. Afterwards, the mixture was heated at 100 °C for 10 min to stop the reaction. The concentration of glucose was measured as described in Section 2.3.2. An acarbose calibration curve was done for quantification (0.00–0.20 mM).

For all the inhibitory capacity experiments, a positive control with the addition of PBS instead of tested sample or acarbose was set up. All experiments were performed in triplicate. The results were expressed as mM of acarbose equivalents or by inhibition capacity measured as:

$$\%\ inhibition = \frac{A_{+} - \left(A_{sample} - A_{blank}\right)}{A_{+}} \tag{3}$$

where, A_{+} is the absorbance value for the positive control, A_{sample} is the absorbance of the reaction mixture and A_{blank} is the absorbance of the reaction mixture with PBS instead of enzymatic extract.

2.5.3.2. Glucose Absorption

DBSG and DEBSG glucose absorption in IEC-6 was determined. Tested samples were DBSG (15% *v/v*) and DEBSG (15% *v/v*) and a control containing an equivalent amount of glucose content. The concentration of the intestinal digest was determined by MTT assay [37] to guarantee at least 80% cell viability as shown in (Supplementary Material Figure S1).

Transwell plates with polycarbonate inserts (Transwell®® inserts, 0.4 µm pore size, 1.1 cm^2) were used to estimate glucose transport across IEC-6 cell monolayers. Cells were seeded on 12-well Transwell plates with 7.6 × 10^4 cells/well. Cells were grown (37 °C, 5% CO_2) for 10 days, to differentiate. Five hundred µL and 1500 µL of medium on the apical and basolateral sides of each well were added, respectively. The medium used was formed by DMEM with 4.5 g/L of glucose, 10% *v/v* of FBS, 1% *v/v* of L-glutamine and 1% *v/v* of antibiotics and was changed within 48 h. Transepithelial electrical resistance (TEER) was measured using a Millicell-ERS device (Millipore, Zug, Switzerland) to evaluate monolayer integrity.

On the assay day, cells were washed with PBS 10 mM and pH 7.4. Afterwards, PBS was aspirated, and cells incubated (37 °C, 5% CO_2) for 30 min with PBS in the absence of glucose. Next, PBS was aspirated from both sides of the Transwell and replaced by 500 µL of tested sample prepared in PBS in the apical side and 1500 µL of PBS in the basolateral side. Plates were incubated (37 °C, 5% CO_2). Two hundred µL was removed from the basolateral side at different times (10 min, 30 min, 45 min, 60 min, 75 min, 90 min and 120 min) and the volume in the basolateral side was replaced with PBS.

After finishing the assay, a Lucifer yellow test was performed to evaluate the cell monolayer permeability [41–43]. Wells were washed with PBS and 200 µL of Lucifer yellow reagent (50 µM) was added in the apical side. One and a half mL of PBS was added in the basolateral side and plates were incubated for 1 h at 37 °C. Finally, 200 µL aliquots from the basolateral side were removed and the fluorescence was measured (485 nm/528 nm). The permeability percentage was calculated as the coefficient between the sample fluorescence and Lucifer yellow reagent (50 µM) fluorescence. Permeability less than 5% was considered adequate.

Glucose absorption was determined by measuring glucose concentration (see Section 2.3.2) in the basolateral side at different times. Glucose contents at different times were plotted graphically and the incremental area under the curve (IAUC) was calculated.

2.5.3.3. Glucose Transport Inhibition

The inhibition capacity of DBSG and DEBSG over glucose transporters present in the IEC-6 cells was tested in the presence and absence of sodium. Assays were based on Fernández et al. [44] with modifications.

The procedure was the same as that described in Section 2.5.3.2 with slight differences. Two buffer solutions: PBS and potassium phosphate buffer (PK), were used for making the assay in the presence or absence of sodium, respectively. PBS (10 mM, pH 7.4) was prepared by dissolving in 1 L of distilled water, 8 g of NaCl, 0.2 g of KCl, 1.44 g of Na_2HPO_4 and 0.24 g of KH_2PO_4. PK (10 mM, pH 7.4) was prepared by mixing 10.4 g of KCl, 0.53 g of KH_2PO_4 and K_2HPO_4. Tested samples added to the wells were glucose (25 mM), glucose 25 mM + DBSG (15% *v/v*), glucose 25 mM + DEBSG (15% *v/v*), glucose 25 mM + ferulic acid 0.5 mM and glucose 25 mM with reference inhibitors (0.1 mM phloretin, 0.3 mM phloridzin). Every solution was prepared on PBS or PK depending on whether the experiment was in the presence or absence of sodium.

Glucose absorption was determined by measuring glucose concentration (see Section 2.3.2) in the basolateral side at different times. The glucose contents at different times were plotted graphically and the incremental area under the curve (IAUC) was calculated. The inhibition percentage was calculated as follows:

$$\%\,Glucose\ transport\ inhibition = \frac{IAUC_{Control} - IAUC_{Glucose\ 25\ \mathrm{mM}+\ tested\ sample}}{IAUC_{Control}} \times 100 \quad (4)$$

where, $IAUC_{Control}$ corresponds to the incremental area under the curve of glucose (glucose 25 mM) and $IAUC_{Glucose\ 25\ \mathrm{mM}+tested\ sample}$ is equal to the incremental area under the curve of glucose (25 mM) + DBSG, DEBSG, ferulic acid or reference inhibitors (phloridzin and phloretin).

- Determination of the kinetic mechanism of inhibition

The V_{max} and K_m values for DEBSG and ferulic acid inhibition mechanisms were determined for glucose uptake and transport in IEC-6 based on Manzano et al. [45].

After seeding and growing cells as described in Section 2.5.3.2, cells were washed with PBS. Five hundred µL and 1500 µL of PBS were added to the apical and basolateral sides, respectively. Cells were incubated for 30 min (37 °C, 5% CO_2). Afterwards, PBS was aspirated from both sides of the well and different treatments were applied to the cells: 0.5–40 mM glucose, 0.5–40 mM glucose + DEBSG (15% *v/v*) and 0.5–40 mM glucose + ferulic acid (0.5 mM). Five hundred µL of each treatment was added in the apical side and 1500 µL of PBS in the basolateral side. Plates were incubated for 30 min (37 °C, 5% CO_2). Finally, sampling was carried out by removing 200 µL from the apical and basolateral sides.

Glucose content was measured in the basolateral and apical aliquots by the glucose oxidase-peroxidase method (Section 2.3.2). Glucose uptake was calculated as the difference between glucose concentration load and the remaining glucose concentration in the apical and basolateral sides. Lineweaver–Burk plots were constructed for glucose uptake and transport to estimate V_{max} and K_m.

2.6. Statistical Data Analysis

All data are reported as mean ± standard deviation. T-student tests and one-way analysis of variance (ANOVA), applying Tukey's test, were used to determine significant differences between samples ($\alpha \leq 0.05$). Analyses were performed using XLSTAT Version 2011 (Addinsoft 1995–2010, France). Pearson correlation coefficients were obtained with a significance level of 0.05.

3. Results and Discussion

3.1. Duodenal Bioaccessibility of Nutrients

Data on bioaccessible amino acids, glucose, gluten and soluble protein are presented in Table 1. Although in vitro static models are over-simplistic and do not reproduce all dynamic aspects of the gastrointestinal tract, they have been shown to be useful in predicting macronutrient digestion [46].

Table 1. Bioaccessibility of glucose, gluten, free amino acids and soluble protein in duodenal digests.

	Glucose (mM)	Soluble Protein (mg BSA/mL)	Gluten (ppm)	Amino Acids (mM of Equivalent *N*-acetyl lysine)
DBSG	4.654 ± 0.347 [a]	17.089 ± 0.069 [a]	<3 [a]	55.781 ± 3.516 [a]
DEBSG	4.788 ± 0.217 [a]	18.942 ± 0.012 [b]	<3 [a]	69.776 ± 6.389 [b]

Different letters within the same column show significant differences ($p < 0.05$). BSA: bovine serum albumin; AGEs: advanced glycation end products; DSG: brewers' spent grain intestinal digest; DEBSG: extruded brewers' spent grain intestinal digest.

Regarding glucose bioaccessibility, no significant differences ($p > 0.05$) were observed in the BSG intestinal digest (DBSG) and extruded BSG intestinal digest (DEBSG) bioaccessible glucose levels (Table 1). The glucose released during human oral-gastro-intestinal digestion depends on the free glucose content of the food matrix and on the enzymatic digestion of nutritional polysaccharides. α-Amylase found in the saliva and duodenum catalyzes the α-1,4-glucan bonds forming starch, maltodextrins and malto-oligosaccharides, followed by hydrolytic reactions catalyzed by α-glucosidase which liberate glucose residues [47]. The values obtained were within the range considered normal (4.4–6.1 mM) in the blood of humans under fasting conditions [48].

Protein digestibility is considered the most important determinant of protein quality in adults [10]. Extrusion increased BSG protein digestibility as a higher content ($p < 0.05$) of soluble protein and free amino acids was found in DEBSG than in DBSG (Tables 1 and 2). This might be caused by denaturation of proteins and inactivation of antinutritional factors, such as trypsin inhibitors, during extrusion [10,11]. The EBSG protein digestibility percentage (Equation (1)) was 48.3% while that for BSG was 40.3%. Soluble protein was determined using a bicinchoninic acid (BCA) method, in which tripeptides and larger peptides react [49]. Although, BCA is a method for protein quantification, the results may be influenced by soluble peptides present in the digests due to enzymatic activity. In fact, the method has previously been used for peptide quantification [50]. Furthermore, BSG and EBSG contain gluten [8]. However, gluten was not found in the intestinal digests (Table 1). This data suggests that gluten was completely digested and implies that the extrusion process did not affect the digestibility of this protein.

The amino acid profiles for the BSG and EBSG intestinal digests are shown in Table 2. Both food matrices contained serine and alanine as the major non-essential amino acids (NEAAs), while leucine, valine and phenylalanine were the most abundant essential amino acids (EAAs) present in its intestinal digests. However, aspartic acid, cysteine histidine and methionine showed the lowest bioaccessibility. Protein nutritional value is dependent on the quantity, digestibility, and availability of EAAs [10]. The intestinal EBSG digests presented higher ($p < 0.05$) EAA content than the BSG intestinal digests (Table 2), suggesting EBSG is of better nutritional quality than BSG. Intestinal EBSG digests presented higher content ($p < 0.05$) of leucine, isoleucine and threonine than intestinal BSG digests. The differential muscle protein synthetic response is largely dependent on the post-prandial availability of EAA, particularly leucine [51]. However, DEBSG presented lower bioaccessible lysine ($p < 0.05$) than DBSG (Table 2). This might be a consequence of advanced glycation end-products (AGEs) formed during the digestion process [27], which were 82.605 ± 1.927 and 83.408 ± 2.349 μM of quinine sulphate for DBSG and DEBSG, respectively. Although no significant difference ($p > 0.05$) was found in the AGEs levels, a significant ($p < 0.05$) negative Pearson correlation was found between lysine and AGEs content present in BSG and EBSG intestinal digests. The reaction of EAA and sugars in the gut may cause a significant reduction in their bioaccessibility and health disorders of different kinds. In addition, the AGEs formed might contribute to the body pool of AGEs and, therefore, to an overall state of chronic oxidative stress [27].

Table 2. Amino acid profile for BSG intestinal digests (DBSG) and EBSG intestinal digests (DEBSG).

	Amino Acid (mM)	DBSG	DEBSG
Non-essential amino acids (NEAA)	Alanine (Ala)	5.759 ± 0.220 [a,J]	6.719 ± 0.353 [b,K]
	Arginine (Arg)	3.442 ± 0.038 [a,G,H]	3.363 ± 0.232 [a,I]
	Aspartic acid (Asp)	0.235 ± 0.025 [a,A]	0.235 ± 0.025 [a,A]
	Cysteine (Cys)	0.502 ± 0.034 [a,A]	0.461 ± 0.075 [a,A,B]
	Glutamic acid (Glu)	1.309 ± 0.070 [a,C,D]	1.328 ± 0.098 [a,C,D,E]
	Glycine (Gly)	3.259 ± 0.075 [a,G]	3.294 ± 0.174 [a,I]
	Proline (Pro)	1.535 ± 0.332 [a,D]	1.578 ± 0.351 [a,D,E,F]
	Serine (Ser)	8.762 ± 0.242 [a,L]	10.498 ± 0.382 [b,N]
	Tyrosine (Tyr)	1.600 ± 0.098 [a,D]	1.788 ± 0.096 [b,E,F]
Total NEAA		26.282 ± 0.717 [a,*]	28.729 ± 1.788 [b]
Essential amino acids (EAA)	Histidine (His)	1.139 ± 0.034 [a,B,C]	1.158± 0.455 [a,C,D]
	Isoleucine (Ile)	2.683 ± 0.092 [a,F]	2.870 ± 0.133 [b,H,I]
	Leucine (Leu)	7.360 ± 0.249 [a,K]	8.729 ± 0.471 [b,M]
	Lysine (Lys)	2.744 ± 0.060 [b,F]	2.580 ± 0.154 [a,H,G]
	Methionine (Met)	0.876 ± 0.021 [a,B]	0.892 ± 0.068 [a,B,C]
	Phenylalanine (Phe)	3.632 ± 0.189 [a,H]	3.909 ± 0.295 [a,J]
	Threonine (Thr)	1.975 ± 0.049 [a,E]	2.079 ± 0.081 [b,F,G]
	Tryptophan (Trp)	n.d.	n.d.
	Valine (Val)	4.541 ± 0.132 [a,I]	4.824 ± 0.400 [a,L]
Total EAA		24.947 ± 0.661 [a]	27.041 ± 1.570 [b]
Total		51.457 ± 1.230 [a]	56.233 ± 3.074 [b]

Different letters within the same line (lower case) and column (capital letters) show significant differences ($p < 0.05$). * shows significant differences between Total NEAA and EAA content for each sample. DSG: brewers' spent grain intestinal digest; DEBSG: extruded brewers' spent grain intestinal digests.

3.2. Duodenal Bioaccessibility of Non-Nutrients

Many factors influence the bioaccessibility of polyphenols, such as the food matrix and interactions with other components. [52]. Table 3 shows the bioaccessible total phenolic content in the duodenum (DBSG and DEBSG) and in the colon (FBSG and FEBSG). The bioaccessible compound concentration in the BSG and EBSG intestinal digests were 2.492 ± 0.107 and 2.787± 0.122 mM of FAeq ($p < 0.05$), respectively.

Table 3. Total polyphenolic content for BSG and EBSG intestinal- and colonic-fermented digests.

Total Phenolic Content (mg FAeq/g of Digested Sample)		
	BSG	EBSG
Duodenal bioaccessibility	3.221 ± 0.116 [a,A]	3.604 ± 0.111 [b,A]
Colon bioaccessibility	4.743 ± 0.235 [a,B]	5.290 ± 0.072 [b,B]

Different letters (lower case for rows and capital letter for columns) within the same column show significant differences ($p < 0.05$). BSG: brewers' spent grain; EBSG: extruded brewers' spent grain. FA: ferulic acid.

A higher ($p < 0.05$) duodenal bioaccessibility of phenolic compounds was found after the extrusion process for BSG, when measured by the Folin–Ciocalteu method. Extrusion might have released bound phenolics due to breaking of conjugated moieties [53]. A higher bioaccessibility of phenolic acids after extrusion has previously been reported for mango bagasse [54]. However, the bioaccessibility of phenolic compounds after extrusion of brown rice and oat was decreased, while it remained unchanged for wheat [55].

These results are consistent with the phenolic profile presented in Table 4, which shows an increase in the duodenal bioaccessibility of ferulic acid, dihydrocaffeic acid, and particularly of benzoic acid, in DEBSG compared to DBSG. Ferulic acid was quantified as 0.007 ± 0.003 mg/100 g of the digested sample in DBSG and 0.021 ± 0.004 mg/100 g of the digested sample in DEBSG ($p < 0.05$). However, similar phenolic profiles were found for DBSG and DEBSG, with 2-(3-hydroxyphenyl) propionic acid the most abundant compound. Polyphenols are highly sensitive to mild alkaline conditions in the small intestine and most dietary polyphenols are degraded or transformed into other compounds at this stage [52]. Thus, it is relevant to identify which phenolic compounds are bioaccessible after undergoing oral-gastro-intestinal digestion.

Table 4. Bioaccessibility of individual polyphenols in duodenal and colonic digests.

Proposed Compound	Molecular Formula	Molar Mass (g/mol)	Retention Time (min)	Duodenal Bioaccessibility			Colon Bioaccessibility		
				DBSG [1] (%)	DEBSG [1] (%)	Variation after Extrusion (%) [2]	FBSG [1] (%)	FEBSG [1] (%)	Variation after Extrusion (%) [2]
2-(3-hydroxyphenyl) propionic acid	$C_9H_{10}O_3$	165.1	16.7	95.0	94.3	−3	96	96	14
Ferulic acid	$C_{15}H_{18}O_8$	193.1	20.0	0.5	0.5	3	0	0	0
Dihydrocaffeic acid	$C_9H_{10}O_4$	181.1	6.7	2.8	2.8	2	0	0	0
Benzoic acid	$C_7H_6O_2$	121.0	11.5	1.8	2.4	31	4	4	0

([1]) Percentage of chromatogram peak area respect total chromatogram area, ([2]) Percentage of variation due to extrusion process. DSG: brewers' spent grain intestinal digest; DEBSG: extruded brewers' spent grain intestinal digest; FBSG: brewers' spent grain colonic fermented digest; FEBSG: extruded brewers spent grain colonic fermented digest.

3.3. Colonic Bioaccessibility of Metabolites Formed by Microbial Fermentation of Nutrients: Short-Chain Fatty Acids (SCFAs)

The gut microbiota, consisting of numerous microbial species, is considered a vital organ that plays a pivotal role in the host's health [19]. Most of the contributions made by the gut microbiota to human physiology are related to microbial metabolism [56]. The fermentation of exogenous or endogenous substrates is of major importance for the host's health through the production of a wide variety of metabolites [57]. To establish if BSG and EBSG are fermented by gut microbiota after oral-gastro-intestinal digestion, an in vitro colonic fermentation was applied using human feces.

Fecal inoculum was obtained using a pool of fecal samples from seven European volunteers, six women and one man, with a normal body mass index and between 18 and 54 years of age. Table 5 shows the characterization of the viable microbial populations in the fecal inoculum. Similar microbial distributions were found in the culture-dependent characterization of fecal inoculum made by Tamargo et al. [58]. Nevertheless, an analysis considering genomic sequences could have been performed to better understand the microorganisms present in the fecal inoculum. The dominant gut microbial phylum are Firmicutes, Bacteroidetes, Actinobacteria, Proteobacteria, Fusobacteria, and Verrucomicrobia, with the first two phyla listed representing 90% of gut microbiota [14,59]. *Clostridium* spp. and *Bifidobacterium* appeared to be the predominant orders in the fecal inoculum (Table 5), belonging to the Firmicutes and Actinobacteria phyla, respectively. Firmicutes spore-forming bacteria typically account for 64–78% of Western human microbiota [60]. Although the Actinobacteria phylum is proportionally less abundant [59], it has been found that subjects adopting a Mediterranean diet had a higher abundance of *Bifidobacterium* [61]. A lower proportion of bacteria belonging to the Proteobacteria phylum was identified, corresponding to the *Enterobacteriaceae*. Moreover, lower counts of total aerobes were expected, as gut microorganisms live in an anaerobic environment. However, some *Enterobacteriaceae* are anaerobic facultative and may have grown under aerobic conditions.

Table 6 presents the short-chain fatty acid (SCFA) content after in vitro fecal fermentation of BSG and EBSG undigested material. SCFAs are essential gut microbiota metabolites, produced mainly from the fermentation of non-digestible carbohydrates that become available to the gut microbiota. SCFAs impact on host health at cellular, tissue and organ levels by mechanisms related to gut barrier function, glucose homeostasis, immunomodulation and obesity [61].

The degradation of undigested polysaccharides involves a variety of hydrolytic enzymes that are not produced by the host. This is essential for providing bacteria with carbon and energy from released sugars [57]. BSG dietary fiber is mainly formed of cellulose, hemicellulose and lignin [4], which are substrates for microbial degradation. Furthermore, resistant starch, whose presence has been previously reported in BSG [8], is a potential source of energy for gut microbes leading to the production of gas and SCFAs [62]. Amounts of 0.839 ± 0.107 μM of glucose and 0.742 ± 0.082 μM of glucose ($p > 0.05$) were found in BSG colonic-fermented digest (FBSG) and EBSG colonic-fermented digest (FEBSG), which might be associated with cellulose or resistant starch degradation. Moreover, arabinose and xylose monosaccharides present in BSG hemicellulose were quantified as arabinose by 0.802 ± 0.007 μM in FBSG and 0.650 ± 0.096 μM in FEBSG ($p > 0.05$), as both coelute in the chromatogram. In fact, the digestibility of cellulose and hemicellulose in a group of seven women was estimated at 70% and 72%, respectively, showing that there is extensive degradation of these polysaccharides in dietary plant cell wall material [63]. Cellulolytic strains isolated from human feces have been classified as *Ruminococcus* spp., *Clostridium* spp., *Eubacterium* spp. and *Bacteroides* spp. [63], some of which were identified in the fecal inoculum. Hemicellulose degrading activities were associated with *Bacteroides* species [63]. No fructose content was identified in FBSG and FEBSG.

Table 5. Microbial count in fecal inoculum.

	log (CFU/mL)	**Relative Percentage to Total Anaerobes**
Total anaerobes	9.575 ± 0.077 [d]	100.00%
Total aerobes	7.504 ± 0.035 [a,b]	0.842%
Enterobacteriaceae	7.435 ± 0.094 [a,b]	0.728%
Staphylococcus spp.	7.155 ± 0.045 [a]	0.377%
Lactic acid bacteria	8.354 ± 0.039 [b,c]	5.965%
Lactobacillus spp.	7.110 ± 0.067 [a]	0.342%
Clostridium spp.	9.357 ± 0.157 [d]	62.281%
Entereococcus spp.	7.385 ± 0.027 [a,b]	0.640%
Bifidobacteria	9.290 ± 0.061 [d,c]	51.754%

Different letters within the same column show significant differences (Δlog > 1). CFU: colony forming units.

Table 6. Short-chain fatty acids and organic acids present in FBSG and FEBSG.

	Short-Chain Fatty Acids (mM)	
	FBSG	FEBSG
Acetic acid	9.471 ± 0.139 [b,D]	8.185 ± 0.674 [a,C]
Propionic acid	5.808 ± 0.287 [a,C]	5.739 ± 0.158 [a,B]
Butyric acid	5.747 ± 0.142 [a,C]	5.551 ± 0.216 [a,B]
Isobutyric acid	0.302 ± 0.012 [a,A]	0.284 ± 0.001 [a,A]
Isovaleric acid	1.084 ± 0.048 [a,B]	1.017 ± 0.090 [a,A]
Valeric acid	1.008 ± 0.128 [a,B]	0.837 ± 0.208 [a,A]
Caproic acid	0.328 ± 0.036 [a,A]	0.298 ± 0.085 [a,A]
Total	23.748 ±0.181 [b]	21.895 ± 1.035 [a]

Different letters within the same line (lower case) and column (capital letters) show significant differences ($p < 0.05$). FSG: brewers' spent grain colonic fermented digest; FEBSG: extruded brewers' spent grain colonic-fermented digest.

Sugars fermentation results in acetate, propionate and butyrate as the main fermentation end-products [57,61]. These were the major SCFAs obtained for the fermentation of BSG and EBSG (Table 6). Acetate, propionate and butyrate are expected to exist in a molar ratio of 60:20:20, respectively [61], while our studies showed a ratio of 45:28:27 (1:0.62:0.60) in FBSG and 42:29:29 (1:0.69:0.69) in FEBSG. The relative proportion of each SCFA depends on the substrate, microbiota composition and gut transit time [61]. Significant differences ($p < 0.05$) for acetate content in FBSG and FEBSG may be related to the higher content of insoluble dietary fiber in unextruded BSG (48.57 ± 0.15% in dry weight basis) [8]. However, it has been found that extrusion increased the metabolism of dietary fibers from grains by gut microbiota in vitro using fecal microbiota and in vivo in rat feeding studies [64]. Furthermore, in vitro fermentation in a batch human fecal model showed ratios of acetic:propionic:butyric equal to 65:23:11, 1:0.4:0.1 and 1:0.3:0.3 for BSG arabinoxylans. The results confirmed that arabinose was consumed during fermentation of BSG-derived products [65,66]. The abundance of *Bacteroides* spp. is associated with the production of propionate and acetate, while butyrate is produced mainly by the Firmicutes phylum [61]. Moreover, several studies have reported the ability of selected strains of *Lactobacillus* and *Bifidobacterium* to ferment arabinoxylans, resulting in propionate production [65]. Different physiological effects have been associated with these SCFAs. Acetic acid can induce apoptosis of cancerous cells and prevent DNA oxidative damage caused by hydrogen peroxide in epithelial cells of the distal colon. Propionate has been shown to increase satiety and improve glucose homeostasis. The presence of butyrate has been related to anti-inflammatory and anticarcinogenic effects [14,63].

Different pathways are involved in the biosynthesis of these major SCFAs from dietary fiber fermentation once pyruvate is produced from sugars. Propionate can be synthesized by two pathways involving succinate and lactate as intermediates [56,57,61]. Although lactate or succinate presence was not identified in FBSG and FEBSG, these intermediates may have been further metabolized into major metabolites, such as propionate [57]. In addition, in previous studies, lactic acid production was noted during the first twelve fermentation hours, which was then depleted over time [65]. Propionate could also have been formed through the propanendiol pathway after microbial transformation of deoxy-sugars [61]. Although traces of rhamnose have been found in BSG [4], this is not the major type of monosaccharide present.

The digestibility of proteins by the host is more variable than that of carbohydrates which leads to different amino acid composition available to the gut microbiota [14,56]. Similar to fiber fermentation, protein fermentation produces SCFAs accompanied by branch-chained fatty acids (BCFAs), ammonia, amines, hydrogen sulfide, phenols and indoles [67]. BCFAs (isobutyric and isovaleric) were found in FBSG and FEBSG (Table 6) to a low extent. BCFAs are reliable markers of proteolytic fermentation as they are produced exclusively through the fermentation of branched-chain amino acids [67]. A significant amount of these amino acids, especially leucine, was previously identified in BSG and EBSG [8]. In fact, a significantly higher ($p < 0.05$) amount of isovaleric acid (formed by leucine fermentation) than isobutyric acid was found in FBSG. Little is known about the effect of BCFAs on the host's health. Previous studies suggested their capacity to modulate glucose and lipid metabolism in the liver as SCFAs. However, protein catabolism in the gut has potential negative implications, as toxic compounds can be released [56]. Nevertheless, carbohydrate presence alters amino acid utilization by microbes, reducing the uptake of some amino acids, such as tyrosine, and increasing the uptake of others, such as valine [67].

3.4. Colonic Bioaccessibility of Non-Nutrients

The results for bioaccessible polyphenols in the colon are presented in Table 3. The molar concentrations of phenolic compounds bioaccessible in the colon were 1.903 ± 0.101 and 2.009 ± 0.059 mM FA eq. As well as phenolic duodenal bioaccessibility, extrusion favored ($p < 0.05$) the release of bioactive compounds in the colon after microbial fermentation. To the best of our knowledge, there are no previous studies evaluating the effect of extrusion on phenolic colon bioaccessibility. Nevertheless, the extrusion of barley and oat increased the total tract bioaccessibility of dietary polyphenols in pigs [68].

The gut microbiota biotransforms polyphenols into metabolites that may have a greater biological activity than their precursor structures [15]. Microbiota enzymes can perform hydrolysis, dehydroxylation, demethylation and decarboxylation [52]. In the phenolic profile (Table 4), only benzoic acid and 2-(3-hydroxyphenyl) propionic acid were detected in FBSG and FEBSG, which may have been a result of microbial metabolism. Hydroxyphenyl propionic acids are a metabolite of phenolic acid microbial fermentation and are later transformed into benzoic acid [52]. Nevertheless, new phenolic metabolites might also have been formed after microbial metabolism, depending on the structure of polyphenols, that might have not been identified in this analysis.

Comparing the bioaccessibility at different digestion stages (Table 5), it was 32% higher in the colon than in the duodenum stage for both BSG and EBSG. The bioaccessibility of polyphenols present in cereal grains was 68% in the colon and 28% in the small intestine [69]. Most phenolic acids in BSG are bound to arabinoxylans by ester bonds [70], so microbial esterases are required during digestion to improve their bioaccessibility [15].

3.5. *Bioactivity of Bioaccessible Compounds*

3.5.1. Antioxidant Capacity

The antioxidant capacity of BSG and EBSG digests, after in vitro oral-gastro-intestinal digestion and colonic fermentation, is shown in Table 7. EBSG presented higher ($p < 0.05$) antioxidant activity measured by ABTS and ORAC after oral-gastro-intestinal digestion. This may have been a consequence of the higher ($p < 0.05$) content of phenolic compounds present in DEBSG than in DBSG (Table 3). A slight significant correlation was found for antioxidant capacity measured by ABTS and the total phenolic content of intestinal digests ($p < 0.05$, 0.72), while no correlation was found between their total phenolic content and antioxidant capacity as measured by ORAC. Furthermore, extruded apple pomace and sorghum have also shown higher antioxidant capacity after in vitro gastrointestinal digestion than the untreated material, measured by ORAC and ABTS, respectively [71,72]. Regarding the colonic-fermented digests, significant differences ($p < 0.05$) were observed for the antioxidant capacity of FBSG and FEBSG measured by ABTS, but not by ORAC. Additionally, colonic-fermented digests presented higher ($p < 0.05$) antioxidant capacity than intestinal digests.

Table 7. Antioxidant capacity for BSG and EBSG intestinal and colonic fermented digests.

	ABTS (mM FAeq)		**ORAC (mM FAeq)**	
	Duodenal	Colonic	Duodenal	Colonic
BSG	2.899 ± 0.158 [a]	1.967 ± 0.081 [a]	10.809 ± 0.109 [a]	1.624 ± 0.044 [a]
EBSG	3.232 ± 0.089 [b]	2.132 ± 0.088 [b]	13.000 ± 0.080 [b]	1.785 ± 0.053 [a]
	ABTS (µmol FAeq/g of digested sample)		ORAC (µmol FAeq/g of digested sample)	
	BSG	EBSG	BSG	EBSG
Duodenal	19.916 ± 1.170 [A]	23.900 ± 0.673 [A]	12.487 ± 0.344 [A]	16.950 ± 0.195 [A]
Colonic	25.354 ± 0.920 [B]	28.184 ± 1.157 [B]	21.262 ± 0.262 [B]	21.632 ± 0.210 [B]

Different letters (lower case for results in mM FA and capital letter for results in µmol FA/g) within the same column show significant differences ($p < 0.05$). BSG: brewers' spent grain; EBSG: extruded brewers' spent grain; FA: ferulic acid.

Furthermore, intestinal digests (DBSG and DEBSG) were tested for their ability to inhibit intracellular ROS formation in IEC-6 (Figure 1). Cells treated with DEBSG showed a significant reduction ($p < 0.05$) in the formation of intracellular ROS under physiological conditions, i.e., in DEBSG-treated cells, ROS formation was 18% less than in untreated cells (control). This effect was not observed in cells treated with DBSG. This is in accordance with the higher amount of phenolic compounds (Table 3) with antioxidant capacity (Table 7) in DEBSG than DBSG. However, addition of 10 µM ferulic acid resulted in higher ($p < 0.05$) inhibition (70%) than for DBSG and DEBSG, although a dose-dependent effect was not observed for the tested concentrations (10–100 µM). Nevertheless, the results might also have been due to a higher content of soluble protein in DEBSG (Table 1). It has been found previously that BSG protein hydrolysates inhibited oxidative stress by reducing intracellular ROS formation in a Caco-2 cell line [73].

3.5.2. Anti-Inflammatory Properties

NO formation in RAW264.7 macrophages was determined as an inflammation biomarker after stimulation with LPS and coadministration with DBSG, DEBSG and ferulic acid (Figure 2). While DEBSG showed anti-inflammatory capacity under assay conditions, DBSG did not. In line with previous studies, greater anti-inflammatory properties have been recorded for extruded sorghum than unextruded sorghum on LPS-stimulated

RAW264.7 [74]. Therefore, it may be inferred that the extrusion process favors the bioaccessibility of compounds with anti-inflammatory properties. This might be due to the greater ($p < 0.05$) release of phenolic compounds (Table 3) and soluble proteins (Table 1) in EBSG than in BSG after oral-gastro-intestinal digestion.

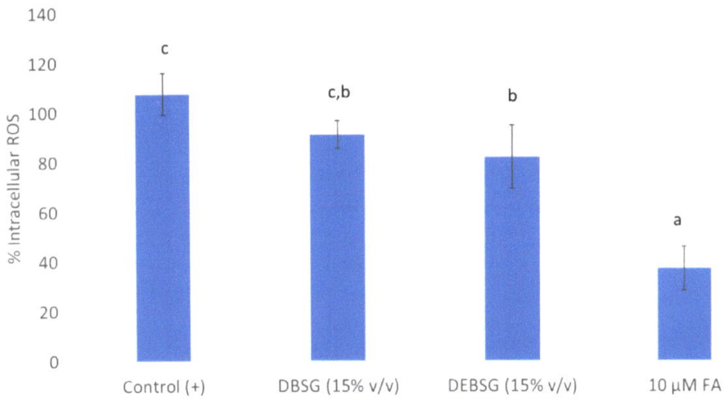

Figure 1. Intracellular ROS formation under physiological conditions in IEC-6 for untreated cells (Control (+)), cells treated with DBSG, cells treated with DEBSG and an antioxidant standard (FA). Percentage of ROS formation was calculated with respect to the Control (+). Different letters within columns show significant differences ($p < 0.05$). DBSG: brewers' spent grain intestinal digest; DEBSG: extruded brewers' spent grain intestinal digest; FA: ferulic acid.

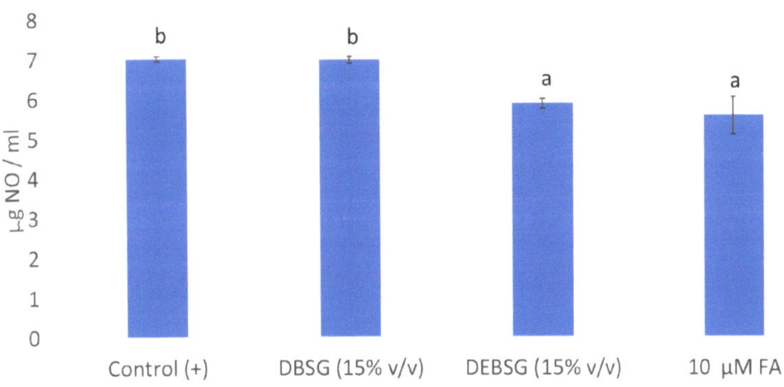

Figure 2. NO formation (μg/mL) in RAW 264.7 induced by LPS (1 μg/mL). Different letters within columns show significant differences ($p < 0.05$). DBSG: brewers' spent grain intestinal digest; DEBSG: extruded brewers' spent grain intestinal digest; FA: ferulic acid.

On the one hand, phenolic compounds are essential for the suppression of inflammation and their potent anti-inflammatory capacity has been shown [75]. A negative and significant correlation ($p < 0.001$, -0.94) was found between NO levels in cells treated with DBSG and DEBSG and their phenolic content measured by the Folin–Ciocalteu method. Furthermore, DEBSG exhibited statistically the same ($p > 0.05$) anti-inflammatory capacity than ferulic acid at the tested concentrations. Ferulic acid has already been reported to show anti-inflammatory effects inhibiting inducible nitric oxide synthase expression in LPS stimulated RAW264.7 cells, as well as NO production [76].

On the other hand, the anti-inflammatory effect of BSG protein hydrolysates has been previously described [77]. A negative significant correlation ($p < 0.05$, -0.73) was established between the soluble protein content of DBSG and DEBSG and NO levels. It was expected that, due to protein enzymatic hydrolysis during oral-gastro-intestinal digestion simulation, peptides might be solubilized into bioaccessible fractions which might be reflected in the BCA method results. Regarding mechanisms of action, studies have shown that protein hydrolysates limited proinflammatory cytokines and augmented anti-inflammatory IL-10 levels of spleen cells stimulated with LPS and Concanavalian A. Moreover, the involvement of both TLR2 and TLR4, and the main role of NFKB, in the immunomodulation by BSG hydrolysates has been demonstrated [78].

3.5.3. Antidiabetic Properties

Carbohydrase Activity

Dietary starch hydrolysis, by α-amylase and α-glucosidase, is the major source of glucose in blood. Inhibition of these enzymes can significantly decrease the postprandial increase of blood glucose levels and therefore can be an important strategy in the management of hyperglycemia linked to type II diabetes [39].

Table 8 shows results obtained for DBSG and DEBSG α-amylase, α-glucosidase and sucrase activity of rat intestine acetone powder, as well as the acarbose concentration, which caused 50% inhibition of these enzymes' activity (IC_{50}). To our knowledge, this is the first investigation that has characterized α-amylase activity of rat acetone intestine powder. Previous studies suggested that a considerable amount of amylase seems to be formed in the rat small-intestinal mucosa itself [79]. The acarbose IC_{50} value for α-amylase activity was of the same order as that found in previous studies for human α-amylase (34.1 ± 0.8 μg/mL) [44] under similar experimental conditions. BSG and EBSG exhibited low α-amylase inhibitory activity which corresponded to 6.12% ± 0.31% and 12.44% ± 0.08%, respectively. These findings are consistent with previous studies which showed 13.35% ± 1.21% inhibition for porcine pancreatic α-amylase for BSG protein hydrolysates [80]. Additionally, different types of BSG acetone extracts showed porcine pancreatic α-amylase inhibitory capacity of 23.1% ± 4.2% and 49.7% ± 12.3% [47]. Regarding DBSG and DEBSG's α-glucosidase inhibition capacity, it was significantly higher ($p < 0.05$) than that for α-amylase. BSG and EBSG intestinal digests showed strong α-glucosidase inhibition capacity equal to 57.71% ± 0.66% and 63.70% ± 1.04%, respectively. These results are in agreement with previous findings for BSG protein hydrolysates [80]. However, DBSG and DEBSG did not show sucrase inhibitory capacity under the assay experimental conditions.

Table 8. Inhibition of digestive carbohydrases.

	IC50 Acarbose	DBSG	DEBSG
α-amylase (μM Acarbose)	10.522	0.435 ± 0.010 ^{a,A}	0.689± 0.004 ^{b,A}
α-glucosidase (mM Acarbose)	1.002	3.005 ± 0.123 ^{a,B}	7.829 ± 1.560 ^{b,B}
Sucrase (μM Acarbose)	15.152	n.i.d.	n.i.d.

Different letters within the same row (lower case) and column (capital letters) show significant differences ($p < 0.05$). n.i.d means no inhibition determined; DBSG: brewers' spent grain intestinal digest; DEBSG: extruded brewers' spent grain intestinal digest.

The extrusion process of BSG showed a significant positive effect ($p < 0.05$) on its carbohydrase inhibition capacity. This might be due to the higher bioaccessibility ($p < 0.05$) of phenolic compounds (Table 3) and soluble protein (Table 1) present in EBSG. Previous findings suggested that ferulic acid, caffeic acid and isoferulic acid inhibited rat intestinal maltase activity in a mixed-inhibition manner, while no pancreatic α-amylase inhibition was found [81]. Furthermore, α-amylase and α-glucosidase inhibition effects of BSG protein hydrolysate have previously been reported [80,82]. Pearson correlation tests between α-glucosidase inhibition capacity and phenolic compounds or protein content in intestinal digests both resulted in a strong (Pearson coefficient ≥ 0.95) significant ($p < 0.05$) positive correlation. On the other hand, only intestinal digest protein content showed a significant

($p < 0.05$) correlation with α-amylase inhibition capacity, i.e., no correlation with phenolic content and α-amylase inhibition capacity of DBSG and DEBSG.

Glucose Absorption

Figure 3 shows glucose 0.8 mM DBSG and DEBSG glucose transport in IEC-6 over time. Glucose control absorption showed an acute peak at 45 min which decreased from 75 min onwards. Both, DBSG and DEBSG showed steadier glucose release into the basolateral side. Maximum glucose absorption for DBSG was registered after 90 min, while maximum glucose absorption for DEBSG was at 120 min. Therefore, use of BSG and EBSG may have implications for the prevention and management of chronic diseases such as type 2 diabetes. Carbohydrate-rich foods that promote sustained but low glucose levels, lower the glucose and insulin response and may have benefits for the metabolic control of diabetes and its complications [83]. Therefore, incorporating BSG as a food ingredient may contribute to satiety; glucose may play a role on the short-term satiety signal as maintained glucose plasma levels activate neuronal satiety pathways and contribute to satiety mechanisms by inhibiting gastric emptying [48].

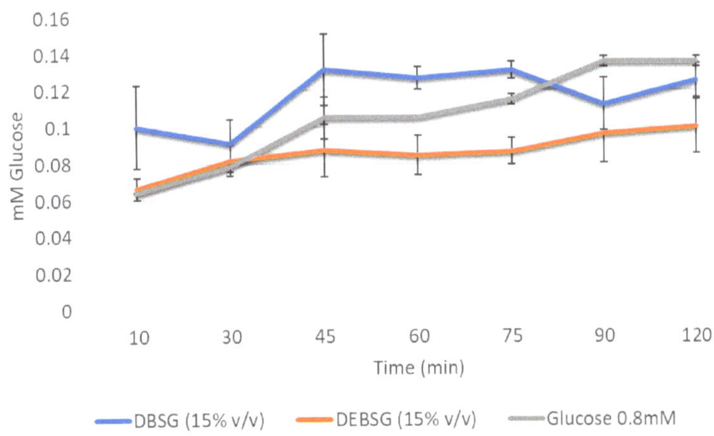

Figure 3. Brewers' spent grain intestinal digests (DBSG) and extruded brewers' spent grain intestinal digests (DEBSG) glucose absorption in IEC-6 cells over time.

Furthermore, the incremental areas under the curves (IAUC) presented in Figure 3 were 11.960 ± 0.058, 13.583 ± 0.386 and 10.129 ± 0.195 (mM × min) for glucose 0.8 mM, DBSG and DEBSG, respectively. There were significant differences ($p < 0.05$) between DEBSG's IAUC and those for DBSG and glucose control, perhaps due to the higher bioaccessibility of phenolic compounds with glucose transporter inhibition capacity in DEBSG than in DBSG.

Glucose Absorption Inhibition

After polysaccharides are digested by pancreatic and intestinal enzymes, the resulting glucose cannot penetrate the lipidic intestinal cell membrane, so specific transporters are required. Two types of glucose transporter have been identified in human intestinal cells, as well as in rat enterocytes: sodium-dependent glucose transporters (SGLT1), which mediate glucose uptake into the cell at a 1:2 ratio, with sodium ions in favor of its electrochemical gradient, and a facilitated diffusion system (GLUT2). Classic theories have suggested that SGLT1 located in the brush border membrane mediates the uptake of glucose inside the cell, while GLUT2 transports it across the basolateral membrane to the blood. However, recent studies of the rat jejunum have suggested that, at high concentrations, luminal glucose

(≥30 mM) saturates SGLT1 causing the translocation of GLUT2 into the apical membrane, increasing the capacity for glucose transport in the enterocyte [48,84–89].

Disturbed regulation of blood glucose levels can lead to hyperglycemia, which is a central problem in the pathophysiology of metabolic diseases, such as obesity, metabolic syndrome and type II diabetes [48]. Hence, controlling glucose levels after meals is a relevant strategy for the prevention and/or treatment of these chronic diseases [44].

Glucose transport levels in IEC-6 over time in the presence of different inhibitors (phloridzin, phloretin), ferulic acid (500 μM), DBSG and DEBSG are presented in Figure 4 (sodium-dependent conditions) and Figure 5 (sodium-free conditions). Furthermore, the incremental areas under the curve (IAUCs) are presented in Table 9.

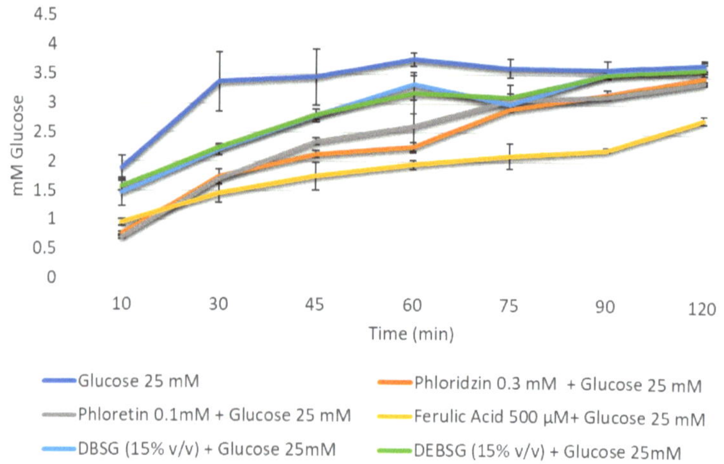

Figure 4. Glucose absorption in IEC-6 cells over time against reference inhibitors of glucose transporters (phloridzin and phloretin), ferulic acid, brewers' spent grain intestinal digest (DBSG) and extruded brewers' spent grain intestinal digest (DEBSG) under sodium-dependent conditions.

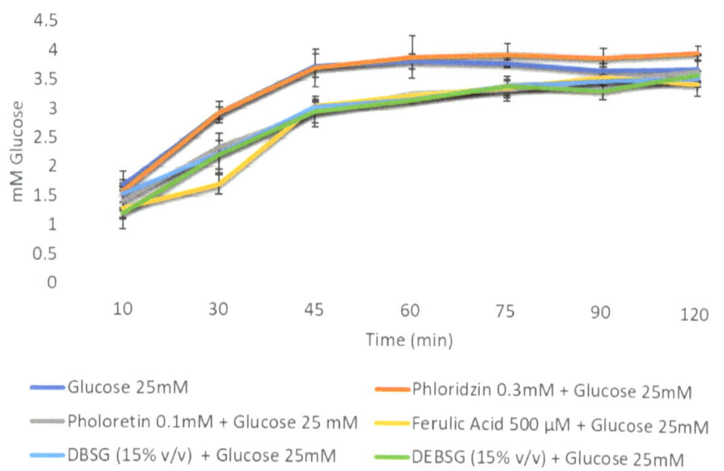

Figure 5. Glucose absorption in IEC-6 cells over time against reference inhibitors of glucose transporters (phloridzin and phloretin), ferulic acid, brewers' spent grain intestinal digest (DBSG) and extruded brewers' spent grain intestinal digest (DEBSG) under sodium-free conditions.

Table 9. Incremental area under the curve for glucose absorption inhibition models.

	IAUC Sodium Dependent Conditions (mM × min)	IAUC Sodium Free Conditions (mM × min)
Glucose 25 mM	377.153 ± 21.049 ^{d,A}	380.010 ± 8.862 ^{b,A}
Phloridzin 3 mM + Glucose 25 mM	272.741 ± 7.324 ^{b,A}	381.344 ± 20.569 ^{b,B}
Phloretin 1 mM + Glucose 25 mM	274.627 ± 6.602 ^{b,A}	334.700 ± 7.394 ^{a,B}
Ferulic acid 500μM + Glucose 25 mM	213.868 ± 11.630 ^{a,A}	321.482 ± 7.037 ^{a,B}
DBSG (15% *v/v*) + Glucose 25 mM	322.757 ± 1.854 ^{c,A}	336.658 ± 9.084 ^{a,A}
DEBSG (15% *v/v*) + Glucose 25 mM	324.793 ± 4.320 ^{c,A}	324.747 ± 10.224 ^{a,A}

Different letters within the same row (capital letters) or column (lower case) show significant differences ($p < 0.05$). DBSG: brewers' spent grain intestinal digest; DEBSG: extruded brewers' spent grain intestinal digest.

Under sodium-dependent conditions, it was expected that SGLT1 and GLUT2 transporters would be active, while under sodium-free conditions, only GLUT2 would be active. Phloridzin and phloretin, phloridzin aglycone, were used as standard inhibitors to check this. As calculated from Table 9, using Equation (4), phloretin inhibited glucose transport under sodium-dependent and sodium-free conditions by 27.3% and 12.1%, respectively. However, consistent with previous studies [90,91], no inhibition was observed for phloridzin under sodium-free conditions (Table 9), as this compound is supposed to inhibit SGLT1 transporter. Regarding ferulic acid, it presented a stronger inhibition of glucose transport under sodium-dependent conditions than under sodium-free conditions: 43.1% and 15.5%, respectively. Welsch et al. [90] reported a 38% ± 2.6% inhibition of glucose uptake under sodium-dependent conditions in rat intestinal brush border membrane vesicles, and Malunga et al. [92] suggested that ferulic acid inhibited glucose uptake in Caco-2 cells by interfering with GLUT2. Finally, to the best of our knowledge, this is the first time that brewers' spent grain and extruded brewers' spent grain intestinal digest glucose transport inhibition capacity has been studied. The results showed that DBSG and DEBSG exhibited the same glucose transport inhibition capacity as each other and under different experimental conditions (Table 9). Therefore, GLUT2 appears to have been the principal glucose transporter inhibited by DBSG and DEBSG. Antidiabetic properties have been associated with foods containing flavonoids and other phenolic compounds [88]. Li et al. [93] suggested that the mild glucose transport inhibition in Caco-2 cells caused by oat food digests may be due to a synergistic effect of all compounds present in the digests, including phenolic acids.

To better elucidate the inhibition mechanism, transporter kinetic parameters for IEC-6 glucose uptake and transport were determined for DEBSG and ferulic acid (Table 10). Regarding glucose uptake at the apical side, the glucose Km value was similar to that found by Zheng et al. [89] after 1 h of glucose starvation (18.5 ± 4.6 mM). Only ferulic acid showed a glucose uptake inhibition due to an increase of 61% in the Km and no statistical ($p > 0.05$) change in Vmax, owing to a competitive type of inhibition. However, both DEBSG and ferulic acid showed a non-competitive inhibition pattern for the transport of glucose in the basolateral side, the same type of inhibition as that reported for strawberry extracts [45]. Hence, different inhibition kinetics were found for DEBSG and ferulic acid. Further investigation is needed to establish which compounds present in DEBSG may be responsible for glucose transport inhibition.

Table 10. Glucose uptake and transport kinetics in IEC-6 cells in presence of EBSG and ferulic acid.

	Km (mM)	Vmax (mM/min)	Inhibition
	Uptake		
Control (Glucose)	19.490 ± 3.156	0.171 ± 0.027	n.a.
DEBSG (15% *v/v*)	16.273 ± 0.557	0.180 ± 0.029	No
Ferulic acid 500 μM	31.493 ± 3.646 *	0.336 ± 0.056	Competitive

Table 10. *Cont.*

	Km (mM)	Vmax (mM/min)	Inhibition
	Transport		
Control (Glucose)	128.427 ± 13.830	0.632 ± 0.003	n.a.
DEBSG (15% *v/v*)	97.726 ± 1.509	0.478 ± 0.015 *	Non-competitive
Ferulic acid 500 μM	133.578± 19.526	0.546 ± 0.021 *	Non-competitive

* Different letters within a column shows significant differences ($p < 0.05$) with the control value. n.a means not applicable; DEBSG: extruded brewers' spent grain intestinal digest.

4. Conclusions

In general, extruded BSG showed improved nutritional value and health-promoting properties compared to untreated BSG. The technological process did not affect glucose bioaccessibility or gluten digestibility ($p > 0.05$), while it favored amino acid release during digestion ($p < 0.05$). Acetate, propionate and butyrate, in a ratio of 45:28:27 and 42:29:29, were the major short-chain fatty acids generated by colonic fermentation of untreated BSG and extruded BSG dietary fiber, respectively. On the other hand, a higher ($p < 0.05$) gastrointestinal and colonic bioaccessibility of brewers' spent grain phenolic compounds was achieved with the extrusion process. Extruded brewers' spent grain intestinal digests inhibited glucose transport and showed higher ($p < 0.05$) α-amylase and α-glucosidase inhibition than untreated BSG. Moreover, lower ($p < 0.05$) glucose absorption was observed for extruded brewers' spent grain than for untreated brewers' spent grain. Furthermore, extruded brewers' spent grain intestinal digests inhibited intracellular ROS formation and had anti-inflammatory properties, while untreated brewers' spent grain did not. Therefore, extruded brewers' spent grain appears to represent an interesting sustainable ingredient with potential biological properties that might help treating and preventing non-communicable diseases, especially those dependent on energetic metabolism.

Supplementary Materials: The following supporting information can be downloaded at: https://www.mdpi.com/article/10.3390/nu14173480/s1. Figure S1: IEC-6 viability for DBSG and DEBSG. * shows significant differences with positive control (C+) ($p < 0.05$). Figure S2: IEC-6 viability for ferulic acid. Figure S3: RAW 264.7 viability for DBSG and DEBSG. * shows significant differences with positive control (C+) ($p < 0.05$). Figure S4: RAW264.7 viability for ferulic acid. Figure S5: Maltose released from potato starch due to α-amylase activity of rat intestinal extract. Figure S6: p-Nitrophenyl released due to α-glucosidase activity of rat intestinal extract. Figure S7: Glucose released from sucrose due to sucrose activity of rat intestinal extract.

Author Contributions: M.B.G.-B. performed the investigation, formal analysis, data curation, methodology, visualization, interpretation and global integration of the results, and writing of the original draft of the manuscript, as well as review and editing; S.C. contributed to writing review and the editing and interpretation of results; P.A. contributed to writing review and editing; M.D.d.C. contributed to methodology, the interpretation of the global results, supervision, writing review and editing, conceptualization and funding acquisition. All authors have read and agreed to the published version of the manuscript.

Funding: This research was funded by Agencia Nacional de Investigación e Innovación from Uruguay (POS_EXT_2018_1_154447) and by Ministerio de Ciencia e Innovación of Spain (PID2019-111510RB-I00).

Institutional Review Board Statement: The study was conducted according to the guidelines of the Declaration of Helsinki and approved by the Institutional Review Board (or Ethics Committee) of CSIC (protocol code 174/21, approved 2 February 2022).

Informed Consent Statement: Informed consent was obtained from all subjects involved in the study.

Data Availability Statement: The data presented in this study are available in the present article and in the supplementary material provided.

Acknowledgments: We are grateful to the Analysis Service Unit facilities of ICTAN and to the Servicio de Química de Proteinas of CIB for the analysis of chromatography and mass spectrometry. We thank the National Center of Biotechnology for gluten determination. Thanks are due to all the volunteers who have provided us their feces samples to be used in the in vitro colonic fermentation. We also want to thank our working colleagues from the Instituto de Investigacion en Ciencias de la Alimentacion Elena Bercianos for helping us with in vitro simulation of colonic fermentation, Alba Tamargo for doing the microbiota characterization and help during the writing and editing, and Amaia Iriondo-DeHond for the final writing revisions of the present study. Finally, thanks to Fábricas Nacionales de Cerveza for providing the raw material.

Conflicts of Interest: The authors declare no conflict of interest.

References

1. UN General Assembly, Transforming Our World: The 2030 Agenda for Sustainable Development, 21 October 2015, A/RES/70/1. Available online: https://www.refworld.org/docid/57b6e3e44.html (accessed on 22 August 2022).
2. De Foidmont-Goertz, I.; Faure, U.; Gajdzinska, M.; Haent Jens, W.; Krommer, J.; Lizaso, M.; Lutzeyer, H.-J.; Mangan, C.; Markakis, M.; Schoumacher, C.; et al. *Food 2030 Pathways for Action*; Fabbri, K., Ndongosi, I., Eds.; Publications Office: Luxembourg, 2020; ISBN 9789276181224.
3. WHO Noncommunicable Diseases. Available online: https://www.who.int/news-room/fact-sheets/detail/noncommunicable-diseases#:~{}:text=Keyfacts,-andmiddle-incomecountries (accessed on 8 June 2022).
4. Lynch, K.M.; Steffen, E.J.; Arendt, E.K. Brewers ' Spent Grain: A Review with an Emphasis on Food and Health. *Inst. Brew. Distill.* **2016**, *122*, 553–568. [CrossRef]
5. ReGrained. Available online: https://www.regrained.com/pages/about-us (accessed on 29 June 2022).
6. Grainstone. Available online: https://www.grainstone.com.au/ (accessed on 29 June 2022).
7. Jimenez, B.; Gopi, A.; Fitting, A.; Jimenez, E. Rise. Available online: https://www.riseproducts.co/ (accessed on 29 June 2022).
8. Gutiérrez-Barrutia, M.B.; del Castillo, M.D.; Arcia, P.; Cozzano, S. Feasibility of Extruded Brewer's Spent Grain as a Food Ingredient for a Healthy, Safe, and Sustainable Human Diet. *Foods* **2022**, *11*, 1403. [CrossRef] [PubMed]
9. AdedejAdekola, K.A. Engineering Review Food Extrusion Technology and Its Applications. *J. Food Sci. Eng.* **2016**, *6*, 149–168. [CrossRef]
10. Singh, S.; Gamlath, S.; Wakeling, L. Nutritional Aspects of Food Extrusion: A Review. *Int. J. Food Sci. Technol.* **2007**, *42*, 916–929. [CrossRef]
11. Björck, I.; Asp, N.G. The Effects of Extrusion Cooking on Nutritional Value—A Literature Review. *J. Food Eng.* **1983**, *2*, 281–308. [CrossRef]
12. Obradović, V.; Babic, J.; Šubarić, D.; Ačkar, D.; Jozinović, A. Improvement of Nutritional and Functional Properties of Extruded Food Products Improvement of Nutritional and Functional Properties of Extruded Food Products. *J. Food Nutr. Res.* **2014**, *53*, 189–206.
13. Grundy, M.M.; Edwards, C.H.; Mackie, A.R.; Gidley, M.J.; Butterworth, P.J.; Ellis, P.R. Re-Evaluation of the Mechanisms of Dietary Fi Bre and Implications for Macronutrient Bioaccessibility, Digestion and Postprandial Metabolism. *Br. J. Nutr.* **2016**, *116*, 816–833. [CrossRef]
14. Williams, B.A.; Grant, L.J.; Gidley, M.J.; Mikkelsen, D. Gut Fermentation of Dietary Fibres: Physico-Chemistry of Plant Cell Walls and Implications for Health. *Int. J. Mol. Sci.* **2017**, *18*, 2203. [CrossRef]
15. Ozdal, T.; Sela, D.A.; Xiao, J.; Boyacioglu, D.; Chen, F. The Reciprocal Interactions between Polyphenols and Gut Microbiota and Effects on Bioaccessibility. *Nutrients* **2016**, *8*, 78. [CrossRef]
16. Nignpense, B.E.; Francis, N.; Blanchard, C.; Santhakumar, A. Bioaccessibility and Bioactivity of Cereal Polyphenols: A Review Borkwei. *Foods* **2021**, *10*, 1595. [CrossRef]
17. Anson, N.M.; Mateo, N.; Havenaar, R.; Bast, A.; Haenen, G.R.M.M. Antioxidant and Anti-Inflammatory Capacity of Bioaccessible Compounds from Wheat Fractions after Gastrointestinal Digestion Antioxidant and Anti-Inflammatory Capacity of Bioaccessible Compounds from Wheat Fractions after Gastrointestinal Digestion. *J. Cereal Sci.* **2010**, *51*, 110–114. [CrossRef]
18. Xu, L.; Li, Y.; Dai, Y.; Peng, J. Natural Products for the Treatment of Type 2 Diabetes Mellitus: Pharmacology and Mechanisms. *Pharmacol. Res.* **2018**, *130*, 451–465. [CrossRef] [PubMed]
19. Zhao Zhang, X.Z.; Chen Yang, F.H.; Huang, Q. In Vitro Digestion and Fermentation by Human Fecal Microbiota of Polysaccharides from Flazseed. *Molecules* **2020**, *25*, 4354. [CrossRef]
20. Hollebeeck, S.; Borlon, F.; Schneider, Y.J.; Larondelle, Y.; Rogez, H. Development of a Standardised Human in Vitro Digestion Protocol Based on Macronutrient Digestion Using Response Surface Methodology. *Food Chem.* **2013**, *138*, 1936–1944. [CrossRef] [PubMed]
21. Martinez-Saez, N.; Hochkogler, C.M.; Somoza, V.; del Castillo, M.D. Biscuits with No Added Sugar Containing Stevia, Coffee Fibre and Fructooligosaccharides Modifies α-Glucosidase Activity and the Release of GLP-1 from HuTu-80 Cells and Serotonin from Caco-2 Cells after In Vitro Digestion. *Nutrients* **2017**, *9*, 694. [CrossRef] [PubMed]
22. Smith, P.; Krohn, R.; Hermanson, G.; Mallia, A.; Gartner, F.; Provenzano, M.; Fujimoto, E.K.; Olson, B.J.; Klenk, D. Measurement of Protein Using Bicinchoninic Acid. *Anal. Biochem.* **1985**, *150*, 76–85. [CrossRef]
23. Ketnawa, S.; Ogawa, Y. In Vitro Protein Digestibility and Biochemical Characteristics of Soaked, Boiled and Fermented Soybeans. *Sci. Rep.* **2021**, *11*, 14257. [CrossRef]

24. Mena, M.C.; Lombardía, M.; Hernando, A.; Méndez, E.; Albar, J.P. Comprehensive Analysis of Gluten in Processed Foods Using a New Extraction Method and a Competitive ELISA Based on the R5 Antibody. *Talanta* **2012**, *91*, 33–40. [CrossRef]
25. Shene, C.; Asenjo, J.A.; Paredes, P.; Flores, L.; Leyton, A.; Chisti, Y. Dynamic Flux Balance Analysis of Biomass and Lipid Production by Antarctic Thraustochytrid *Oblongichytrium* sp. RT2316—13. *Biotechnol. Bioeng.* **2020**, *117*, 3006–3017. [CrossRef]
26. Spackman, D.; Stein, W.; Moore, S. Automatic Recording Apparatus for Use in the Chromatography of Amino Acids. *Anal. Chem.* **1958**, *30*, 1190–1206. [CrossRef]
27. Martinez-Saez, N.; Fernandez-Gomez, B.; Cai, W.; Uribarri, J.; del Castillo, M.D. In Vitro Formation of Maillard Reaction Products during Simulated Digestion of Meal-Resembling Systems. *Food Res. Int.* **2019**, *118*, 72–80. [CrossRef] [PubMed]
28. Singleton, V.L.; Rossi, J.A. Colorimetry of Total Phenolics with Phosphomolybdic-Phosphotungstic Acid Reagents. *Am. J. Enol. Vitic.* **1965**, *16*, 144–158. [CrossRef]
29. Iriondo-Dehond, A.; García, N.A.; Fernandez-Gomez, B.; Guisantes-Batan, E.; Velázquez, F.; Patricia, G.; Ignacio, M.; Andres, S.; Sanchez-Fortun, S.; Dolores, M. Validation of Coffee By-Products as Novel Food Ingredients. *Innov. Food Sci. Emerg. Technol.* **2019**, *51*, 194–204. [CrossRef]
30. Navajas-Porras, B.; Pérez-Burillo, S.; Valverde-Moya, Á.; Hinojosa-Nogueira, D. Effect of Cooking Methods on the Antioxidant Capacity of Foods of Animal Origin Submitted to In Vitro. *Antioxidants* **2021**, *10*, 445. [CrossRef] [PubMed]
31. Pérez-Burillo, S.; Rufián-Henares, J.A.; Pastoriza, S. Towards an Improved Global Antioxidant Response Method (GAR+): Physiological-Resembling in Vitro Digestion-Fermentation Method Q. *Food Chem.* **2018**, *239*, 1253–1262. [CrossRef] [PubMed]
32. Pérez-Burillo, S.; Hinojosa-Nogueira, D.; Pastoriza, S.; Rufián-Henares, J. Plant Extracts as Natural Modulators of Gut Microbiota Community Structure and Functionality. *Heliyon* **2020**, *6*, e05474. [CrossRef]
33. Tamargo, A.; Cueva, C.; Laguna, L.; Moreno-arribas, M.V. Understanding the Impact of Chia Seed Mucilage on Human Gut Microbiota by Using the Dynamic Gastrointestinal Model Simgi ®®. *J. Funct. Foods* **2018**, *50*, 104–111. [CrossRef]
34. Hartemink, R.; Domenech, V.R.; Rombouts, F.M. Methods LAMVAB—A New Selective Medium for the Isolation of Lactobacilli from Faeces. *J. Microbiol. Methods* **1997**, *29*, 77–84. [CrossRef]
35. Re, R.; Pellegrini, N.; Proteggente, A.; Pannala, A.; Yang, M.; Rice-Evans, C. Antioxidant Activity Applying an Improved ABTS Radical Cation Decolorization Assay. *Free Radic. Biol. Med.* **1999**, *26*, 1231–1237. [CrossRef]
36. Iriondo-DeHond, A.; Ramírez, B.; Velazquez Escobar, F.; del Castillo, M.D. Antioxidant Properties of High Molecular Weight Compounds from Coffee Roasting and Brewing Byproducts. *Bioact. Compd. Health Dis.* **2019**, *2*, 48–63. [CrossRef]
37. Bakondi, E.; Gonczi, M.; Szabós, É.; Bai, P.; Pacher, P.; Gergely, P.; Kovács, L.; Hunyadi, J.; Szabó, C.; Csernoch, L.; et al. Role of Intracellular Calcium Mobilization and Cell-Density-Dependent Signaling in Oxidative-Stress-Induced Cytotoxicity in HaCaT Keratinocytes. *J. Investig. Dermatol.* **2003**, *121*, 88–95. [CrossRef] [PubMed]
38. Benayad, Z.; Martinez-Villaluenga, C.; Frias, J.; Gomez-Cordoves, C.; Es-Safi, N.E. Phenolic Composition, Antioxidant and Anti-Inflammatory Activities of Extracts from Moroccan Opuntia Ficus-Indica Flowers Obtained by Different Extraction Methods. *Ind. Crops Prod.* **2014**, *62*, 412–420. [CrossRef]
39. Lordan, S.; Smyth, T.J.; Soler-vila, A.; Stanton, C.; Ross, R.P. The a -Amylase and a -Glucosidase Inhibitory Effects of Irish Seaweed Extracts. *Food Chem.* **2013**, *141*, 2170–2176. [CrossRef] [PubMed]
40. Li, W.; Chuang, Y.; Hsieh, J. Characterization of Maltase and Sucrase Inhibitory Constituents from Rhodiola Crenulata. *Foods* **2019**, *8*, 540. [CrossRef]
41. Puthia, M.K.; Sio, S.W.S.; Lu, J.; Tan, K.S.W. Blastocystis Ratti Induces Contact-Independent Apoptosis, F-Actin Rearrangement, and Barrier Function Disruption in IEC-6 Cells. *Infect. Innmunity* **2006**, *74*, 4114–4123. [CrossRef]
42. Miller, R.D.; Monsul, N.T.; Vender, J.R.; Lehmann, J.C. NMDA- and Endothelin- 1 -Induced Increases in Blood-Brain Barrier Permeability Quantitated with Lucifer Yellow. *J. Neurol. Sci.* **1996**, *136*, 37–40. [CrossRef]
43. Jennis, M.; Cavanaugh, C.; Leo, G.; Mabur, J.; Lenhard, J.; Hornby, P. Derived Tryptophan Indoles Increase after Gastric Bypass Surgery and Reduce Intestinal Permeability in Vitro and in Vivo. *Neurogastroentology Motil.* **2017**, *30*, e13178. [CrossRef]
44. Fernández, A.; Iriondo-dehond, A.; Nardin, T.; Larcher, R.; Dellacassa, E.; Medrano, A.; del Castillo, M.D. In Vitro Bioaccessibility of Extractable Compounds from Tannat Grape Skin Possessing Health Promoting Properties with Potential to Reduce the Risk of Diabetes. *Foods* **2020**, *9*, 1575. [CrossRef]
45. Manzano, S.; Williamson, G. Polyphenols and Phenolic Acids from Strawberry and Apple Decrease Glucose Uptake and Transport by Human Intestinal Caco-2 Cells. *Mol. Nutr. Food Res.* **2010**, *54*, 1773–1780. [CrossRef]
46. Bohn, T.; Carriere, F.; Day, L.; Deglaire, A.; Egger, D.; Freitas, D.; Golding, M.; Le Feuntenun, S.; Macierzanka, A.; Menard, O.; et al. Correlation between in Vitro and in Vivo Data on Food Digestion. What Can We Predict with Static in Vitro Digestion Models? *Crit. Rev. Food Sci. Nutr.* **2017**, *58*, 2239–2261. [CrossRef]
47. Becker, D.; Bakuradze, T.; Hensel, M.; Beller, S.; Corral, C.; Richling, E. Influence of Brewer's Spent Grain Compounds on Glucose Metabolism Enzymes. *Nutrients* **2021**, *13*, 2696. [CrossRef] [PubMed]
48. Gromova, L.V.; Fetissov, S.O.; Gruzdkov, A.A. Mechanisms of Glucose Absorption in the Small Intestine in Appetite Regulation. *Nutrients* **2021**, *13*, 2474. [CrossRef] [PubMed]
49. ThermoFishcer Scientific Kit de Ensayo de Proteínas BCA PierceTM. Available online: https://www.thermofisher.com/order/catalog/product/23225 (accessed on 20 July 2022).
50. Kapoor, K.N.; Barry, D.T.; Rees, R.C.; Dodi, I.A.; Mcardle, S.E.B.; Creaser, C.S.; Bonner, P.L.R. Estimation of Peptide Concentration by a Modified Bicinchoninic Acid Assay. *Anal. Biochem.* **2009**, *393*, 138–140. [CrossRef] [PubMed]

51. Gorissen, S.H.M.; Crombag, J.J.R.; Senden, J.M.G.; Waterval, W.A.H.; Bierau, J. Protein Content and Amino Acid Composition of Commercially Available Plant—Based Protein Isolates. *Amino Acids* **2018**, *50*, 1685–1695. [CrossRef] [PubMed]
52. Wojtunik-Kulesza, K.; Oniszczuk, A.; Oniszczuk, T.; Combrzynski, M.; Nowakowska, D.; Matwijczuk, A. Influence of In Vitro Digestion on Composition, Bioaccessibility and Antioxidant Activity of Food. *Nutrients* **2020**, *12*, 1401. [CrossRef]
53. Acosta-Estrada, B.A.; Gutiérrez-Uribe, J.A.; Serna-Saldívar, S.O. Bound Phenolics in Foods, a Review. *FOOD Chem.* **2014**, *152*, 46–55. [CrossRef]
54. Herrera-Cazares, L.A.; Luzardo-Ocampo, I.; Ramírez-Jiménez, A.K.; Gutiérrez-Uribe, J.A.; Campos-Vega, R.; Gaytán-Martínez, M. Influence of Extrusion Process on the Release of Phenolic Compounds from Mango (*Mangifera indica* L.) Bagasse-Added Confections and Evaluation of Their Bioaccessibility, Intestinal Permeability, and Antioxidant Capacity. *Food Res. Int.* **2021**, *148*, 110591. [CrossRef]
55. Zeng, Z.; Liu, C.; Luo, S.; Chen, J.; Gong, E. The Profile and Bioaccessibility of Phenolic Compounds in Cereals Influenced by Improved Extrusion Cooking Treatment. *PLoS ONE* **2016**, *11*, e161086. [CrossRef]
56. Oliphant, K.; Allen-Vercoe, E. Macronutrient Metabolism by the Human Gut Microbiome: Major Fermentation by- Products and Their Impact on Host Health. *Microbiome* **2019**, *7*, 91. [CrossRef]
57. Bernalier-Donadille, A. Fermentative Metabolism by the Human. *Gastroentérologie Clin. Biol.* **2010**, *34*, S16–S22. [CrossRef]
58. Tamargo, A.; Cueva, C.; Silva, M.; Molinero, N.; Miralles, B.; Moreno-arribas, M.V. Gastrointestinal Co-Digestion of Wine Polyphenols with Glucose/Whey Proteins Affects Their Bioaccessibility and Impact on Colonic Microbiota Bego N. *Food Res. Int.* **2022**, *155*, 111010. [CrossRef] [PubMed]
59. Rinninella, E.; Raoul, P.; Cintoni, M.; Franceschi, F.; Abele, G.; Miggiano, D.; Gasbarrini, A.; Mele, M.C. What Is the Healthy Gut Microbiota Composition? A Changing Ecosystem across Age, Environment, Diet, and Diseases. *Microorganisms* **2019**, *7*, 14. [CrossRef] [PubMed]
60. Havensone, G.; Meija, L.; Lejnieks, A. Perspective: Physiological Benefits of Short-Chain Fatty Acids from Cereal Grain Fibre Fermentation and Metabolic Syndrome. *Proc. Latv. Acad. Sci.* **2020**, *74*, 65–67. [CrossRef]
61. Portincasa, P.; Bonfrate, L.; Vacca, M.; De Angelis, M.; Farella, I.; Lanza, E.; Khalil, M.; Wang, D.Q.; Sperandio, M.; De Ciaula, A. Gut Microbiota and Short Chain Fatty Acids: Implications in Glucose Homeostasis. *Int. J. Mol. Sci.* **2022**, *23*, 1105. [CrossRef]
62. Bui, A.T.; Williams, B.A.; Murtaza, N.; Lisle, A.; Mikkelsen, D.; Morrison, M.; Gidley, M.J. Wheat-Based Food Form Has a Greater Effect than Amylose Content on Fermentation Outcomes and Microbial Community Shifts in an in Vitro Fermentation Model. *Food Hydrocoll.* **2021**, *114*, 106560. [CrossRef]
63. Flint, H.J.; Scott, K.P.; Duncan, S.H.; Louis, P.; Forano, E. Microbial Degradation of Complex Carbohydrates in the Gut. *Gut Microbes* **2012**, *3*, 289–306. [CrossRef]
64. Smith, C.; Haute, M.J. Van Processing Has Differential Effects on Microbiota-Accessible Carbohydrates in Whole Grains during In Vitro Fermentation. *Appl. Environ. Microbiol.* **2020**, *86*, 1705–1720. [CrossRef]
65. Sajib, M.; Falck, P.; Sardari, R.R.R.; Mathew, S.; Grey, C.; Karlsson, E.N.; Adlercreutz, P. Valorization of Brewer' s Spent Grain to Prebiotic Oligosaccharide: Production, Xylanase Catalyzed Hydrolysis, in-Vitro Evaluation with Probiotic Strains and in a Batch Human Fecal Fermentation Model. *J. Biotechnol.* **2018**, *268*, 61–70. [CrossRef]
66. Reis, S.F.; Gullón, B.; Gullón, P.; Ferreira, S.; Maia, C.J.; Alonso, J.L.; Domingues, F.C.; Abu-ghannam, N. Evaluation of the Prebiotic Potential of Arabinoxylans from Brewer ' s Spent Grain. *Appl. Microb. Cell Physiol.* **2014**, *98*, 9365–9373. [CrossRef]
67. Diether, N.E.; Willing, B.P. Microbial Fermentation of Dietary Protein: An Important Factor in Diet—Microbe—Host Interaction. *Microorganisms* **2019**, *7*, 19. [CrossRef]
68. Hole, A.S.; Kjos, N.P.; Grimmer, S.; Kohler, A.; Lea, P.; Rasmussen, B.; Lima, L.R.; Narvhus, J.; Sahlstrøm, S. Extrusion of Barley and Oat Improves the Bioaccessibility of Dietary Phenolic Acids in Growing Pigs. *J. Agric. Food Chem.* **2013**, *61*, 2739–2747. [CrossRef] [PubMed]
69. Chitindingu, K.; Benhura, M.A.N.; Muchuweti, M. LWT—Food Science and Technology In Vitro Bioaccessibility Assessment of Phenolic Compounds from Selected Cereal Grains: A Prediction Tool of Nutritional Ef Fi Ciency. *LWT Food Sci. Technol.* **2015**, *63*, 575–581. [CrossRef]
70. Lynch, K.M.; Strain, C.R.; Johnson, C.; Patangia, D.; Stanton, C.; Koc, F.; Gil, J.; Patrick, M.; Aylin, O.R.; Paul, W.S.R.; et al. Extraction and Characterisation of Arabinoxylan from Brewers Spent Grain and Investigation of Microbiome Modulation Potential. *Eur. J. Nutr.* **2021**, *60*, 4393–4411. [CrossRef] [PubMed]
71. Liu, G.; Ying, D.; Guo, B.; Cheng, J.; May, B. Extrusion of Apple Pomace Increases Antioxidant Activity upon in Vitro Digestion. *Food Funct.* **2019**, *10*, 951–963. [CrossRef] [PubMed]
72. Salazar-López, N.J.; González-Aguilar, G.; Rouzaud-Sanchez, O.; Robles-Sánchez, M. Bioaccessibility of Hydroxycinnamic Acids and Antioxidant Capacity from Sorghum Bran Thermally Processed during Simulated in Vitro Gastrointestinal Digestion. *J. Food Sci. Technol.* **2018**, *55*, 2021–2030. [CrossRef]
73. Vieira, E.F.; Dias, D.; Carmo, H.; Ferreira, I.M.P.L.V.O. Protective Ability against Oxidative Stress of Brewers ' Spent Grain Protein Hydrolysates. *Food Chem.* **2017**, *228*, 602–609. [CrossRef] [PubMed]
74. Salazar Lopez, N.; Loarca-Piña, G.; Campos-Vega, R.; Martínez, M.G.; Moralez Sánchez, E.; Esquerra-Brauer, J.M.; Gonzalez-aguilar, G.A.; Robles-Sánchez, R.M. The Extrusion Process as an Alternative for Improving the Biological Potential of Sorghum Bran: Phenolic Compounds and Antiradical and Anti-Inflammatory Capacity. *Evid. Based Complement. Alterantive Med.* **2016**, *2016*, 8387975. [CrossRef]

75. Shahidi, F.; Yeo, J. Bioactivities of Phenolics by Focusing on Suppression of Chronic Diseases: A Review. *Int. J. Mol. Sci.* **2018**, *19*, 1573. [CrossRef]
76. Alam, A. Anti-Hypertensive Effect of Cereal Antioxidant Ferulic Acid and Its Mechanism of Action. *Front. Nutr.* **2019**, *6*, 121. [CrossRef]
77. Mccarthy, A.L.; Callaghan, Y.C.O.; Connolly, A.; Piggott, C.O.; Fitzgerald, R.J.; Brien, N.M.O. In Vitro Antioxidant and Anti-Inflammatory Effects of Brewers ' Spent Grain Protein Rich Isolate and Its Associated Hydrolysates. *Food Res. Int.* **2013**, *50*, 205–212. [CrossRef]
78. Cian, R.; Hernández-Chirlaque, C.; Gámez-Belmonte, R.; Drago, S.; Sánchez de Medina, F.; Martínez-Augustin, O. Molecular Action Mechanism of Anti-Inflammatory Hydrolysates Obtained from Brewers' Spent Grain Running. *J. Sci. Food Agric.* **2020**, *100*, 2880–2888. [CrossRef] [PubMed]
79. Dahlqvist, A.; Thomson, D. Separation and Characterization of Two Rat-Intestinal Amylases. *Biochem. J.* **1963**, *89*, 272–277. [CrossRef] [PubMed]
80. Connolly, A.; Connolly, A.; Piggott, C.O.; Fitzgerald, R.J. In Vitro α-Glucosidase, Angiotensin Converting Enzyme and Dipeptidyl Peptidase-IV Inhibitory Properties of Brewers' Spent Grain Protein Hydrolysates Food Research International In Vitro α - Glucosidase, Angiotensin Converting Enzyme and Dipeptidyl Pept. *Food Res. Int.* **2014**, *56*, 100–107. [CrossRef]
81. Adisakwattana, S.; Chantarasinlapin, P.; Thammarat, H.; Adisakwattana, S.; Chantarasinlapin, P.; Thammarat, H. A Series of Cinnamic Acid Derivatives and Their Inhibitory Activity on Intestinal α-Glucosidase A Series of Cinnamic Acid Derivatives and Their Inhibitory Activity on Intestinal α-Glucosidase. *J. Enzym. Inhib. Med. Chem.* **2009**, *6366*, 1194–1200. [CrossRef] [PubMed]
82. Lin, H.; Li, L.; Tian, Y.; Zhang, X.; Li, B. Protein Hydrolysate from Brewer ' s Spent Grain and Its Inhibitory Ability of α-Glucosidase. *Adv. Mater. Res.* **2012**, *582*, 138–141. [CrossRef]
83. Guo, Y.; Huang, Z.; Sang, D.; Gao, Q.; Li, Q. The Role of Nutrition in the Prevention and Intervention of Type 2 Diabetes. *Front. Bioeng. Biotechnol.* **2020**, *8*, 575442. [CrossRef] [PubMed]
84. Kellett, G.L.; Helliwell, P.A. Glucose-Induced Recruitment of GLUT2 to the Brush-Border Membrane. *Biochem. J.* **2000**, *162*, 155–162. [CrossRef]
85. Helliwell, P.A.; Richardson, M.; Affleck, J.; Kellett, G.L. Stimulation of fructose transport across the intestinal brush-border membrane by PMA is mediated by GLUT2 and dynamically regulated by protein kinase C. *Biochem J.* **2000**, *154*, 149–154. [CrossRef]
86. Helliwell, P.A.; Rumsby, M.G.; Kellett, G.L. Intestinal Sugar Absorption Is Regulated by Phosphorylation and Turnover of Protein Kinase C βII Mediated by Phosphatidylinositol 3-Kinase- and Mammalian Target of Rapamycin-Dependent Pathways *. *J. Biol. Chem.* **2003**, *278*, 28644–28650. [CrossRef]
87. Nistor, L.A.; Martineau, L.C.; Benhaddou-andaloussi, A.; Arnason, J.T.; Lévy, É.; Haddad, P.S. Inhibition of Intestinal Glucose Absorption by Anti-Diabetic Medicinal Plants Derived from the James Bay Cree Traditional Pharmacopeia. *J. Ethnopharmacol.* **2010**, *132*, 473–482. [CrossRef]
88. Koepsell, H. Glucose Transporters in the Small Intestine in Health and Disease. *Pflügers Arch. Eur. J. Physiol.* **2020**, *472*, 1207–1248. [CrossRef] [PubMed]
89. Zheng, Y.; Jeffrey, S.; Duenes, B.; Sarr, M. Mechanisms of Glucose Uptake in Intestinal Cell Lines: Role of GLUT2. *Natl. Institutes Health* **2013**, *151*, 13–25. [CrossRef]
90. Welsch, C.A.; Lachahce, P.A.; Wasserman, A.P. Nutrient Uptake and Transport Dietary Phenolic Compounds: Inhibition of Na-Dependent o-Glucose Uptake in Rat Intestinal Brush Border Membrane Vesicles. *Am. Inst. Nutr.* **1989**, *119*, 1698–1704. [CrossRef]
91. Johnston, K.; Sharp, P.; Clifford, M.; Morgan, L. Dietary Polyphenols Decrease Glucose Uptake by Human Intestinal Caco-2 Cells.Pdf. *Fed. Eur. Soc. Bioquímicas* **2005**, *579*, 1653–1657. [CrossRef]
92. Malunga, L.N.; Eck, P.; Beta, T. Inhibition of Intestinal-Glucosidase and Glucose Absorption by Feruloylated Arabinoxylan Mono- and Oligosaccharides from Corn Bran and Wheat Aleurone. *J. Nutr. Metab.* **2016**, *2016*, 1932532. [CrossRef]
93. Li, M.; Koecher, K.; Hansen, L.; Ferruzzi, M.G. Phenolics from Whole Grain Oat Products as Modifiers of Starch Digestion and Intestinal Glucose Transport. *J. Agric. Food Chem.* **2017**, *65*, 6831–6839. [CrossRef]

nutrients

MDPI

Article

Shibi Tea (*Adinandra nitida*) and Camellianin A Alleviate CCl_4-Induced Liver Injury in C57BL-6J Mice by Attenuation of Oxidative Stress, Inflammation, and Apoptosis

Ruohong Chen [1,†], Yingyi Lian [2,†], Shuai Wen [1], Qiuhua Li [1], Lingli Sun [1], Xingfei Lai [1], Zhenbiao Zhang [1], Junquan Zhu [3], Linsong Tang [4], Ji Xuan [5], Erdong Yuan [2,*] and Shili Sun [1,*]

1 Tea Research Institute, Guangdong Academy of Agricultural Sciences/Guangdong Key Laboratory of Tea Resources Innovation & Utilization, Guangzhou 510640, China; chenruohong@tea.gdaas.cn (R.C.); wenshuai@tea.gdaas.cn (S.W.); liqiuhua@tea.gdaas.cn (Q.L.); sunlingli@tea.gdaas.cn (L.S.); laixingfei@tea.gdaas.cn (X.L.); zhangzhenbiao@tea.gdaas.cn (Z.Z.)

2 School of Food Science and Engineering, Guangdong Province Key Laboratory for Green Processing of Natural Products and Product Safety, Engineering Research Center of Starch and Plant Protein Deep Processing, Ministry of Education, South China University of Technology, Guangzhou 510641, China; lyylianyingyi@163.com

3 Guangdong Society of Plant Protection, Guangzhou 510640, China; 13535904444@163.com

4 Taihongyuan Agriculture Co., Ltd., Xinyi, Maoming 525000, China; lisotong@sjbutton.com

5 Hospital of South China University of Technology, Guangzhou 510641, China; xuanji@scut.edu.cn

* Correspondence: erdyuan@scut.edu.cn (E.Y.); sunshili@zju.edu.cn (S.S.); Tel.: +86-20-8711-04218 (E.Y.); +86-20-8516-1045 (S.S.)

† These authors contributed equally to this work.

Citation: Chen, R.; Lian, Y.; Wen, S.; Li, Q.; Sun, L.; Lai, X.; Zhang, Z.; Zhu, J.; Tang, L.; Xuan, J.; et al. Shibi Tea (*Adinandra nitida*) and Camellianin A Alleviate CCl_4-Induced Liver Injury in C57BL-6J Mice by Attenuation of Oxidative Stress, Inflammation, and Apoptosis. *Nutrients* **2022**, *14*, 3037. https://doi.org/10.3390/nu14153037

Academic Editors: Maria Digiacomo, Doretta Cuffaro and Pietro Vajro

Received: 28 May 2022
Accepted: 16 July 2022
Published: 24 July 2022

Abstract: Liver injury is a significant public health issue nowadays. Shibi tea is a non-*Camellia* tea prepared from the dried leaves of *Adinandra nitida*, one of the plants with the greatest flavonoid concentration, with Camellianin A (CA) being the major flavonoid. Shibi tea is extensively used in food and medicine and has been found to provide a variety of health advantages. The benefits of Shibi tea and CA in preventing liver injury have not yet been investigated. The aim of this study was to investigate the hepatoprotective effects of extract of Shibi tea (EST) and CA in mice with carbon tetrachloride (CCl_4)-induced acute liver injury. Two different concentrations of EST and CA were given to model mice by gavage for 3 days. Treatment with two concentrations of EST and CA reduced the CCl_4-induced elevation of the liver index, liver histopathological injury score, alanine aminotransferase (ALT), and aspartate aminotransferase (AST). Western blotting and immunohistochemical analysis demonstrated that EST and CA regulated the oxidative stress signaling pathway protein levels of nuclear factor E2-related factor 2 (Nrf2)/heme-oxygenase-1 (HO-1), the expression of inflammatory cytokines, the phosphorylated nuclear factor-kappaB p65 (p-NF-κB)/nuclear factor-kappaB p65 (NF-κB) ratio, the phospho-p44/42 mitogen-activated protein kinase (p-MAPK), and the apoptosis-related protein levels of BCL2-associated X (Bax)/B cell leukemia/lymphoma 2 (Bcl2) in the liver. Taken together, EST and CA can protect against CCl_4-induced liver injury by exerting antioxidative stress, anti-inflammation, and anti-apoptosis.

Keywords: *Adinandra nitida* (Theaceae); Camellianin A; liver injury; oxidative stress; inflammation; anti-apoptosis

1. Introduction

The liver is a key controller of many physiological processes by mediating the synthesis and metabolism of endogenous compounds and participating in biological functions such as storage of liver glycogen, synthesis of secretory proteins, bile secretion, regulation of hematopoiesis, immune responses, and the metabolism of many existing exogenous compounds, including drugs, alcohols, and toxins [1,2]. The liver has a unique ability to regenerate and can fully recover from the most severe nonrecurrent conditions. However, a variety of conditions, including viral hepatitis, non-alcoholic fatty liver disease, chronic alcohol abuse, and long-term drug use, can lead to persistent liver injury, resulting in liver scarring and cirrhosis, which eventually progress to dysfunction and, later, hepatocellular carcinoma [3]. With the deterioration of the environment and the abuse of chemicals, liver injury has become a widespread disease with high morbidity [4].

Reactive oxygen species (ROS) are perceived to be molecular secondary messengers and play an important role in cell signal transduction and physiological processes [5]. However, when ROS are not cleared promptly, they can lead to intracellular protein, lipid, and nucleic acid destruction during pathological processes [6]. Due to the liver tissue usually being susceptible to pathological cascades of oxidative stress, inflammatory apoptotic response and excessive ROS generation have been postulated to trigger the formation of liver injury [7]. CCl_4 is a chemical widely used in the laboratory to induce experimental acute liver injury, and its toxicity is mainly due to the generation of ROS [8] and the p-NF-κB [9], then leading to organ damage. The natural active ingredient flavonoids were found to have some hepatoprotective effects in experimental liver injury models in animals [10]. Silymarin is the most well-studied hepatoprotective flavonoid, which showed good hepatoprotective effects in CCl_4-induced experimental liver injury models and has been applied in the treatment of liver injury [11]. Therefore, the exploration of safe, natural, effective efficacious flavonoid fractions of plant origin with hepatoprotective activity in a model for liver injury induced by intraperitoneal injection of CCl_4 is essential to develop measures to elucidate the mechanism of hepatoprotective action.

Shibi tea, also known as Shiya tea, is a traditional non-*Camellia* Chinese tea prepared from *Adinandra nitida* leaves, which are high in flavonoids. Shibi tea is a unique wild tea that has been used as a health tea and herbal medicine in Southeast Asia, including China, for hundreds of years [12]. It is mainly grown in the cool, moist, and high-altitude (above 500 m) cliffs in southern China, and in recent years, it has also been cultivated on a large scale in Guangdong and Guangxi [13]. Shibi tea is not only tasty and sweet, but also rich in flavonoids (up to 28.4%), and it has been found to have antioxidant and antibacterial effects and prevent peptic ulcers [14,15]. Shibi tea is one of the plants with the highest flavonoid content. Among them, CA, a flavonoid glycoside with apigenin as its parent nucleus, is its main flavonoid (nearly 60%) [16]. However, the advantages of Shibi tea and CA in the prevention of liver injury are still unknown. In this study, we successfully established CCl_4-induced liver injury in C57BL-6J mice and explored the anti-inflammatory, anti-apoptosis, and antioxidative effects of EST and the main flavonoid, CA, in repairing acute liver injury. We have established the hepatoprotective effects of Shibi tea and the functional ingredient CA in vivo. Consequently, we propose that Shibi tea might serve as a functional beverage with hepatoprotective effects, and emphasize CA as a novel natural plant-souse hepatoprotective ingredient that plays a significant role in the alleviation of liver injury.

2. Materials and Methods

2.1. Chemicals and Reagents

Shibi tea was obtained from Zhengqi Agricultural Development Co., Ltd. in Yingde, and from Taihongyuan Agriculture Co., Ltd. in Xinyi, Maoming, Guangdong, China. The EST was prepared and the component analysis of EST was realized in our previous study [17]. CA (HPLC $\geq$ 98%) was obtained according to our previous study [16]. The 4% paraformaldehyde (BL539A) was purchased from Biosharp (Anhui, China).

2.2. Establishment of Murine Liver Injury Model

Male C57BL/6 mice (7 weeks old) were purchased from the Beijing Huafukang Bioscience Co. Ltd. (Beijing, China). The mice were housed at room temperature (22 $\pm$ 2 °C) with 60% $\pm$ 15% humidity on a 12 h light/dark cycle, with free access to deionized water and basic feed. The mice were acclimatized for 1 week, and then randomly divided into the following seven groups (n = 7 each): control group (Control), untreated CCl_4 model (Model),

CCl_4 + 100 mg/kg silymarin (Positive), CCl_4 + 30 mg/kg CA (L-CA), CCl_4 + 100 mg/kg CA (H-CA), CCl_4 + 200 mg/kg EST (L-EST), and CCl_4 + 700 mg/kg EST (H-EST). Silymarin, EST, and CA were dissolved in a 0.5% CMC-Na solution. Except for the Control group, the acute liver injury model was induced by intraperitoneal injection of 0.2 mL/kg CCl_4 (dissolved in maize oil) to each animal in the remaining groups, and the control group was injected with an equal amount of maize oil as a negative control. Two hours after intraperitoneal injection, gavage treatment was performed according to the dose administered to each group for three days. The Control and Model groups received the same volume of 0.5% CMC-Na solution.

2.3. Tissue Processing

Two hours after the 3-day treatment, the body weights of mice were measured. Then, the mice were anaesthetized with 40 mg/kg pentobarbital and euthanized by cervical dislocation. Whole blood was collected in heparinized tubes, and the sera were separated by centrifuging at 3000 rpm for 10 min. The wet weight of liver tissue was measured and collected for further analysis. The liver and spleen index were calculated as follows: liver or spleen index (%) = liver or spleen wet weight (mg)/mouse body weight (mg) × 100%. After weighing, the liver tissue was divided into two parts; the intact liver lobules were fixed in 4% paraformaldehyde solution for the preparation of liver tissue sections, and the remaining liver tissue was stored at −80 °C for subsequent analysis.

2.4. Biochemical Analysis

The serum levels of AST (C010-2-1) and ALT (C009-2-1) were measured using commercially available kits from Nanjing Jiancheng Bioengineering Institute (Nanjing, China) according to the manufacturer's instructions. The liver tissues were fully homogenized by a homogenizer (OMNI Bead Ruptor 24) after adding saline in the ratio of weight (g) to volume (mL) of 1:9, and the homogenate was centrifuged at 4 °C for 10 min at 2500 rpm. The supernatant was taken and the protein content in the supernatants was measured using the Pierce BCA protein assay kit (Thermo VK312556). The ROS levels in liver were measured using an ELISA kit (MM-43700MA) purchased from Jiangsu Meimian (Yancheng, China). The levels of malondialdehyde (MDA) (A003-1-2), glutathione peroxidase (GPx) (H545-1-1), catalase (CAT) (A007-1-1), glutathione (GSH) (A006-2-1), and superoxide dismutase (SOD) (A001-3-2) in the liver were measured using specific assay kits according to the manufacturer's instructions (Nanjing Jiancheng Bioengineering Institute).

2.5. Histological Evaluation

The fixed liver tissues were dehydrated with 70% ethanol for 24 h and further embedded in paraffin. Paraffin-embedded liver samples were sectioned (thickness approximately 2 μm) and stained with hematoxylin and eosin (HE stain) using a commercial kit (C0105S) from Beyotime (Shanghai, China) as the standard protocol. The sections were sealed with neutral gum and observed under a microscope (Olympus, Tokyo, Japan, 100 X), and a histopathological assessment was performed as previously described [18].

The severity of the liver injury is graded on a scale of 1 to 5 depending on the degree of cellular necrosis, coagulum, central area, and surrounding inflammatory infiltrate: a score of 0 indicates normal, 1 indicates very low (<1%), 2 indicates mild (1–25%), 3 indicates moderate (26–50%), 4 indicates moderate/severe (51–75%), and 5 indicates severe/high (76–100%) [19,20]. The injury score was averaged for each group of animals.

2.6. Western Blotting

The liver tissues (20 mg) were homogenized by a homogenizer in 180 μL Radio-Immunoprecipitation assay (RIPA) lysis buffer (P0013B, Beyotime, Shanghai, China) supplemented with 2 μL phenylmethanesulfonyl fluoride (PMSF). The tissue lysates further lysed were incubated on ice for 1 h and then centrifuged at 13,200 rpm at 4 °C for 5 min. The protein content was measured using the Pierce BCA protein assay kit (Thermo VK312556).

An equal amount of protein per sample was collected and boiled for Western blotting analysis. Western blotting was performed as previously described [21]. The antibodies used in this work included Nrf2 (12721S, CST, Danvers, MA, USA), HO-1 (43966S, CST), β-actin (A1978, Sigma, MO, USA), NF-κB (8242S, CST), p-NF-κB (Ser536) (13346, CST), p-MAPK (3510, CST), Bcl2 (ab117115, Abcam, Cambridge, U.K.), and Bax (ab32503, Abcam). Positive signals were visualized using a chemiluminescence (ECL) analysis kit (170-5061, Bio-Rad, Hercules, CA, USA) and recorded with the Chemi Doc system (Bio-Rad, USA). The positive bands were quantified by densitometry using ImageJ software (Bethesda, Rockville, MD, USA) and normalized to the density of β-actin.

2.7. Immunohistochemical (IHC)

Immunohistochemical analysis was performed using the 3,3N-Diaminobenzidine Tertrahydrochloride (DAB) horseradish peroxidase color development kit (P0203, Beyotime) as previously described [22]. The sections were incubated with primary antibodies targeting tumor necrosis factor-α (TNF-α) (ab6671, Abcam, Cambridge, U.K.), interleukin-6 (IL-6) (SC-1265, SantaCruz, Dallas, TX, USA), Interleukin-1beta (IL-1β) (bs-0812R, Bioss, Woburn, MA, USA). After being mounted with neutral resin, the sections were observed under a light microscope (Olympus, Japan) and analyzed using the ImageJ software.

2.8. Statistical Analysis

The experimental results are presented as mean ± standard deviation (mean ± SD). Experimental data and graphs were statistically analyzed and plotted using GraphPad Prism 7 software for Windows (GraphPad Software Inc., San Diego, CA, USA). One-way analysis of variance (ONE WAY-ANOVA) was used to compare the significant differences between all groups, and all experiments were performed with at least three independent replications. $p < 0.05$ (* compared with the Model group/# compared with the Control group) and $p < 0.01$ (** compared with the Model group/ ## compared with the Control group) were considered statistically significant.

3. Results

3.1. EST and CA Ameliorated the CCl_4-Induced Liver Injury

As shown in Figure 1a,b, the liver and spleen indices of mice in the model group were significantly higher ($p < 0.01$) and decreased in mice treated with silymarin, CA, and EST. Among them, the liver index of L-CA and L-EST was comparable to that of the control group. To assess the effect of EST and CA on liver injury, the levels of AST and ALT in serum and liver tissues were detected, respectively. Compared with the control group, the AST and ALT of the model group were significantly increased ($p < 0.01$). Consistent with the gross observations, the AST (Figure 1c,e) and ALT (Figure 1d,f) levels of the mice treated with EST and CA were significantly reduced.

Histological evaluation was performed to visualize the extent of liver tissue injury. As shown in Figure 2a, the hepatocytes of the control mice had normal morphology, regular and tight arrangement, and the structure of liver lobules was clear. After the CCl_4 was induced, the liver tissues of mice in the model group showed obvious pathological changes, which were manifested by obvious inflammatory cell infiltration and disorder of hepatocyte arrangement. In comparison to the model group, treatment with the positive drugs silymarin, EST (200 and 700 mg/kg), and CA (30 and 100 mg/kg) alleviated the inflammatory cell infiltration induced by CCl_4.. Furthermore, the liver injury score showed that EST and CA significantly reduced liver injury ($p < 0.01$) and restored normal cell morphology (Figure 2b). In addition, we detected the collagen fiber content of liver tissues in different treatment groups by using Masson trichrome staining. Masson staining of the model group showed increased collagen staining. Both CA and EST ameliorated the histopathological lesions in the CCL4-induced model group of mice (Figure S1).

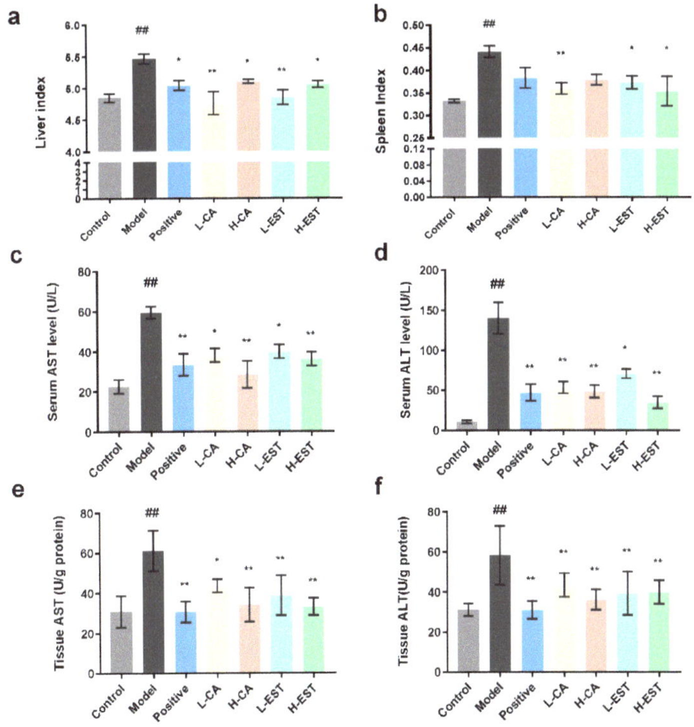

Figure 1. EST and CA ameliorated the CCl_4-induced liver injury: (**a**) the liver index, (**b**) the spleen index of different groups, (**c**) serum AST, (**d**) serum ALT, (**e**) liver tissue AST, and (**f**) liver tissue ALT level of each group. Data are expressed as mean $\pm$ SD of at least three independent experiments ($n \geq 5$). ## $p < 0.01$ compared with the control group; * $p < 0.05$, ** $p < 0.01$ compared with the model group.

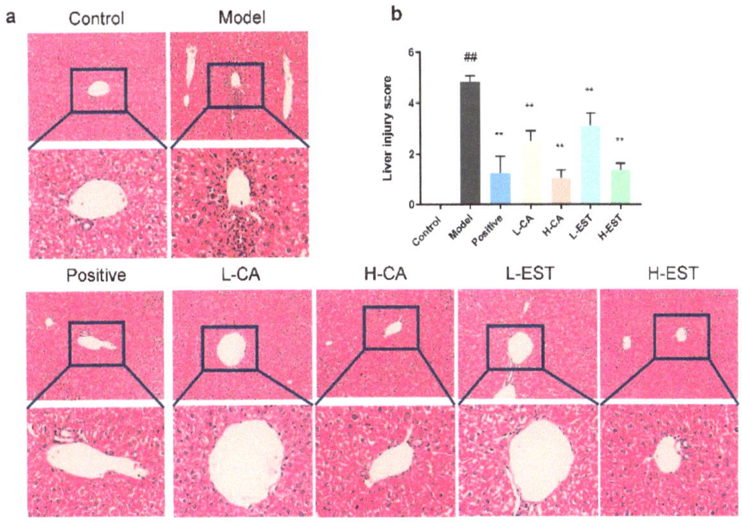

Figure 2. (**a**) Histopathological liver tissue (100×), (**b**) the liver injury score of mice in different groups. Data are expressed as mean $\pm$ SD of at least three independent experiments ($n \geq 5$). ## $p < 0.01$ compared with the control group, and ** $p < 0.01$ compared with the model group.

3.2. EST and CA Decrease the CCl4-Induced Oxidative Stress by Activating the Nrf2/HO-1 Pathway

Oxidative stress is an important factor contributing to CCl_4-induced liver injury. To evaluate the antioxidant effect of the EST and CA, we analyzed the levels of MDA, ROS, SOD, CAT, GSH-Px, and GSH in the liver tissues. As shown in Figure 3, EST and CA significantly reduced MDA ($p < 0.01$) and ROS ($p < 0.05$) levels and increased those of endogenous antioxidants such as SOD, CAT, GSH-Px, and GSH by varying degrees. These data indicate that both EST and CA restored antioxidant enzyme activity in the liver at different degrees compared with the model group.

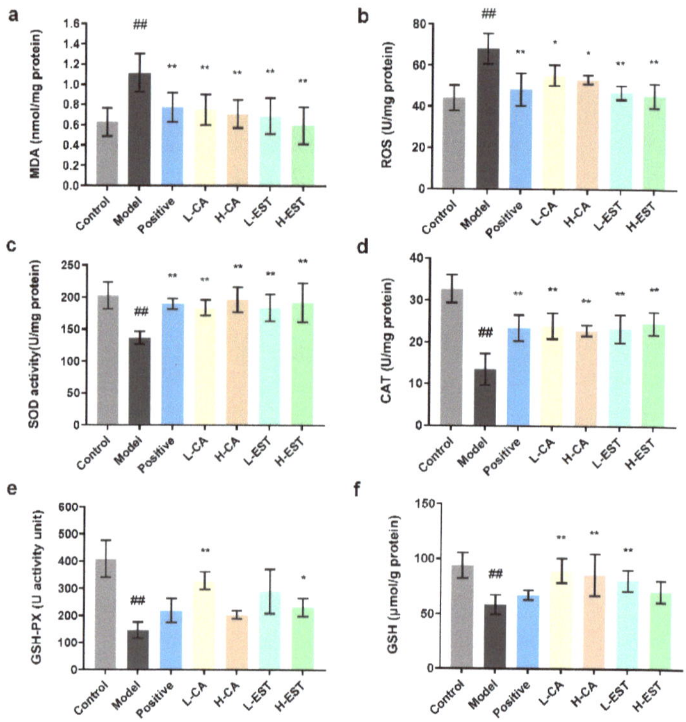

Figure 3. (**a**) MDA, (**b**) ROS level, (**c**) SOD, (**d**) CAT, (**e**) GSH-Px activities, and (**f**) GSH content of liver tissues in each group. Data are expressed as mean ± SD of at least three independent experiments ($n \geq 5$). ## $p < 0.01$ compared with the control group; * $p < 0.05$, ** $p < 0.01$ compared with the model group.

To explore whether the Nrf2/HO-1 signaling pathway was involved in the hepatoprotective effects of EST and CA, we examined the protein expression levels of Nrf2 and HO-1 in liver tissues of different treatment groups by Western blot analysis. As shown in Figure 4, both Nrf2 ($p < 0.01$) and its downstream HO-1 protein expression were increased in the model group compared to the normal group, and the expression was further increased after EST or CA treatment. These results indicate that EST and its main active component CA activate the Nrf2/HO-1 signaling pathway to exert hepatic antioxidant protective effects.

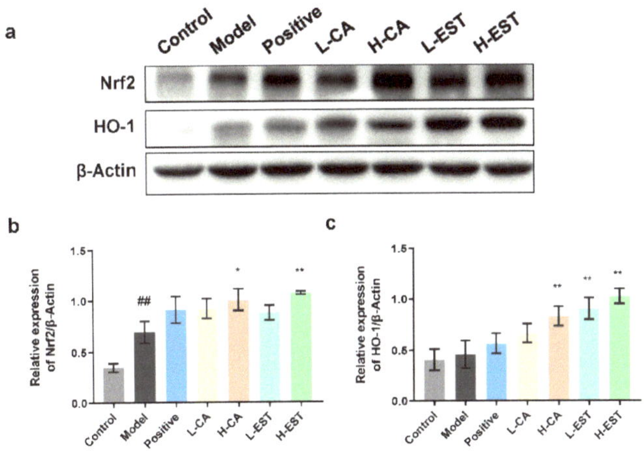

Figure 4. EST and CA decrease the CCl_4-induced oxidative stress by activating the Nrf2/HO-1. (**a**) Western blotting analysis of Nrf2 and HO-1 expression in liver tissues and densitometric quantification of Nrf 2 (**b**) and HO-1 (**c**) ($n = 3$). Data are expressed as mean ± SD. ## $p < 0.01$ compared with the control group; * $p < 0.05$, ** $p < 0.01$ compared with the model group.

3.3. EST and CA Suppress the CC_{14}-Induced Pro-Inflammatory Cytokines Expression via Inhibiting the NF-κB Pathway

In the early stages of oxidative stress, liver immune cells secrete pro-inflammatory factors, such as TNF-α, IL-6, and IL-1β, to trigger inflammatory responses and aggravate liver injury [23]. Immunohistochemical results showed that the protein expression levels of TNF-α, IL-6, and IL-1β were significantly increased in the CCl_4-induced model group compared with the control group ($p < 0.01$), while both EST and CA could inhibit the expression of these three pro-inflammatory factors in a gradient manner, and the effect was comparable to that of the positive drug, silymarin, indicating that EST and CA could inhibit the expression of CCl_4-induced pro-inflammatory factors and exerted anti-inflammatory effects (Figure 5).

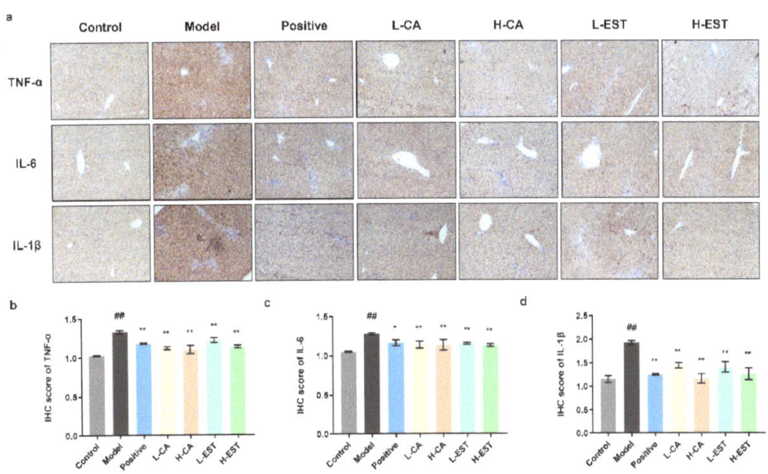

Figure 5. Effects of EST and CA on liver TNF-α, IL-6, and IL-1β expression. (**a**) Immunohistochemical staining of gastric TNF-α, IL-6, and IL-1β. (**b–d**) Densitometric quantification ($n = 3$). Data are expressed as mean ± SD. ## $p < 0.01$ compared with the control group; * $p < 0.05$, ** $p < 0.01$ compared with the model group.

To further examine the effect of EST and CA on inflammatory signaling pathways, the expression of NF-κB and p-MAPK (Figure 6a) in liver tissues, which is vital in inflammatory responses, were determined by Western blot analysis. We found a significant 2-fold upregulation ($p < 0.01$) of p-NF-κB/NF-κB in the liver of the model group compared with the normal group. In contrast, silymarin, EST (200 and 700 mg/kg), and CA (30 and 100 mg/kg) significantly ($p < 0.01$) inhibited the CCl_4-induced p-NF-κB/NF-κB upregulation (Figure 6b). The effect of different treatment groups of p-MAPK was basically consistent with the ratio of p-NF-κB/NF-κB (Figure 6c). Together, these results suggest that EST and CA alleviate the CCl_4-induced hepatic inflammatory response by inhibiting the NF-kB signaling pathway and the expression of pro-inflammatory cytokines.

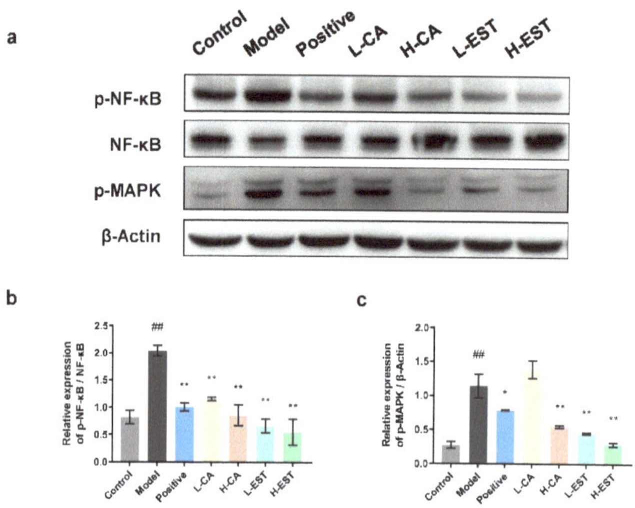

Figure 6. Effects of EST and CA on liver p-NF-κB, NF-κB, p-MAPK expression. (**a**) Western blotting analysis of p-NF-κB, NF-κB, and p-MAPK expression in liver tissues and densitometric quantification of p-NF-κB/NF-κB (**b**) ($n = 3$). (**c**) Western blotting analysis of p-MAPK expression in liver tissues and densitometric quantification of p-MAPK ($n = 3$). Data are expressed as mean ± SD. ## $p < 0.01$ compared with the control group; * $p < 0.05$, ** $p < 0.01$ compared with the model group.

3.4. EST and CA Alleviate CCl_4-Induced Hepatocyte Apoptosis by Regulating the Expression of Bax/Bcl-2

Bax, a typical pro-apoptotic factor in cytoplasmic lysis, can be transferred to mitochondria to induce apoptosis, while Bcl-2 can inhibit Bax-induced apoptosis and is an anti-apoptotic factor [24]. Since hepatocyte apoptosis plays an important role in models of liver injury, we investigated the protein expression levels of Bax/Bcl-2 in liver tissues of different groups of mice. As shown in Figure 7, Western blot analysis showed that the expression of the pro-apoptotic protein Bax was significantly upregulated ($p < 0.01$), and the expression of the anti-apoptotic protein Bcl-2 was significantly downregulated ($p < 0.05$) in the model mice compared to the control mice. As with silymarin, CA (30 and 100 mg/kg) and EST (200 and 700 mg/kg) inhibited the CCl_4-induced increase in Bax expression and decrease in Bcl-2 expression, and the effect was more pronounced in the EST-treated group. We also performed TUNEL analysis to detect the degree of apoptosis in hepatocytes of different treatment groups in Figure S2. These results suggest that EST and CA inhibited the apoptotic signaling pathway by regulating the expression of apoptosis-related proteins Bax and Bcl-2.

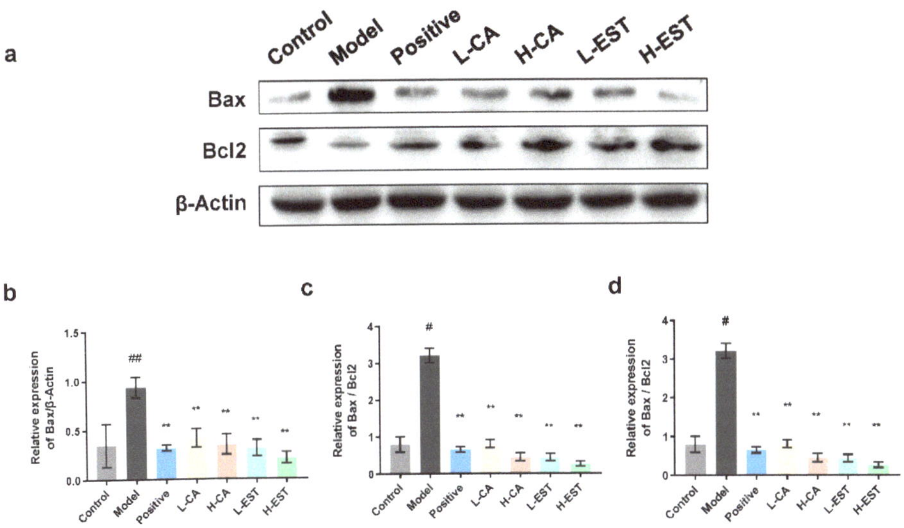

Figure 7. Effects of EST and CA on liver Bax and Bcl2 expression. (**a**) Western blotting analysis of Bax and Bcl2 expression in liver tissues and densitometric quantification of Bax (**b**), Bcl2 (**c**), and Bax/Bcl2 ratio (**d**), (n = 3). Data are expressed as mean ± SD. # $p < 0.05$, ## $p < 0.01$ compared with the control group; ** $p < 0.01$ compared with the model group.

4. Discussion

Making tea from plants for drinking has been a tradition in China since ancient times. According to the origin of the plants, tea can be divided into traditional tea and non-*Camellia* tea. Traditional tea is a tea beverage made from processed leaves of *Camellia sinensis* plants. In contrast, non-*Camellia* tea refers to tea beverages made from leaves, flowers, and roots of plants that do not belong to *Camellia* [25]. Non-*Camellia* tea usually accumulated certain application experience within a certain region or ethnic group, and, together with traditional tea, forms the colorful tea culture of China [26]. Shibi tea is a kind of non-*Camellia* tea made from the dried leaves of *Adinandra nitida*, which has been reported to have various health benefits such as lowering blood pressure and antibacterial, antioxidant, and analgesic effects, and is commonly used in food and medicine [27,28]. Shibi tea is rich in flavonoids and other active ingredients. The research on the health activity of Shibi tea is mainly focused on the active efficacy of the flavonoid components of Shibi tea, but there is relatively little research on the active efficacy of its major flavonoid monomers, especially CA. In one study, 3T3-L1 cell lines were used as an in vitro model of obesity, combined with nuclear magnetic resonance (NMR) and liquid chromatography (LC)–MS techniques to identify four triterpenoid saponins in Shibi tea that inhibit adipogenesis [29]. In our previous study, we identified CA as the predominant flavonoid in Shibi tea and established a method to prepare CA from Shibi tea [30]. We also investigated for the first time that Shibi tea and CA alleviated alcoholic gastric injury by attenuating HCl/EtOH-induced oxidative stress in the stomach and inhibiting the NF-kB signaling pathway to suppress the expression of inflammatory factors in vivo [17]. However, the mitigating effects of EST and CA on liver injury have not yet been studied.

Liver injury is a common pathological basis of various liver diseases, and clinical study has found that long-term liver injury could lead to mild hepatic inflammation and even liver fibrosis, cirrhosis, and hepatocellular carcinoma [31,32]. Liver injury is still a global health issue, and the hunt for novel hepatoprotective methods is extremely important. Experimental animal models are useful models for the study of drug-induced liver injury and its pathogenesis. Among them, acetaminophen (Acetaminophen, APAP) and CCl_4 are the two most common model inducers in the study of endogenous drug-induced liver injury [33]. Oxidative stress pathways, inflammatory cytokines, and apoptosis are the main regulatory targets of liver injury [34]. When liver injury occurs, hepatocytes and endothelial cells are damaged, hepatic Kupffer cells (KC) and hepatic stellate cells (HSC) are activated, and inflammatory factors accumulate in the liver. As the therapeutic effects of flavonoids on liver injury have been studied recently, the significant protective effects of flavonoids on liver injury have been confirmed. Flavonoids, such as silymarin [35,36], can reduce the production of oxidative free radicals and increase the activity of antioxidant enzymes, effectively protecting the liver from oxidative damage. Studies have shown that flavonoids protect hepatocytes and endothelial cells from oxidative stress and inhibit the proliferation and activation of activated HSC and promote their apoptosis; in addition, flavonoids inhibit TNF-α production and KC activation and reduce the accumulation of inflammatory factors [37]. In this study, we demonstrated that the EST (200 and 700 mg/kg) and its major flavonoid CA (30 and 100 mg/kg) significantly ameliorated CCl_4-induced acute liver injury (Figures 1 and 2) in a mouse model with intraperitoneal injection of CCl_4, using silymarin as a positive drug control, and that its hepatoprotective mechanism was related to the antioxidant, anti-inflammatory, and anti-apoptotic effects of EST and CA. Given that CA is the main flavonoid component of EST and the hepatoprotective effect is comparable to that of EST, we propose that CA is potentially the main active substance contributing to the hepatoprotective effect in EST.

Oxidative stress is an important cause of liver injury, and the metabolites of CCl_4, CCl3- and CCl3OO-, both of which can cause oxidative stress in hepatocytes and consequently hepatocellular injury [38]. Flavonoids improve oxidative stress by activating the Nrf2/HO-1 pathway, and Nrf2 is a key transcription factor that upregulates the antioxidant gene HO-1 [39]. Plant-derived flavonoids such as cyanidin flavonoids [40] and alpinetin [41] have also been reported to effectively alleviate CCl_4-induced acute liver injury in mice by modulating the Nrf2 signaling pathway, reducing ROS and MDA levels, and increasing antioxidant enzyme activity. In this study, we found that EST and CA increased the activities of antioxidant enzymes SOD, CAT, and GSH-Px in liver tissues; increased the level of antioxidant GSH; inhibited the CCl_4-induced increase in MDA and ROS oxidative stress levels (Figure 3); and significantly activated antioxidant pathway targets Nrf2 and HO-1 to perform antioxidant effects (Figure 4).

Abnormal expression of inflammatory cytokines, which mediate the interference of various immune cells, can directly affect the immune response, and occur frequently during liver injury. Morin [42] significantly inhibited lipopolysaccharide (LPS)-induced production of serum AST, ALT, IL-6, and TNF-α, which could exert antioxidant and anti-inflammatory protective effects by activating the Nrf2 antioxidant signaling pathway and inhibiting the NF-κB inflammatory pathway. Breviscapine [8] inhibited significantly elevated levels of serum TNF-α, IL-6, IL-1β, and monocyte chemotactic protein-1 (MCP-1) in a CCl_4 model group of mice and suppressed the expression of downstream IκBα and NF-κB inflammatory pathways, acting as hepatoprotective agents. Similarly, EST and CA significantly inhibited CCl_4-induced liver injury by targeting the MAPK and NF-κB pathways (Figure 6) and the downstream inflammatory cytokines, such as TNF-α, IL-6, and IL-1β (Figure 5).

Flavonoids can also reduce liver injury by regulating apoptosis; for example, dihydromyricetin significantly inhibited the expression of the pro-apoptotic protein Bax and upregulated the expression of the anti-apoptotic protein Bcl-2, thereby inhibiting chronic liver injury [43]. Baicalin played an important hepatic repair role in oxidative stress-induced liver injury by regulating mitochondria-related apoptosis [44]. Wogonin significantly increased the Bax/Bcl-2 ratio in T6 cells and regulated the activation and apoptosis of hepatic stellate cells to reduce liver fibrosis [45]. We found that EST and CA regulate apoptosis by regulating Bax/Bcl-2 expression to alleviate CCl_4-induced liver injury (Figure 7).

In this study, we found that for the CA and EST treatment groups, the high-dose group generally showed better results than the low-dose group, but did not reach a significant difference in some of the results. Therefore, the current study provides a basis for further research on the active dosage and underlying mechanisms of CA and EST hepatoprotective effects. In addition, we used 30 and 100 mg/kg CA treatments and 200 and 700 mg/kg EST treatments, which are much lower than the reported flavonoid monomer concentrations (>500 mg/kg) that cause toxicity or side effects. Orally administrated dosages of 200–700 mg/kg EST are in accordance with the dosage taken by drinking tea (3–10 g) per day. There were no effects of CA and EST on animal behavior during the experiment and no significant visual damage to other organs of mice in all groups.

5. Conclusions

Our study demonstrates that EST and its major flavonoid component, CA, can effectively alleviate CCl_4-induced acute liver injury in mice. More importantly, the hepatoprotective effects of EST and CA are primarily attributed to the reduction of oxidative stress, inhibition of inflammation, and regulation of apoptosis. Additionally, these results provide a theoretical basis for the further investigation of EST and CA as potential agents for the treatment and prevention of liver disease.

Supplementary Materials: The following supporting information can be downloaded at: https://www.mdpi.com/article/10.3390/nu14153037/s1, Figure S1: Masson trichrome staining of the liver sections of different groups; Figure S2. Histological changes (TUNEL assay) in the liver of different groups.

Author Contributions: Conceptualization, R.C. and E.Y.; Data curation, Y.L.; Formal analysis, Y.L. and L.S.; Funding acquisition, R.C. and S.S.; Investigation, Y.L., S.W. and Q.L.; Methodology, S.W., Q.L. and X.L.; Project administration, E.Y. and S.S.; Resources, Q.L., J.Z., J.X. and L.T.; Software, L.S. and Z.Z.; Supervision, E.Y. and S.S.; Validation, S.W. and X.L.; Visualization, Z.Z.; Writing—original draft, R.C., Y.L. and S.S. All authors have read and agreed to the published version of the manuscript.

Funding: This study was funded by the 14th Five-Year Plan team-building projects of the Guangdong Academy of Agricultural Sciences (grant no. 202126TD); Guangzhou Science and Technology Plan Projects (grant no. 202102020047, 202002030202); the Guangdong Basic and Applied Basic Research Foundation (grant no. 2020A1515011266, 2021A1515010958); the Key-Area Research and Development Program of Guangdong Province (grant no. 2020B0202080003); Innovation Fund projects of Guangdong Academy of Agricultural Sciences (grant no. 202115, 202035); the special fund for scientific innovation strategy-construction of high level of the Academy of Agriculture Science (grant no. R2019PY-JX004); the Innovation Fund projects of Guangdong Key Laboratory of Tea Plant Resources Innovation and Utilization (grant no. 2021CX02); the Qingyuan Science and Technology Program (grant no. DZXQY021); and the special fund project for the introduction of scientific and technological talents of the Guangdong Academy of Agricultural Sciences (project no. R2021YJ-YB3014). The funders did not have any role in the study design, data collection, or data analysis.

Institutional Review Board Statement: All experimental procedures were conducted following the Animal Care and Use Guidelines of Tea Research Institute, Guangdong Academy of Agricultural Sciences, and approved by the Institutional Animal Care and Use Committee (serial number: 2020010).

Informed Consent Statement: Not applicable.

Data Availability Statement: Not applicable.

Acknowledgments: The authors thank Zhengqi Agricultural Development Co., Ltd. in Yingde and Taihongyuan Agriculture Co., Ltd. in Xinyi, Maoming (tea brand: Dawutengyun) for providing Shibi tea samples.

Conflicts of Interest: The authors declare no conflict of interest.

References

1. Trefts, E.; Gannon, M.; Wasserman, D.H. The liver. *Curr. Biol. CB* **2017**, *27*, R1147–R1151. [CrossRef] [PubMed]
2. Kubes, P.; Jenne, C. Immune Responses in the Liver. *Annu. Rev. Immunol.* **2018**, *36*, 247–277. [CrossRef] [PubMed]
3. Bernal, W.; Auzinger, G.; Dhawan, A.; Wendon, J. Acute liver failure. *Lancet* **2010**, *376*, 190–201. [CrossRef]
4. Kisseleva, T.; Brenner, D. Molecular and cellular mechanisms of liver fibrosis and its regression. *Nat. Rev. Gastroenterol. Hepatol.* **2021**, *18*, 151–166. [CrossRef]

5. Zhang, C.; Wang, N.; Xu, Y.; Tan, H.Y.; Li, S.; Feng, Y. Molecular Mechanisms Involved in Oxidative Stress-Associated Liver Injury Induced by Chinese Herbal Medicine: An Experimental Evidence-Based Literature Review and Network Pharmacology Study. *Int. J. Mol. Sci.* **2018**, *19*, 2745. [CrossRef]
6. Yang, S.; Lian, G. ROS and diseases: Role in metabolism and energy supply. *Mol. Cell. Biochem.* **2020**, *467*, 1–12. [CrossRef]
7. Feng, M.; Ding, J.; Wang, M.; Zhang, J.; Zhu, X.; Guan, W. Kupffer-derived matrix metalloproteinase-9 contributes to liver fibrosis resolution. *Int. J. Biol. Sci.* **2018**, *14*, 1033–1040. [CrossRef]
8. Liu, Y.; Wen, P.H.; Zhang, X.X.; Dai, Y.; He, Q. Breviscapine ameliorates CCl4induced liver injury in mice through inhibiting inflammatory apoptotic response and ROS generation. *Int. J. Mol. Med.* **2018**, *42*, 755–768. [CrossRef]
9. Wang, K.; Fang, S.; Liu, Q.; Gao, J.; Wang, X.; Zhu, H.; Zhu, Z.; Ji, F.; Wu, J.; Ma, Y.; et al. TGF-beta1/p65/MAT2A pathway regulates liver fibrogenesis via intracellular SAM. *EBioMedicine* **2019**, *42*, 458–469. [CrossRef]
10. Zhang, J.; Chen, Y.; Luo, H.; Sun, L.; Xu, M.; Yu, J.; Zhou, Q.; Meng, G.; Yang, S. Recent Update on the Pharmacological Effects and Mechanisms of Dihydromyricetin. *Front. Pharmacol.* **2018**, *9*, 1204. [CrossRef]
11. Federico, A.; Dallio, M.; Loguercio, C. Silymarin/Silybin and Chronic Liver Disease: A Marriage of Many Years. *Molecules* **2017**, *22*, 191. [CrossRef] [PubMed]
12. Jie, Z.; Yang, J.; Duan, J.; Zhen, L.; Zhang, L.; Huo, Y.; Zhang, Y. Quantitative and qualitative analysis of flavonoids in leaves of Adinandra nitida by high performance liquid chromatography with UV and electrospray ionization tandem mass spectrometry detection. *Anal. Chim. Acta* **2005**, *532*, 97–104.
13. Liu, B.G.; Ma, Y.X.; Liu, Y.; Yang, Z.; Zhang, L.P. Ultrasonic-Assisted Extraction and Antioxidant Activity of Flavonoids from Adinandra nitida Leaves. *Trop. J. Pharm. Res.* **2013**, *12*, 1045–1051. [CrossRef]
14. Chen, Y.; Ma, X.; Fu, X.; Yan, R. phytochemical content, cellular antioxidant activity and antiproliferative activity of adinandra nitida tea (shiyacha) infusion subjected to in vitro gastrointestinal digestion. *RSC Adv.* **2017**, *7*, 50430–50440. [CrossRef]
15. Zhang, W.; Lian, Y.; Li, Q.; Sun, L.; Chen, R.; Lai, X.; Lai, Z.; Yuan, E.; Sun, S. Preventative and Therapeutic Potential of Flavonoids in Peptic Ulcers. *Molecules* **2020**, *25*, 4626. [CrossRef]
16. Yuan, E.; Liu, B.; Ning, Z.; Chen, C. Preparative separation of flavonoids in Adinandra nitida leaves by high-speed counter-current chromatography and their effects on human epidermal carcinoma cancer cells. *Food Chem.* **2009**, *115*, 1158–1163. [CrossRef]
17. Yuana, E.; Liana, Y.; Li, Q.; Lai, X.; Chen, R.; Wen, S.; Zhu, J.; Zhang, W.; Sun, S. Roles of Adinandra nitida (Theaceae) and camellianin A in HCl/ethanol-induced acute gastric ulcer in mice. *Food Sci. Hum. Wellness* **2022**, *11*, 1053–1063. [CrossRef]
18. Wen, S.; Sun, L.; An, R.; Zhang, W.; Xiang, L.; Li, Q.; Lai, X.; Huo, M.; Li, D.; Sun, S. A combination of Citrus reticulata peel and black tea inhibits migration and invasion of liver cancer via PI3K/AKT and MMPs signaling pathway. *Mol. Biol. Rep.* **2020**, *47*, 507–519. [CrossRef]
19. Hao, L.; Liu, M.W.; Gu, S.T.; Huang, X.; Deng, H.; Wang, X. Sedum sarmentosum Bunge extract ameliorates lipopolysaccharide- and D-galactosamine-induced acute liver injury by attenuating the hedgehog signaling pathway via regulation of miR-124 expression. *BMC Complement. Med.* **2020**, *20*, 88. [CrossRef]
20. Hart, N.A.; van der Plaats, A.; Leuvenink, H.G.; Wiersema-Buist, J.; Olinga, P.; van Luyn, M.J.; Verkerke, G.J.; Rakhorst, G.; Ploeg, R.J. Initial blood washout during organ procurement determines liver injury and function after preservation and reperfusion. *Am. J. Transpl.* **2004**, *4*, 1836–1844. [CrossRef]
21. Chen, R.; Lai, X.; Xiang, L.; Li, Q.; Sun, L.; Lai, Z.; Li, Z.; Zhang, W.; Wen, S.; Cao, J.; et al. Aged green tea reduces high-fat diet-induced fat accumulation and inflammation via activating the AMP-activated protein kinase signaling pathway. *Food Nutr. Res.* **2022**, *66*, 7923. [CrossRef] [PubMed]
22. Zhang, Y.; Sun, L.; Lai, X.; Peng, X.; Wen, S.; Zhang, Z.; Xie, Y.; Li, Q.; Chen, R.; Zheng, X.; et al. Gastroprotective effects of extract of Jasminum grandiflorum L. flower in HCl/EtOH-induced gastric mucosal ulceration mice. *Biomed. Pharm.* **2021**, *144*, 112268. [CrossRef] [PubMed]
23. Kim, S.H.; Oh, D.S.; Oh, J.Y.; Son, T.G.; Yuk, D.Y.; Jung, Y.S. Silymarin Prevents Restraint Stress-Induced Acute Liver Injury by Ameliorating Oxidative Stress and Reducing Inflammatory Response. *Molecules* **2016**, *21*, 443. [CrossRef] [PubMed]
24. Rahmani, M.; Nkwocha, J.; Hawkins, E.; Pei, X.; Parker, R.E.; Kmieciak, M.; Leverson, J.D.; Sampath, D.; Ferreira-Gonzalez, A.; Grant, S. Cotargeting BCL-2 and PI3K Induces BAX-Dependent Mitochondrial Apoptosis in AML Cells. *Cancer Res.* **2018**, *78*, 3075–3086. [CrossRef] [PubMed]
25. Long, P.; Cui, Z.; Wang, Y.; Zhang, C.; Zhang, N.; Li, M.; Xiao, P. Commercialized non-Camellia tea: Traditional function and molecular identification. *Acta Pharm. Sin. B* **2014**, *4*, 227–237. [CrossRef]
26. Jiang, B.; Lv, Q.; Wan, W.; Le, L.; Xu, L.; Hu, K.; Xiao, P. Transcriptome analysis reveals the mechanism of the effect of flower tea Coreopsis tinctoria on hepatic insulin resistance. *Food Funct.* **2018**, *9*, 5607–5620. [CrossRef]
27. Wang, Y.; Chen, S.B.; Jie, N.I.; Yao, X.; Wen-Cai, Y.E. Chemical Studies on the Adinandra nitida. *J. China Pharm. Univ.* **2003**, *34*, 407–409.
28. Liu, B.; Yang, J.; Ma, Y.; Yuan, E.; Chen, C. Antioxidant and angiotensin converting enzyme (ACE) inhibitory activities of ethanol extract and pure flavonoids from Adinandra nitida leaves. *Pharm. Biol.* **2010**, *48*, 1432–1438. [CrossRef]
29. Yuan, C.; Huang, L.; Suh, J.H.; Wang, Y. Bioactivity-Guided Isolation and Identification of Antiadipogenic Compounds in Shiya Tea (Leaves of Adinandra nitida). *J. Agric. Food Chem.* **2019**, *67*, 6785–6791. [CrossRef]
30. Gao, H.; Liu, B.; Liu, F.; Chen, Y. Anti-proliferative effect of camellianin A in Adinandra nitida leaves and its apoptotic induction in human Hep G2 and MCF-7 cells. *Molecules* **2010**, *15*, 3878–3886. [CrossRef]

31. Singh, S.; Osna, N.A.; Kharbanda, K.K. Treatment options for alcoholic and non-alcoholic fatty liver disease: A review. *World J. Gastroenterol.* **2017**, *23*, 6549–6570. [CrossRef] [PubMed]
32. Yang, J.D.; Hainaut, P.; Gores, G.J.; Amadou, A.; Plymoth, A.; Roberts, L.R. A global view of hepatocellular carcinoma: Trends, risk, prevention and management. *Nat. Rev. Gastroenterol. Hepatol.* **2019**, *16*, 589–604. [CrossRef] [PubMed]
33. McGill, M.R.; Jaeschke, H. Animal models of drug-induced liver injury. *Biochim. Biophys. Acta Mol. Basis Dis.* **2019**, *1865*, 1031–1039. [CrossRef]
34. Tsuchida, T.; Friedman, S.L. Mechanisms of hepatic stellate cell activation. *Nat. Rev. Gastroenterol. Hepatol.* **2017**, *14*, 397–411. [CrossRef] [PubMed]
35. Clichici, S.; Olteanu, D.; Nagy, A.L.; Oros, A.; Filip, A.; Mircea, P.A. Silymarin inhibits the progression of fibrosis in the early stages of liver injury in CCl(4)-treated rats. *J. Med. Food* **2015**, *18*, 290–298. [CrossRef]
36. Kalopitas, G.; Antza, C.; Doundoulakis, I.; Siargkas, A.; Kouroumalis, E.; Germanidis, G.; Samara, M.; Chourdakis, M. Impact of Silymarin in individuals with nonalcoholic fatty liver disease: A systematic review and meta-analysis. *Nutrition* **2021**, *83*, 111092. [CrossRef]
37. Wan, L.; Jiang, J.G. Protective effects of plant-derived flavonoids on hepatic injury. *J. Funct. Foods* **2018**, *44*, 283–291. [CrossRef]
38. Teschke, R.; Danan, G. Molecular Research on Drug Induced Liver Injury. *Int. J. Mol. Sci.* **2018**, *19*, 216. [CrossRef]
39. Motohashi, H.; Yamamoto, M. Nrf2-Keap1 defines a physiologically important stress response mechanism. *Trends Mol. Med.* **2004**, *10*, 549–557. [CrossRef]
40. Xie, J.; Wang, W.; Dong, C.; Huang, L.; Wang, H.; Li, C.; Nie, S.; Xie, M. Protective effect of flavonoids from Cyclocarya paliurus leaves against carbon tetrachloride-induced acute liver injury in mice. *Food Chem. Toxicol.* **2018**, *119*, 392–399. [CrossRef]
41. Zhu, Z.; Hu, R.; Li, J.; Xing, X.; Chen, J.; Zhou, Q.; Sun, J. Alpinetin exerts anti-inflammatory, anti-oxidative and anti-angiogenic effects through activating the Nrf2 pathway and inhibiting NLRP3 pathway in carbon tetrachloride-induced liver fibrosis. *Int. Immunopharmacol.* **2021**, *96*, 107660. [CrossRef] [PubMed]
42. Tian, Y.; Li, Z.; Shen, B.; Zhang, Q.; Feng, H. Protective effects of morin on lipopolysaccharide/d-galactosamine-induced acute liver injury by inhibiting TLR4/NF-kappaB and activating Nrf2/HO-1 signaling pathways. *Int. Immunopharmacol.* **2017**, *45*, 148–155. [CrossRef] [PubMed]
43. Cheng, Q.C.; Fan, J.; Deng, X.W.; Liu, H.C.; Ding, H.R.; Fang, X.; Wang, J.W.; Chen, C.H.; Zhang, W.G. Dihydromyricetin ameliorates chronic liver injury by reducing pyroptosis. *World J. Gastroenterol.* **2020**, *26*, 6346–6360. [CrossRef] [PubMed]
44. Yu, Z.; Li, Q.; Wang, Y.; Li, P. A potent protective effect of baicalein on liver injury by regulating mitochondria-related apoptosis. *Apoptosis* **2020**, *25*, 412–425. [CrossRef]
45. Du, X.S.; Li, H.D.; Yang, X.J.; Li, J.J.; Xu, J.J.; Chen, Y.; Xu, Q.Q.; Yang, L.; He, C.S.; Huang, C.; et al. Wogonin attenuates liver fibrosis via regulating hepatic stellate cell activation and apoptosis. *Int. Immunopharmacol.* **2019**, *75*, 105671. [CrossRef] [PubMed]

nutrients

MDPI

Review

Food Anthocyanins: Malvidin and Its Glycosides as Promising Antioxidant and Anti-Inflammatory Agents with Potential Health Benefits

Anna Merecz-Sadowska [1,*], Przemysław Sitarek [2], Tomasz Kowalczyk [3], Karolina Zajdel [4], Mariusz Jęcek [1], Paweł Nowak [1] and Radosław Zajdel [1]

1 Department of Economic and Medical Informatics, University of Lodz, 90-214 Lodz, Poland; mariusz.jecek@uni.lodz.pl (M.J.); pawel.nowak@uni.lodz.pl (P.N.); radoslaw.zajdel@uni.lodz.pl (R.Z.)
2 Department of Medical Biology, Medical University of Lodz, 90-151 Lodz, Poland; przemyslaw.sitarek@umed.lodz.pl
3 Department of Molecular Biotechnology and Genetics, Faculty of Biology and Environmental Protection, University of Lodz, 90-237 Lodz, Poland; tomasz.kowalczyk@biol.uni.lodz.pl
4 Department of Medical Informatics and Statistics, Medical University of Lodz, 90-645 Lodz, Poland; karolina.smigiel@umed.lodz.pl
* Correspondence: anna.merecz-sadowska@uni.lodz.pl

Abstract: Anthocyanins are flavonoid compounds that are abundantly present in fruits and vegetables. These compounds contribute to the color of these foods and offer various health benefits to consumers due to their biological properties. There are more than 1000 types of anthocyanins in nature, all derived from 27 anthocyanidin aglycones that have different glycosylations and acylations. Malvidin is one of the most well-known anthocyanidins. Several studies, including those conducted on cell lines, animals, and humans, have suggested that malvidin and its glycosides possess anti-carcinogenic, diabetes-control, cardiovascular-disease-prevention, and brain-function-improvement properties. These health benefits are primarily attributed to their antioxidant and anti-inflammatory effects, which are influenced by the molecular mechanisms related to the expression and modulation of critical genes. In this article, we review the available information on the biological activity of malvidin and its glycosides concerning their health-promoting effects.

Keywords: malvidin; malvidin glycosides; antioxidant activity; anti-inflammatory activity; health benefits

Citation: Merecz-Sadowska, A.; Sitarek, P.; Kowalczyk, T.; Zajdel, K.; Jęcek, M.; Nowak, P.; Zajdel, R. Food Anthocyanins: Malvidin and Its Glycosides as Promising Antioxidant and Anti-Inflammatory Agents with Potential Health Benefits. *Nutrients* **2023**, *15*, 3016. https://doi.org/10.3390/nu15133016

Academic Editors: Maria Digiacomo and Doretta Cuffaro

Received: 23 May 2023
Revised: 23 June 2023
Accepted: 28 June 2023
Published: 1 July 2023

1. Introduction

Flavonoids are a class of polyphenolic compounds found ubiquitously in plants and encompass a diverse range of chemical structures and characteristics. They consist of a 15-carbon atom skeleton (C6-C3-C6) comprising three rings: two phenolic rings (A and B) and one pyran ring (C) [1,2]. Anthocyanins are water-soluble flavonoids that are differentiated based on the nature, number, and location of sugars attached to the molecule; the number of aliphatic or aromatic acids attached to sugars; the number of hydroxyl groups present; and the degree of methylation of hydroxyl groups. Malvidin, peonidin, cyanidin, pelargonidin, petunidin, and delphinidin are some of the most abundant anthocyanin aglycones. Anthocyanins aglycones are glycosidically linked to sugars in the C-3 position of the anthocyanidin. The glycone part of anthocyanins provides chemical stability and solubility, while the conjugated double bonds in the anthocyanidin moiety are responsible for light absorption and unique color production. Generally, methoxylation imparts a red color, while increased hydroxylation produces a blue pigment [3,4].

It has been reported that more than 1000 distinct anthocyanins have been identified to date [5]. These molecules are prevalent in plants and are responsible for the red, purple, and blue pigments observed in a variety of fruits, vegetables, and their derivatives [6]. The composition of anthocyanins may differ based on various factors, such as the source of the food and its variety [7], as well as seasonal changes and environmental conditions [8]. Additionally, factors such as the state of the food, whether fresh, frozen, or dried [9]; and the manner in which the food is stored and prepared, for instance, whether it is peeled or unpeeled, can also play a role in the concentration of anthocyanins [7,10]. The quantity of anthocyanins found in fruits and vegetables can vary significantly, with levels ranging from 30 to 1500 mg/100 g [11]. These compounds exhibit a wide range of biological activities, including antioxidant and anti-inflammatory properties, which are associated with their potential to provide anti-carcinogenic, cardioprotective, antidiabetic, and neuroprotective properties [12]. Moreover, due to their beneficial effects on human health, anthocyanins may be used in food supplements [13].

Malvidin (PubChem CID: 159287), named 3′,5′-dimethoxy-3,4′,5,7-tetrahydroxyflavylium acid anion, is a type of anthocyanidin cation whose chemical structure is similar to delphinidin but has methyl groups attached to positions 3′ and 5′. This compound is present in various fruits, vegetables, and their derivatives, and it accounts for approximately 7% of the pigment distribution in edible plant parts. Malvidin is commonly linked to various sugar moieties in the C-3 position. This molecule has four hydrogen-bond donors, making it a potent scavenger of reactive oxygen species (ROS) [14]. Malvidin and its glycosides play an important role as antioxidant and anti-inflammatory agents [15].

There has been a significant increase in the frequency of articles published on the properties of anthocyanins. However, malvidin, a potent antioxidant and anti-inflammatory agent, has been relatively less studied among anthocyanidins. Therefore, this review article aims to discuss the biological activity and potential health benefits of malvidin and its glycosides.

2. Study Design

Published data were explored using databases such as NCBI-PubMed, Google Scholar, Scopus, and ScienceDirect. The following keywords were used: malvidin, malvidin glycosides, in vitro models, in vivo models, antioxidant potential of malvidin and its glycosides, anti-inflammatory potential of malvidin and its glycosides, anticancer potential of malvidin and its glycosides, antidiabetic potential of malvidin and its glycosides, cardioprotective potential of malvidin and its glycoside, and neuroprotective potential of malvidin and its glycosides. This review of the literature encompassed studies investigating the effects of malvidin and its glycosides on cellular systems, animal models, and human models.

3. Exploring the Characteristic and Food Sources of Malvidin and Other Anthocyanins

Anthocyanins are a group of water-soluble pigments that can be found in all higher plant tissues, such as leaves, stems, roots, flowers, and fruits. These compounds are produced via the phenylpropanoid pathway, which is part of the plant's secondary metabolism. Anthocyanins play several important functions that are essential for the plant's survival. The primary role of anthocyanins is to attract animals and pollinating insects, aiding in the dissemination of seeds or the spread of pollen. Moreover, it has been demonstrated that the synthesis of anthocyanins is induced during the establishment of adverse conditions, suggesting their involvement in both biotic and abiotic stresses [16,17].

Anthocyanins are a class of flavonoid derivatives characterized by a flavylium cation backbone, which is hydroxylated at various positions, typically on carbons C3, C5, C6, C7, C3′, C4′, and C5′. This results in a large class of approximately 1000 different anthocyanin derivatives, each consisting of one of 27 aglycones, known as anthocyanidins. The anthocyanidins usually contain one sugar moiety, which is commonly conjugated to the C3 hydroxyl group in the C-ring, making them glycosides. Sugar moieties such as glucose, galactose, rhamnose, arabinose, and xylose are commonly found in anthocyanins. With the exception of 3-deoxyanthocyanins, anthocyanins exist almost exclusively in a glycosylated form. Their aglycone counterparts are not stable and are rarely found in nature. Six anthocyanidins, namely cyanidin, pelargonidin, delphinidin, petunidin, peonidin, and malvidin, are particularly prevalent in nature and account for approximately

90% of all anthocyanins [18,19]. Isolated anthocyanins are highly unstable and prone to degradation. Their stability can be affected by various factors, such as the pH; storage temperature; chemical structure; concentration; light; oxygen; solvents; and presence of enzymes, flavonoids, proteins, and metallic ions [20].

Anthocyanins are widely distributed in nature and are primarily responsible for the red, blue, and purple colors seen in vegetables, fruits, and their derivatives [21]. Berries, red wine, vegetables, and other fruits are the primary sources of anthocyanins [22], with levels varying considerably among different species. These levels are primarily influenced by plant genotypes and, to a lesser extent, by agricultural practices, growing area, seasonal variability, climatic conditions, temperature and light exposure, ripening stage, and harvesting time, as well as the methods adopted for processing and storage [23]. The daily average intake of anthocyanins is estimated to range from several milligrams to hundreds of milligrams; however, its evaluation is imprecise and depends on factors such as diet, gender, food intolerances in individuals, and the quantities of anthocyanins present in foods [24]. The relative bioavailability of anthocyanins has recently been suggested to be approximately 12% [25].

Anthocyanin pigments have a significant impact on the sensory properties of food, making them desirable for use as food colorants [3]. However, beyond their colorant capacity, anthocyanins may offer numerous health benefits to consumers. There is a growing body of epidemiological and clinical evidence supporting the notion that anthocyanins, as part of a diet rich in fruits and vegetables, play an important role in countering the onset and progression of several disease pathologies, particularly cardiovascular, metabolic, and neurodegenerative diseases, as well as certain types of cancers. The health benefits of anthocyanins are attributed to their free-radical scavenging capacity and their ability to modulate inflammatory cytokine signaling, which is associated with maintaining homeostatic balance in the body [6,26,27].

Malvidin is one of the six most prominent anthocyanins and a member of the *O*-methylated anthocyanidin family. In nature, it is mainly found in its glycosylated form, such as malvidin-3-glucoside and malvidin-3-galactoside, with a sugar moiety attached at position 3 on the C-ring (Figure 1). The solubility of malvidin in water is higher than in methanol and ethanol, likely due to its higher static dipole moment in water [28]. However, this solubility decreases as the degree of acylation increases [29]. Syringic and 4-hydroxybenzoic acids are the primary metabolites of malvidin [30]. When subjected to high pH conditions (pH > 7), acylated malvidin breaks down and releases syringic acid [31].

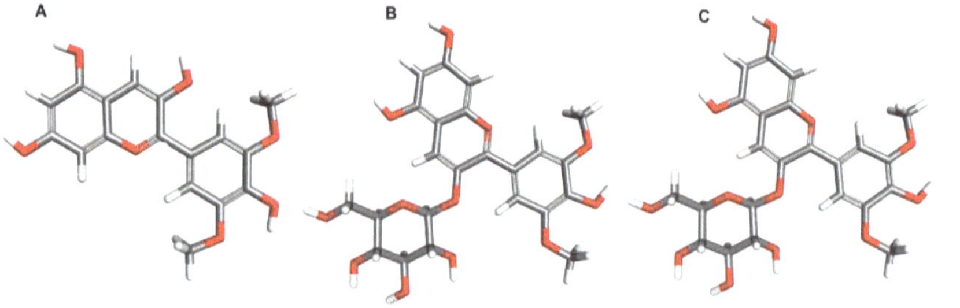

Figure 1. The structure of malvidin (**A**) and its selected glycosides: malvidin-3-glucoside (**B**) and malvidin-3-galactoside (**C**) (https://pubchem.ncbi.nlm.nih.gov/ (accessed on 22 March 2023)).

Malvidin exhibits a visible purple color, is abundantly found in blue-colored flowers [32], and serves as the major red pigment in red wine [33]. The distribution of malvidin glycosides varies across different plant species, with malvidin 3-glucoside and malvidin 3-galactoside being most abundant in fruits and malvidin 3,5-diglucoside in flowers [34]. Grapes predominantly contain malvidin-3-glucoside [25]. The sugar moieties of malvidin glycosides are found at approximately 7% in foods. Malvidin plays a predominant role in regulating both short- and long-term cellular activities and exhibits significant antioxidant properties [6]. Rahman et al. conducted a study on individual anthocyanins and found that delphinidin, isolated from blueberry extracts, has the highest ability to scavenge superoxide species, followed by petunidin, malvidin, cyanidin, peonidin, and pelargonidin, all at concentration of 1 μM. The same trend was observed in their ability to capture peroxynitrite radicals at the same concentration [35]. Table 1 provides a list of the dietary sources of malvidin, which is one of the six major anthocyanins.

Table 1. Types of anthocyanins in fruits, vegetables, and their derivatives, with special emphasis in malvidin and its glycosides.

Source	Dominant Anthocyanins	Concentration	Refs.
Berry (*Berberis lycium* Royle)	Cyanidin-3,5-dihexoside, cyanidin-3-galatoside, cyanidin-3-glucoside, cyanidin-3-lathyroside, cyanidin-3-rut, delphinidin-3-glucoside, malvidin-3,5-dihexoside, pelargonidin-3,5-diglucoside, pelargonidin-3-pentoxilhexoside, pelargonidin-3-rutinoside, pelargonidin-hexoside, peonidin-3-rutinoside	Malvidin-3,5-dihexoside: 4.21%	[36]
Bilberry (*Vaccinium myrtillus* L.)	Cyanidin-3-arabinoside, cyanidin-3-galatoside, cyanidin-3-glucoside, delphinidin-3-arabinoside, delphinidin-3-glucoside, delphinidin-3-galatoside, malvidin-3-arabinoside, malvidin-3-galatoside, malvidin-3-glucoside, peonidin-3-arabinoside, peonidin-3-galatoside, peonidin-3-glucoside, petunidin-3-arabinoside, petunidin-3-galatoside, petunidin-3-glucoside	Malvidin-3-arabinoside: 175–295 mg/100 g dry weight; malvidin-3-galatoside: 82–127 mg/100 g dry weight; malvidin-3-glucoside: 344–506 mg/100 g dry weight	[37]
Blackberry (*Rubus fruticosus* L.)	Cyanidin, cyanidin-3-glucoside, cyanidin-3-arabinoside, malvidin-3-galactoside, malvidin-3-glucoside	-	[38,39]
Blueberry (*Vaccinium corymbosum* L.)	Cyanidin-3-arabinoside, cyanidin-3-galatoside, cyanidin-3-glu, delphinidin-3-arabinoside, delphinidin-3-galatoside, delphinidin-3-glucoside, malvidin-3-arabinoside, malvidin-3-galatoside, malvidin-3-glucoside, peonidin-3-galatoside, peonidin-3-glucoside, petunidin-3-arabinoside, petunidin-3-galatoside, petunidin-3-glucoside	Malvidin-3-arabinoside: 147.6–697 mg/100 g dry weight Malvidin-3-galatoside: 137–330 mg/100 g dry weight Malvidin-3-glucoside: 3–200.4 mg/100 g dry weight	[37,40,41]
Dabai (*Canarium odonthophyllum* Miq.)	Delphinidin, cyanidin, cyanidin-3-glucoside, cyanidin-3-rutinoside, malvidin-3,5-diglucoside, pelargonidin, peonidin-3-glucoside	Malvidin-3,5-diglucoside: 0.07–0.20 mg/100 g dry weight	[42]
Red grape (*Vitis vinifera* L.)	Cyanidin-3-glucoside, delphinidin-3-glucoside, malvidin-3-acetylglucoside, malvidin-3-glucoside, malvidin-3-*p*-coumarylglucoside, peonidin-3-acetylglu, peonidin-3-glucoside, peonidin-3-*p*-coumarylglucoside, petunidin-3-glucoside	Malvidin-3-acetylglucoside: 1.51–29.22% Malvidin-3-glucoside: 32.40–58.96% Malvidin-3-*p*-coumarylglucoside: 7.29–18.46%	[43]
Red wine	Cyanidin-3-glucoside, delphinidin 3-glucoside, malvidin-3-acetylglucoside, malvidin-3-coumarylglucoside, malvidin 3-glucoside, peonidin-3-acetylglucoside, peonidin 3-glucoside, peonidin-3-ρ-coumarylglucoside, petunidin 3-glucoside	Malvidin-3-acetylglucoside: 2.42–144.27 mg/L Malvidin-3-coumarylglucoside: 2.48–19.96 mg/L Malvidin 3-glucoside: 125.44–353.13 mg/L	[44,45]
Sweet cherries (*Prunus avium* L.)	Cyanidin 3-rutinoside, cyanidin-3-5-diglucoside, cyanidin-3-arabinoside, cyanidin-3-coumaroyl-diglucoside, cyanidin-3-glucoside, cyanidin-3-glucoside, cyanidin-3-rutinoside, cyanidin-3-rutinoside, cyanidin-3-sambubioside, cyanidin-3-sophoroside, delphinidin 3-rutinoside, malvidin-3-glucoside-acetaldehyde, pelargonidin 3-rutinoside, pelargonidin-3-glucoside, pelargonidin-3-rutinoside, pelargonidin-3-rutinoside, peonidin 3-rutinoside, peonidin-3-glucoside, peonidin-3-rutinoside, peonidin-3-rutinoside	Malvidin-3-glucoside-acetaldehyde: 0.08–011 mg/100 g fresh weight	[46–48]
Tomato (*Solanum lycopersicum* L.)	Delphinidin-glycoside, delphinidin-rutinoside, malvidin 3-glucoside, malvidin-glycoside, malvidin-ρ-coumaroyl-rutinoside-glycoside, petunidin rutinoside, petunidin ρ-coumaroyl-rutinoside, petunidin ρ-coumaroyl-rutinoside-glycoside	Malvidin 3-glucoside: 54.77–298.57 μg/1 g dry weight	[49,50]

4. Antioxidant Properties of Malvidin and Their Glycosides

Free radicals are reactive species that can be produced through natural metabolic processes or external sources. They can be derived from oxygen (such as hydroxyl, peroxyl, and superoxide) or nitrogen (such as nitric oxide and peroxynitrite). In addition, there are even-numbered free radical species, such as H_2O_2 and lipid peroxide. The accumulation of these radicals can be toxic to cells and trigger reactions such as the oxidation of cellular components, including nucleic acids, proteins, and lipids. This condition is known as oxidative stress, resulting from an imbalance between free radical production and neutralization [51,52]. Anthocyanins have been shown to possess the ability to scavenge free radicals, particularly harmful oxidants such as reactive oxygen and nitrogen species (ROS and RNS) [34,53].

The unique structure of the flavylium cation (AH+) gives anthocyanins distinct antioxidant properties [34,54]. These compounds can neutralize reactive radical species through a single electron transfer (SET) reaction or by a hydrogen atom transfer (HAT) mechanism. In SET, the antioxidant donates an electron to the free radical to neutralize it, while in HAT, the antioxidant donates a hydrogen atom to the free radical that then stabilizes the radical. Both mechanisms usually occur simultaneously, and the reaction mechanism is determined by the structure, solubility, partition coefficient, and solvent polarity of the antioxidant [51,55,56]. Molecules such as anthocyanins have been reported to have the ability to modulate oxidative stress [34,57,58]. In vitro studies demonstrating the antioxidant potential of malvidin and its glycosides are presented in Table 2.

Table 2. Antioxidant activities of malvidin and its glycosides evaluated in an in vitro model.

Compounds	Cell Line	Effect	Refs.
Malvidin Malvidin-3-glucoside Malvidin-3-galactoside	Human umbilical vein endothelial cells	Decreased: ROS levels and xanthine oxidase enzyme activity Increased: superoxide dismutase enzyme activity	[59]
Malvidin 3,5-diglucoside	Human endothelial cells	Decreased: ROS levels	[60]
Malvidin-3-glucoside	Bovine aortic endothelial cells pretreated with peroxynitrite	Decreased: ROS levels Inhibited mitochondrial apoptotic signaling pathways by preventing mitochondrial membrane depolarization, activation of caspase-3 and -9, and reducing the expression of the proapoptotic Bax protein	[61]
Malvidin-3-glucoside, Malvidin-3-galactoside	Human retinal pigment epithelial cells pretreated with H_2O_2	Decreased: ROS and malondialdehyde levels Increased: superoxide dismutase, catalase, and glutathione peroxidase enzymes activity	[62]
Malvidin-3-arabinoside	Human colorectal adenocarcinoma cells pretreated with ethyl carbamate	Decreased: oxidative damages Enhancing autophagy flux	[63]
Malvidin	human fibroblast cells (WI-38) pretreated with H_2O_2	Decreased: lipid peroxidation	[64]

Emandi et al. conducted an in vivo study on rats to investigate the effect of malvidin on renal ischemia–reperfusion injury, which is caused by oxidative stress. The authors found that the administration of malvidin resulted in the increased activity of antioxidant enzymes, such as catalase (CAT) and superoxide dismutase (SOD), as well as decreased levels of the oxidative stress marker malondialdehyde (MDA), compared to the reperfusion injury group. Based on these results, the authors suggest that consumption of malvidin may have a protective effect against acute kidney injury induced by renal reperfusion injury, partly by inhibiting oxidative stress in renal tissues [65].

In summary, malvidin and its glycosides exhibit potent antioxidant activity by donating an electron or hydrogen atom to neutralize free radicals. In addition to this mode of action, they also inhibit enzymes involved in the production of ROS; upregulate or protect antioxidant defenses; and induce antioxidant enzymes such as glutathione peroxidase (GPx), CAT, and SOD, which decompose harmful compounds. Furthermore, studies have shown that malvidin and its glycosides can inhibit the expression of the xanthine oxidase (XO) enzyme. Overall, these findings suggest that malvidin and its glycosides have the potential to block oxidative stress, indicating their usefulness as an antioxidant agents.

5. Anti-Inflammatory Properties of Malvidin and Their Glycosides

Highly reactive free radicals can cause damage to cellular and tissue components by oxidizing nucleic acids, proteins, and lipids, resulting in inflammation. Studies have shown that an increase in ROS production leads to the generation of more pro-inflammatory markers. Additionally, experimental evidence has demonstrated that ROS activate the NF-κB pathway, which triggers the transcriptional activation of genes related to inflammatory response [66,67].

In the human body, inflammation is a crucial response, but it is also involved in the pathogenesis of various diseases, such as microbial and viral infections; exposure to allergens; autoimmune disorders; and chronic diseases. Upon activation, innate immune cells secrete proinflammatory cytokines and chemokines, which trigger the production of ROS/RNS. These molecules initiate signaling cascades that further stimulate the release of more proinflammatory agents. However, prolonged inflammation can lead to cell damage or cellular hyperplasia due to excessive ROS production by inflammatory cells. Additionally, during inflammation, ROS can interact with DNA in mitotic cells, leading to permanent genomic mutations, such as point mutations, gene deletions, or gene rearrangements. In response to inflammation, cellular antioxidant systems are activated to counter the overproduction of free radicals by inducing genes involved in DNA repair. However, in cases of chronic inflammation, the rate of ROS-induced DNA damage is significant due to the depletion of cellular antioxidants. This makes cells more prone to transformation and increases the frequency of mutations caused by inflammatory cells [68,69]. Molecules such as anthocyanins have been reported to have the ability to modulate the inflammatory processes [70,71]. In vitro studies demonstrating the ant-inflammatory potential of malvidin and its glycosides are presented in Table 3.

Table 3. Anti-inflammatory activities of malvidin and its glycosides evaluated in an in vitro model.

Compounds	Cell Line	Effect	Refs.
Malvidin	Human umbilical vein endothelial cells pretreated with tumor necrosis factor-alpha	Decreased: monocyte chemotactic protein-1, intercellular adhesion molecule-1, and vascular cell adhesion molecule-1 production Inhibition: degradation of IκBα and the nuclear translocation of p65	[72]
Malvidin, malvidin-3-glucoside, malvidin-3-galactoside	Human umbilical vein endothelial cells pretreated with tumor necrosis factor-alpha	Decreased: monocyte chemotactic protein1, intercellular adhesion molecule-1, vascular cell adhesion molecule-1 production and angiotensin I-converting enzyme activity Inhibition: degradation of IκBα and the nuclear translocation of p65	[73]
Malvidin-3-glucoside and malvidin-3-galactoside	Human umbilical vein endothelial cells pretreated with tumor necrosis factor-alpha	Decreased: monocyte chemo-tactic protein-1, intercellular adhesion molecule-1, and vascular cell adhesion molecule-1 production Inhibition: degradation of IκBα and the nuclear translocation of p65	[74]
Malvidin-3-glucoside	Bovine arterial endothelial cells pretreated with peroxynitrite	Decreased: inducible nitric oxide synthase activity, cyclooxygenase activity, IL-6 production Increased: eNOS activity and NO production	[75]
Malvidin	RAW264.7 macrophages stimulated by bacterial lipopolysaccharide	Decreased: lipopolysaccharide-induced nuclear factor-kappaB, poly ADP-ribose polymerase and mitogen-activated protein kinase activation, reactive oxygen species production and mitochondrial depolarization Increased: mitogen-activated protein kinase phosphatase-1 expression and Akt activation	[76]
Malvidin-3-glucoside	Rat macrophages stimulated by bacterial lipopolysaccharide	Decreased: tumor necrosis factor-alpha, IL-1, IL-6 and inducible nitric oxide synthase activity	[77]
Malvidin	Human monocytic cells (THP1) stimulated by bacterial lipopolysaccharide	Decreased: IL-6, tumor necrosis factor-α, and IL-1β production Increased: IL-10 production	[78]
Malvidin	Peripheral blood mononuclear cells stimulated by bacterial lipopolysaccharide	Decreased: IL-6, tumor necrosis factor-alpha and IL-1β production, cyclooxygenase 2 activity	[79]
malvidin	Human fibroblast cells (WI-38) pretreated with H_2O_2	Decreased: NF-κB production and cyclooxygenase 2 and inducible nitric oxide synthase activity	[64]

Iban-Arias and colleagues investigated the potential therapeutic effects of malvidin-3-glucoside on inflammation induced by the inflammasome in vitro and in a mouse model of chronic unpredictable stress. Their findings demonstrated that malvidin-3-glucoside targets the inflammasomes NLRP3, NLRC4, and AIM2, leading to a reduction in caspase-1 and IL-1 protein levels in murine primary cortical microglia and the brain. The demonstrated beneficial effect of malvidin includes its ability to counteract anxiety and depression. The study also suggests the potential of malvidin-3-glucoside in mitigating LPS-induced inflammation in vitro, particularly in the context of bacterial-mediated inflammation [80]. Moreover, in a mouse model, malvidin was found to inhibit ROS-dependent NLRP3 inflammasome activation, leading to a decrease in serum pro-inflammatory cytokine secretion and mitochondrial-pathway-mediated apoptosis. Malvidin also increased Bcl-2 protein levels, while decreasing Bax, cytochrome C, and caspase-3 levels. The study demonstrated that malvidin targeted the AMPK-α/UCP2 axis, which restored mitochondrial function and reduced ROS accumulation, ultimately leading to the inhibition of NLRP3 inflammasome activation and mitochondrial apoptosis in a ROS-dependent manner [81].

Dai et al. conducted an in vivo study to examine the effects of malvidin on osteoarthritis in rats induced by monosodium iodoacetate administration. The results showed that malvidin treatment significantly relieved pain in the osteoarthritis rats and reduced the expression of apoptotic markers in chondrocytes. In addition, the upregulation of proinflammatory cytokines IL-1β, IL-6, TNF-α, and matrix metalloproteinases (MMPs) induced by monosodium iodoacetate in cartilage tissues were significantly reversed by malvidin treatment. Malvidin also inhibited the NF-κB pathway through a mechanism independent of NF-κB inhibitor (IκBα) by suppressing p65 nuclear transport in vitro. Overall, the findings suggest that malvidin can effectively attenuate pain and inflammation induced by osteoarthritis by inhibiting the NF-κB signaling pathway, suppressing proinflammatory cytokine expression, and reducing chondrocyte apoptosis [82].

To summarize, malvidin and its glycosides have been shown to possess anti-inflammatory effects through their radical scavenging activities. Various experiments have provided evidence that malvidin and its glycosides can modulate different inflammatory mediators, including protein kinases; transcription factors such as AP-1 and NF-κB; enzymes such as iNOS and COX-2; and cytokines such as IL-1β, IL-6, and TNF-α, which are known to play a central role in the inflammation process. These findings suggest that malvidin and its glycosides could serve as potential therapeutic agents for inflammatory diseases by blocking inflammation.

6. The Roles of Malvidin and Its Glycosides in Oxidative Stress and Inflammation-Mediated Chronic Disorders

Oxidative stress and inflammation are two significant physiological processes with close connections that modulate cellular physiological and pathological responses. The body possesses an endogenous defense system to maintain cellular homeostasis by combating both processes. However, prolonged oxidative stress and inflammation can be detrimental to the body's defense system [83,84]. ROS can activate pro-inflammatory cytokines and the NLRP3 inflammasome, thereby regulating inflammation [85,86]. The findings regarding the antioxidant and anti-inflammatory properties of malvidin and its derivatives provide an insight into their potential molecular mechanisms of action in mitigating oxidative stress and inflammation. Research on malvidin and its glycosides has demonstrated their ability to act as antioxidants through various mechanisms, including scavenging free radicals and anions.

Additionally, studies have found that these molecules can activate the nuclear factor erythroid 2-related factor (Nrf2) pathway, which is sensitive to oxidative stress. Nrf2 is a transcription factor that is normally bound to the actin protein Keap1, but exposure to ROS leads to Nrf2 degradation via the ubiquitin proteasome pathway. However, malvidin and its glycosides stabilize Nrf2, allowing it to accumulate in the nucleus and activate the antioxidant response element (ARE)-regulated target genes. Nrf2 plays a crucial role in regulating endogenous antioxidant defense by upregulating enzymes such as glutathione S-transferase (GST) and peroxidase (GPx), hemeoxygenase-1 (HO-1), CAT, and SOD. The Keap1-Nrf2 system helps protect cellular components from oxidative damage caused by ROS by increasing antioxidant enzyme expression and decreasing sensitivity to oxidative-stress-related inflammatory reactions [87,88]. Additionally, Nrf2 promotes the activation of the pentose phosphate pathway (PPP), leading to NADPH production, which is involved in regenerating reduced glutathione (GSH) from GSH disulfide and maintaining cellular antioxidant levels [89]. Therefore, enhancing Nrf2 induction efficiency provides homeostatic mechanisms for the antioxidant activity of malvidin and its glycosides.

Research has also revealed that these molecules target arachidonic acid. Lipid mediators derived from arachidonic acid, such as prostaglandins produced via COX-2 and leukotrienes via lipoxygenases (LOX), are important targets of anthocyanins. COX enzymes are essential for the conversion of arachidonic acid to prostanoids, which play a significant role in inflammation. LOX uses arachidonic acid as a substrate and catalyzes four different reactions, namely 5*S*, 12*R*, 12*S*, or 15*S* oxygenation. The oxygenated substrates of these enzymes initiate biological reactions, activate cellular signaling through surface receptors, or are metabolized into potent lipid mediators. Previous studies have highlighted the influence of LOX in the inflammatory process [90–92].

According to several studies, malvidin and its glycosides have demonstrated inhibitory effects on the NF-κB pathway which plays a crucial role in regulating the inflammatory response. The canonical NF-κB signaling pathway is activated by various inflammatory stimuli, such as exposure to LPS, TNF-α, or IL-1. Activation of the canonical NF-κB pathway involves phosphorylation-dependent activation of the IKKs complex, leading to the degradation of inhibitory IκB proteins by the ubiquitin–proteasome system. This release allows the κB transcription factor to translocate to the nucleus and activate target genes. NF-κB is responsible for the transcription of a wide range of genes involved in the inflammatory response, including chemokines, cytokines, and adhesion molecules, as well as genes that negatively regulate its activity [93,94].

Several studies have suggested that malvidin and its glycosides possess inhibitory effects on the MAPK pathway. Under normal physiological conditions, MAPK signaling pathways play a crucial role in various processes, including cell proliferation, survival, and differentiation. These pathways are activated by extracellular stimuli, such as cytokines, Toll-like receptors (TLRs), and oxidative stress, which trigger the corresponding receptors to transduce intracellular signaling into the nucleus through three primary MAPK cascades. In humans, the extracellular-signal-regulated kinase (ERK) 1/2, C-jun *N*-terminal kinase (JNK), and p38 MAPKs are the three main kinases. It is hypothesized that intracellular kinases such as MAPKK, MEK, or MKK and MAPK initiate downstream activation of MAPKs (ERK1/2, JNK, and p38 MAPK) by phosphorylating specific residues in the appropriate protein. Once activated, MAPKs move into nuclei and phosphorylate the relevant transcription factors, thereby regulating gene expression. Activation of the MAPK pathway also leads to the activation of the transcription factor AP-1, which, similar to NF-κB, contains transcriptional regulator binding sites for most inflammatory mediators. AP-1 can also bind promoters of inflammatory mediators independently of NF-κB during inflammation [95,96]

The antioxidant and anti-inflammatory mechanisms of action of malvidin and its glycosides are presented in Figure 2. Oxidative stress and inflammation are both key causative factors for the onset of several diseases, including cancer, metabolic and cardiovascular disorders, and neurodegenerative diseases.

6.1. Anti-Cancerogenic Properties of Malvidin and Their Glycosides

Cancer is a significant cause of death worldwide, with approximately 10 million deaths in 2020, accounting for approximately one in six deaths [97]. Breast, lung, colon, rectum, and prostate cancer are among the most prevalent types. Cancer is a complex process involving at least three stages: initiation, promotion, and progression [98]. Cell-cycle dysregulation is a crucial factor in cancer cell growth. The cell cycle comprises a series of tightly regulated events that enable cell growth and division. Cyclin-dependent kinases (CDKs) are essential components of the cell-cycle machinery that, when activated, facilitate progression from one phase to the next. The regulation of CDKs involves positive regulation by cyclins and negative regulation by CDK inhibitors (CDKIs). In cancer, the cell cycle is often dysregulated through the overexpression of cyclins or the absence of CDKIs [99]. Oxidative stress and inflammation are closely linked to cancer progression. ROS may initiate carcinogenesis directly or through the activation of signaling pathways.

Tumor promoters have the ability to attract inflammatory cells and induce them to produce ROS, which is an important characteristic [100,101].

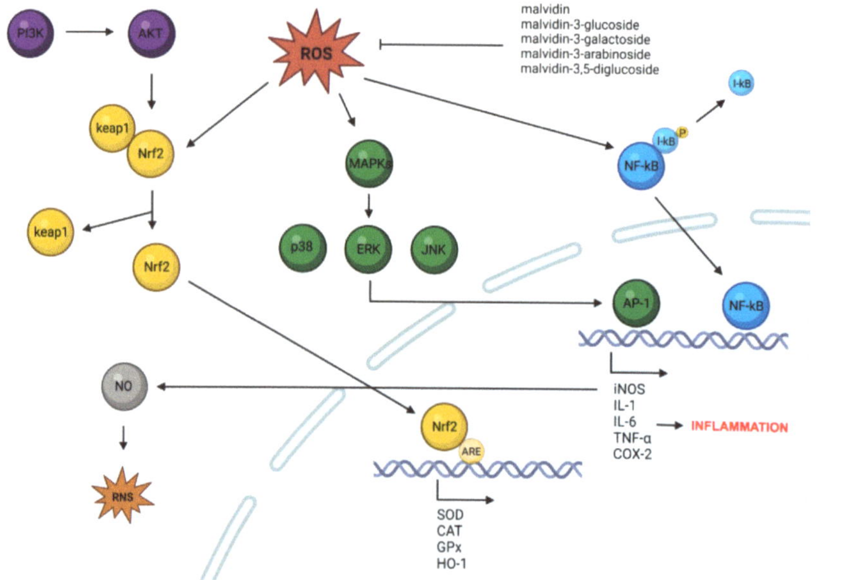

Figure 2. The mechanisms through which malvidin and its glycosides act as antioxidant and anti-inflammatory agents (created by BioRender: https://www.biorender.com/ (accessed on 6 April 2023)).

Cancer cells have a modified metabolism to meet the increased energy requirements due to their rapid growth and proliferation. Consequently, they produce more ROS than normal cells to maintain typical subcellular activities, including signal transduction and gene expression [102,103]. Research has established that ROS contribute to tumorigenesis through the activation of oncogenic cell signaling pathways [104]. The phosphoinositide-3-kinase (PI3K)/AKT and MAPK pathways, which promote cell survival, proliferation, metabolism, nutrient uptake, and inflammation, are commonly associated with cancer [105,106]. H_2O_2 inhibits protein tyrosine phosphatase 1B (PTB1B), which, in turn, prevents the dephosphorylation of epidermal growth factor receptor (EGFR), leading to the activation of downstream PI3K/AKT and RAS-MEK-ERK (ERK/MAPK) pathways [104]. The GTPase RAS is frequently implicated in tumorigenesis by activating MAPK pathways and regulating transcription. Specifically, K-RAS activates JNK, ERK, and p38 signaling pathways, with the latter also contributing to ROS production via NADPH oxidase 1 (NOX1) [107]. It is hypothesized that ROS may have varying effects on tumorigenesis, either enhancing, inhibiting, or regulating it, depending on the activation of MAPK pathways in different cancers [108]. However, the excessive accumulation of ROS is typically detrimental to cancer cells [109]. Despite contributing to cancer progression, high levels of ROS can also induce cell death in cancer cell lines [85,86]. Inflammation is regulated by ROS through the activation of pro-inflammatory cytokines and the NLRP3 inflammasome [56].

Inflammation plays a significant role in the development and progression of cancer by promoting cell proliferation, survival signaling, angiogenesis, invasion, and metastasis. Pro-inflammatory cytokines such as IL-1, IL-6, and TNF-α are elevated in both the plasma and cells of cancer patients [110]. These cytokines activate NF-κB and signal transducers and activators of transcription 3 (STAT3), which are involved in cancer growth [111–113]. TNF-α induces tumorigenesis through ROS production, which can cause DNA damage [114].

Mutant p53 has been shown to alter TNF-α signaling to favor NF-κB activation [115]. NF-κB can increase oncogenic K-RAS levels in a positive feedback loop, further correlating chronic inflammation with cancer progression [116].

Molecules such as malvidin and its glycosides have been reported to be able to modulate the carcinogenesis processes [117,118]. Two separate in vitro studies conducted by Lin et al. and Wang et al. investigated the anticancer effects of malvidin-3-galactoside on hepatocellular carcinoma. The studies found that the compound inhibited cell proliferation and colony formation, induced cell-cycle arrest and apoptosis, and suppressed migration and invasion potential by regulating the expression of MMPs. The anticancer effects of malvidin-3-galactoside were associated with the inhibition of the PI3K/AKT, MAPK, and MMP pathways. These findings suggest that malvidin-3-galactoside may have potential for preventing liver cancer by modulating proliferation, apoptosis, migration, and invasion-related pathways [119,120]. In a study conducted by Xu et al., the anticancer activity of malvidin was evaluated using the human colorectal cancer cell line. Their findings showed that malvidin exhibited cytotoxic effects against the cells by inhibiting colony formation and inducing apoptosis. Malvidin was also found to induce G2/M cell-cycle arrest and inhibit cell-cycle-related proteins [121]. Baba et al. investigated the potential of malvidin to target the transcription factor STAT-3 in the oral cancer cell line. Their results indicated that malvidin acts as a STAT-3 inhibitor, suppressing its phosphorylation and nuclear translocation, which, in turn, induced cell-cycle arrest and apoptosis through mitochondrial-mediated pathways [122]. Dahlawi conducted a study on the effects of malvidin on two human leukemia cell lines. The results revealed that malvidin inhibited cell proliferation and induced apoptosis in both leukemia cell lines, as evidenced by caspase-3 activation. Malvidin also caused cell-cycle arrest at the S phase in both cell lines [123]. Similarly, Hyun and Chung reported on the cytotoxic effect of malvidin on the human monocytic leukemia cell line. Malvidin induced apoptosis and arrested the G2/M phase of the cell cycle [124]. Ouanouki et al. conducted a study to investigate the impact of malvidin on TGF-β-induced epithelial–mesenchymal transition (EMT) and its underlying mechanism. EMT is a process that allows benign tumor cells to infiltrate surrounding tissues. The researchers treated human glioblastoma cell line with malvidin before or together with or after the addition of TGF-β. They found that malvidin inhibited TGF-β-induced EMT by affecting both the TGF-β Smad and non-Smad signaling pathways. This inhibitory effect altered the expression of the EMT mesenchymal markers fibronectin and Snail and significantly reduced the migration of cells [125]. The role of malvidin and its glycosides in the regulation of cell cycles in cancer cells is presented in Figure 3.

The anti-carcinogenic effects observed in vitro have been validated in vivo. In a study by Sakthivel et al., mice with Dalton's lymphoma were treated with malvidin for 10 consecutive days, starting from the day of tumor induction. Treatment with malvidin led to a significant reduction in tumor volume and an increase in white blood cell count. The treatment also maintained body weight and the hemoglobin level, and decreased levels of liver enzymes such as aspartate aminotransferase (AST), alanine aminotransferase (ALT), alkaline phosphatase (ALP), and gamma-glutamyl transferase (GGT) were observed. Additionally, malvidin reduced the levels of inflammatory mediators and cytokines such as TNF-α and IL-6, which are molecular targets for cancer prevention. A decrease in the level of ROS, such as NO, was also observed. A histopathological examination showed altered morphological changes in tumor tissue and the alleviation of hepatic architecture due to Dalton's lymphoma. An immunohistochemical analysis revealed the inhibition of iNOS [126]. Lin et al. reported that malvidin-3-galactoside induced apoptosis in liver tumor cells in mice [119]. In another study, Cheng et al. investigated the effects of malvidin-3-galactoside on gut microbiota in liver-cancer mice. The mice were fed diets supplemented with malvidin-3-galactoside for three weeks. The results showed an increase in the abundance of *Verrucomicrobiaceae* and *Ruminococcus*, as well as anti-inflammatory bacteria such as *Akkermansia* and *Sutterella*. On the other hand, a decrease in the abundance of proinflammatory bacteria such as *Dorea*, *Coprobacillus*, *Clostridium*, *Streptococcus*, and *Oscillospira* was observed. Malvidin-3-galactoside was found to enhance signal transduction, membrane transport, and cell motility, as well as induce cell death. The study suggests that malvidin-3-galactoside may have significant impacts on the structure and metabolic function of gut microbiota in liver-cancer mice [127].

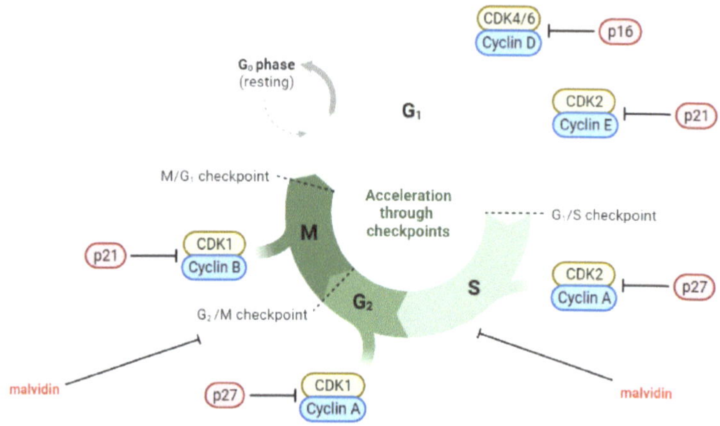

Figure 3. Cell-cycle deregulation in cancer cells and the role of malvidin (created by BioRender: https://www.biorender.com/ (accessed on 10 April 2023)).

In conclusion, abovementioned studies provide evidence of the significant anticancer effects of malvidin and its glycosides, and such effects may be attributed to their antioxidative and anti-inflammatory mechanisms of action. However, there is currently a lack of unequivocal human trials allowing for the assessment of the anticancer effect of malvidin and its glycosides.

6.2. Antidiabetic Properties of Malvidin and Its Glycosides

According to a 2019 report, diabetes and diabetes-related kidney disease were responsible for an estimated two million deaths. Diabetes is a disorder characterized by elevated glucose levels, and it has multiple subtypes [128,129]. Prolonged high levels of glucose and free fatty acids can lead to severe complications, including organ failure and tissue damage [130]. It has been linked to damage in the nervous system, vascular endothelium, and kidneys that can be attributed to stress-activated signaling pathways such as NF-κB and MAPKs and other stress-activated protein kinases.

The production of ROS is triggered by hyperglycemia and is also believed to play a role in the pathogenesis of disease [131]. Elevated glucose levels are known to induce oxidative stress by upregulating mitochondrial ROS, causing protein glycation, and triggering glucose autooxidation. These processes can potentially impair enzyme activity and cellular function. In addition, high levels of free fatty acids can lead to mitochondrial uncoupling and β-oxidation, also resulting in oxidative stress. Moreover, advanced diabetes is associated with reduced levels of important antioxidants such as vitamin E and α-lipoic acid, as well as SOD, an enzyme that plays a crucial role in inactivating the superoxide radical [132]. Superoxide radicals, in turn, activate multiple pathways, including enhanced polyol formation, increased hexosamine pathway flux, and the activation of the PKC isoform [133].

There is a growing body of evidence indicating a correlation between moderate inflammation and the onset of type 2 diabetes mellitus (T2DM). Furthermore, there is a possibility that inflammation may play a role in the development of diabetes. Research has demonstrated that inflammatory cytokines can interfere with insulin-signal transduction, leading to insulin resistance [134]. Additionally, genes such as IL-1α, IL-1β, IL-6, and TNF-α are continually activated in individuals with diabetes. Inflammation primarily targets immune cells and endothelial cells in diabetic patients [135]. In addition, inflammation could potentially trigger the degradation of β-cells. Increased production of IL-1β, IL-6, and IL-8 in pancreatic islets due to glucotoxicity and lipotoxicity in diabetes can lead to a reduction in insulin gene transcription and an increase in macrophages in the pancreas, ultimately leading to apoptosis of β-cells [136].

Molecules such as malvidin and its glycosides have been reported to be able have beneficial effects on diabetes [137,138]. In a study by Herrera-Balandrano et al., it was demonstrated that exposing human hepatocarcinoma cells to high levels of glucose led to a significant increase in hepatic oxidative stress, of up to six times, and a decrease in cell viability. However, pretreatment with malvidin, malvidin-3-glucoside, and malvidin-3-galactoside resulted in a significant reduction of ROS and increase in cell viability. These pretreatments were also able to reduce the expression levels of enzymes involved in gluconeogenesis and lipogenesis and enhance those involved in glycogenolysis and lipolysis via the AMPK signaling pathway in cells, effectively inhibiting hyperglycemia and hyperlipidemia [139]. Huang et al. reported that malvidin, malvidin-3-glucoside, and malvidin-3-galactoside could protect human retinal capillary endothelial cells from high-glucose-induced injury by decreasing ROS and increasing CAT and SOD enzyme activity. The study also revealed that malvidin significantly inhibited NOX4 expression induced by high glucose, while malvidin-3-galactoside downregulated NOX4. Furthermore, malvidin, malvidin-3-glucoside, and malvidin-3-galactoside affected NO levels and inhibited high-glucose-induced intercellular adhesion molecule-1 (ICAM-1) and NF-κB expression. Malvidin-3-galactoside also influenced angiogenesis by decreasing the level of vascular endothelial cell growth factor (VEGF) and inhibiting the PI3K/AKT pathway [140]. In addition, Huang et al. found that pretreatment of human umbilical vein endothelial cells with malvidin, malvidin-3-glucoside, and malvidin-3-galactoside significantly improved high-glucose-induced damage by enhancing endogenous antioxidant SOD and HO-1, reducing ROS generation and NOX4 expression, and increasing cell viability. They also induced vasodilation by increasing the vasodilator NO and its promoters' endothelial NO synthase (eNOS) and peroxisome proliferator-activated receptor (PPAR) levels, while decreasing the vasoconstrictor angiotensin-converting enzyme (ACE), XO-1, and low-density lipoprotein (LDL) levels. These bioactivities were linked to the activation of the PI3K/AKT signaling pathway and the breakdown of the PKC pathway [141].

Clinical trials (https://clinicaltrials.gov (accessed on 14 March 2023)) analyzing the effects of anthocyanins, including malvidin and its glycosides, on patients with T2DM have been reported on the clinicaltrials.gov website. In the first study, patients diagnosed with T2DM were administered Medox, two capsules with anthocyanins, twice daily for 12 weeks. Each capsule contained 80 mg of anthocyanins, including 3.0% of malvidin-3-glucoside, malvidin-3-galactoside, and malvidin-3-arabinoside. The trial showed a significant decrease in serum LDL cholesterol, triglycerides, apolipoprotein B-48, and apo C-III and an increase in high-density lipoprotein (HDL) cholesterol. Moreover, patients in the anthocyanin group had higher total radical-trapping antioxidant parameter and ferric-ion-reducing antioxidant power values than those in the placebo group. Concentrations of 8-iso-prostaglandin F2α, 13-hydroxyoctadecadienoic acid, and carbonylated proteins were significantly lower in patients in the anthocyanin group than in the placebo group. Supplementation with anthocyanin also reduced fasting plasma glucose and homeostasis model assessment for insulin resistance index and increased serum adiponectin and β-hydroxybutyrate concentrations compared with placebo supplementation. The trial was registered as NCT02317211 [142]. In another study, Medox capsules were administered to 160 participants with prediabetes or newly diagnosed T2DM, with a daily supplementation of 320 mg anthocyanins for 12 weeks. The study showed significant increases in serum adipsin and decreases in visfatin between the anthocyanins and placebo groups. Furthermore, improvements in HbA1c, apolipoprotein A-1, and apolipoprotein B were observed in response to the anthocyanin intervention. This study was registered as NCT02689765 [143]. The following study investigated the potential effects of a standardized extract called Mirtoselect® on glucose metabolism in patients with T2DM. Mirtoselect® is derived from the berries of *Vac-*

cinium myrtillus and contains 36% (*w*/*w*) of anthocyanins, including malvidin-3-galactoside, malvidin-3-glucoside, and malvidin-3-arabinoside. Participants with T2DM who were managing their condition through diet and lifestyle alone were given a single oral capsule of either 0.7 g of standardized bilberry extract or a placebo, followed by a polysaccharide drink (equivalent to 75 g of glucose) after a 2-week washout period. The results showed that ingestion of the bilberry extract led to a significant decrease in both glucose and insulin levels compared to the placebo. This trial was registered as NCT01245270 [144]. These findings demonstrate the potential beneficial effects of anthocyanin supplementation in individuals with T2DM.

In conclusion, the abovementioned studies provide evidence of the significant antidiabetic effects of malvidin and its glycosides, which may be attributed to their antioxidative and anti-inflammatory mechanisms of action. The inclusion of diet-induced anthocyanins, such as malvidin and its glycosides, may present a natural alternative for improving glycemic control in individuals with T2DM. However, further research is necessary to fully comprehend the specific ways in which anthocyanins, including malvidin and its glycosides, can contribute to the management of T2DM.

6.3. *Cardioprotective Effect of Malvidin and Their Glycosides*

Cardiovascular diseases (CVDs) are a major cause of mortality worldwide, resulting in an estimated 17.9 million deaths annually. CVDs refer to a group of disorders that affect the heart and blood vessels, including conditions such as myocardial infarction, heart failure, stroke, peripheral arterial disease, arrhythmia, and atrial fibrillation. In addition, CVDs have been linked to the development of dementia and loss of daily living function. Overall, CVDs are a significant cause of mortality, morbidity, disability, and loss of function [145,146].

Oxidative stress is a significant contributor to the development of atherosclerosis, which is the primary cause of cardiovascular diseases. Atherosclerosis involves the formation of fibrofatty lesions in the artery wall. The development of atherosclerotic lesions is likely initiated by the oxidation of LDL, which is responsible for carrying cholesterol through the blood. Oxidized LDL activates the endothelium and triggers an immune system response, leading to the migration of inflammatory cells such as monocytes and T cells into the arterial intima. Phagocytosis of the oxidized LDL by macrophages causes the release of ROS and pro-inflammatory markers, which lead to further LDL oxidation [147]. This cycle perpetuates the progression of atherosclerosis [148].

The dysfunction of the endothelium may be caused by the presence of ROS generated by the electron transport chain of the mitochondria. This includes mitochondrial ROS and ROS generated by NOX, which have been shown to positively regulate each other. CVDs have been linked to ROS produced by NOX enzymes expressed in vascular tissue. Specifically, NOX4 is involved in regulating vascular smooth-muscle cells, fibroblasts, and the differentiation and migration of cardiac cells. Overexpression of NOX4 can lead to negative effects on cells due to the increased production of H_2O_2. Moreover, in a healthy state, NO plays a crucial role in regulating endothelial function. However, under conditions of elevated ROS levels, such as when superoxide ions are present, NO can become oxidized and transform into peroxynitrite, a potent oxidizer that causes further oxidation and cell damage [148–150].

Malfunctions in several enzymes related to ROS, including XO, LOX, SOD, and GPx, may also predispose to CVDs. XO participates in the regulation and dysregulation of the endothelium. The generation of ROS by XO can interact with superoxide ions, resulting in the formation of peroxynitrite and inducing cellular harm. Furthermore, the ROS produced by XO can directly interact with the epidermal growth factor receptor (EGFR), contributing to vascular remodeling and CVDs. LOX also contributes to the development of CVDs [148]. Specifically, 5-LOX is of interest due to its involvement in the activation of inflammatory cells. In response to oxidative stress, 5-LOX can be upregulated; meanwhile, the downregulation of 5-LOX has shown positive effects on myocardial infarction [148,151]. SOD also plays a role in vascular disease under oxidative stress. SOD is responsible for converting superoxide ions into H_2O_2, with SOD1 being crucial in maintaining endothelial function by preserving NO availability [152,153]. GPx activity is frequently used as a marker for disease and ROS levels in individuals with coronary artery disease and atherosclerosis [154]. GPx safeguards erythrocytes from oxidation and subsequent damage [155]. Research has shown an inverse relationship between erythrocyte GPx activity and coronary artery disease, similar to that observed in atherosclerosis [154].

Cytokines are known to play a crucial role in the development and progression of atherosclerosis and CVDs. For instance, interferon (IFN)-γ stimulates the expression of the pro-inflammatory phenotype (M1) of macrophages, leading to the formation of arterial plaques and cell apoptosis, which results in lipid expulsion into adjacent plaque regions. Interleukin IL-1β, in addition to IFN-γ, can also activate the M1 macrophage phenotypes in an auto-inflammatory way or stimulate other pro-inflammatory genes. Additionally, oxidized LDL can increase the expression of IL-1β, resulting in increased inflammation within the atherosclerotic plaque region. The chemokine (C-C motif) ligand 2 (CCL-2) plays a role in the development of atherosclerotic plaques through the activation of PKC, ERK1, and NF-κB signaling pathways. Studies have shown that cells lacking CCL-2 produce smaller plaques [156].

Molecules such as malvidin and its glycosides have been reported to have beneficial effects on CVDs [157,158]. Del Bo et al. conducted a study to investigate the potential of malvidin-3-glucoside to reduce inflammation-driven adhesion of monocytes to endothelial cells and the secretion of cell-adhesion molecules, including E-selectin and vascular cell adhesion molecule 1 (VCAM-1). The adhesion was induced by TNF-α. The results showed that malvidin-3-glucoside reduced monocytes' adhesion to endothelial cells and significantly reduced E-selectin levels, but not VCAM-1 levels. The study concluded that malvidin-3-glucoside had the potential to resolve inflammation-driven adhesion by reducing E-selectin concentrations and that it was effective at physiologically relevant concentrations [159]. Quintieri et al. investigated the effects of malvidin on cardiovascular function, using isolated and Langendorff-perfused rat hearts. The authors observed that malvidin treatment resulted in negative inotropic and lusitropic effects, as well as coronary dilation. The mechanism-of-action analysis revealed that the cardiac effects of malvidin required the activation of the PI3K/NO/cGMP/cGMP-dependent protein kinase pathway and were associated with increased intracellular guanosine-3′,5′-cyclic monophosphate (cGMP) and the phosphorylation of eNOS, PI3K-AKT, ERK1/2, and glycogen synthase kinase 3 beta (GSK-3β). AKT and eNOS phosphorylation were confirmed in endothelial cells. Malvidin acted as a postconditioning agent, being able to elicit cardioprotection against ischemia/reperfusion damages [160].

Wei et al. investigated the effect of malvidin on myocardial infarction induced by isoproterenol in rats. The study demonstrated that malvidin has significant cardioprotective effects by restoring the defensive activities of endogenous antioxidants, such as CAT, SOD, and GSH, and by reducing the levels of lipid peroxidation and serum marker enzymes lactate dehydrogenase (LD) and creatine kinase (CK). Malvidin also improved the histopathological changes and impaired mitochondria in a cardiac necrosis model stimulated by isoproterenol. Additionally, the results showed that the nuclear translocation of Nrf-2 and subsequent HO-1 expression might be associated with the activation of the NF-kB pathway [161].

An ongoing human clinical trial (https://clinicaltrials.gov (accessed on 17 March 2023)) is investigating whether exposure to anthocyanins, including malvidin, can lower the levels of various markers associated with CVDs risk. A study was conducted to investigate the effects of daily consumption of freeze-dried blueberry powder containing *Vaccinium* species—which are a source of anthocyanins, including malvidin [162,163]—on blood pressure and arterial stiffness in postmenopausal women with pre- and stage-one-hypertension. After eight weeks, the blueberry-powder group showed significantly lower systolic and diastolic blood pressure and brachial–ankle pulse wave velocity compared to baseline levels, as well as higher levels of NO. This suggests that daily blueberry consumption may reduce cardiovascular disease risk by reducing blood pressure and arterial stiffness, possibly due to the increased NO production. The trial was registered as NCT03370991 [164]. This finding demonstrates the potential beneficial effects of anthocyanin supplementation in individuals with hypertension.

In conclusion, the abovementioned studies provide evidence of the significant cardioprotective effects of malvidin and its glycosides, and such effects can be attributed to their antioxidative and anti-inflammatory mechanisms of action. The inclusion of diet-induced anthocyanins, such as malvidin and its glycosides, may present a natural alternative for prevention against the development of CVDs. However, further research is necessary to fully comprehend the specific ways in which anthocyanins, including malvidin and its glycosides, can contribute to the protection of CVDs.

6.4. Neuroprotective Effects of Malvidin and Its Derivatives

Dementia, affecting over 55 million individuals globally, arises from a range of brain diseases and injuries. The primary cause is Alzheimer's disease (AD), accounting for 60–70% of cases [165]. Advanced age is a significant risk factor for cognitive impairment and neurodegenerative conditions such as Alzheimer's and Parkinson's diseases [166].

Aging leads to increased oxidative stress in the nervous system, impairing nerve regeneration and function [167]. Specifically, voltage-gated potassium (K+) channel sub-family B member 1 (KCNB1) is shown to be subjected to moderate oxidation in aged people, causing hippocampal functional impairment. In AD or following a trauma, the oxidation of KCNB1 is aggravated, resulting in marked neurodegeneration [148]. Moreover, inflammation is detected in the human central nervous system. Increased levels of inflammatory cytokines, including TNF-α, IFN-γ-induced protein 10 (IP-10), and IL-8, have been observed in the cerebrospinal fluid of aging individuals [168].

Molecules such as malvidin and its glycosides have been reported to be able to impact beneficial effects on neurodegenerative disease [169,170]. In a study by Lin and colleagues, the potential neuroprotective effects of malvidin against hypoxia in glial cells were explored. The cells were initially treated with malvidin under normoxic conditions and subsequently exposed to hypoxia, using sodium dithionite in an anaerobic incubator. The results indicated that preincubation with malvidin led to increased CAT activity and GSH concentration following hypoxia treatment, compared to control samples without anthocyanidin preincubation. Additionally, cells preincubated with malvidin showed higher SOD activities [171]. In another study by Zhao et al., the effects of malvidin on a murine microglial cell line were investigated, showing that it prevented mitochondrial dysfunction and accumulation of ROS, decreased lipid peroxidation, and increased antioxidant enzyme activity in the cerebrum [81]. In the same study, malvidin was also administered to mice with sepsis-associated encephalopathy, resulting in the restoration of neurobehavioral retardation; decreased levels of serum S100 calcium-binding protein β (S100β) and neuron-specific enolase (NSE); sustained cerebrum morphological structure, improved blood–brain barrier integrity, with elevated tight junction proteins; and decreased Evans blue leakage. Ultimately, malvidin protected mice from brain injury [81].

Giliani et al. conducted a study to investigate the antioxidant properties of malvidin against aluminum chloride (AlCl3)-induced neurotoxicity in rats. After 61 days of treatment, the rats' brains were examined using a neurochemical assay. The results showed that malvidin improved the behavioral parameters affected by AlCl3. A biochemical analysis demonstrated that the oral administration of malvidin had neuroprotective effects through the regulation of antioxidant levels and neuroinflammation in the rats exposed to $AlCl_3$. These findings suggest that malvidin possesses antioxidant activity by inhibiting acetylcholinesterase and regulating oxidative stress in neuronal cells [172]. Lapi et al. conducted a study to investigate the protective effects of malvidin against damage caused by bilateral common carotid artery occlusion and reperfusion in rat pial microcirculation. In hypoperfused rats, the occlusion and reperfusion caused a decrease in arteriolar diameter, an increase in microvascular leakage and leukocyte adhesion, and decreased capillary perfusion and red blood cell velocity, as well as marked neuronal damage and ROS generation. However, malvidin administration induced arteriolar dilation in a dose-related manner, reduced microvascular leakage and leukocyte adhesion, protected capillary perfusion and red blood cell velocity, prevented neuronal damage, and decreased ROS generation. The study suggested that malvidin's effects were mediated by eNOS activation and scavenger activity, and NO synthase inhibition significantly attenuated malvidin's effects on arteriolar diameter. Furthermore, an increase in eNOS and *p*-eNOS expression and a decrease in MMP-9 activity were observed after malvidin's administration [173].

Clinical trials on humans (https://clinicaltrials.gov (accessed on 21 March 2023)) have investigated the effects of anthocyanins, including malvidin, on memory disorders and dementia patients. A clinical trial registered as NCT03419039 investigated the effects of anthocyanins, including malvidin, on cognitive functioning in individuals at risk for dementia. The trial included 206 participants with mild cognitive impairment or cardiometabolic disorders who received 320 mg/day of naturally purified anthocyanins, including malvidin, for 24 weeks. Although there was no significant difference in episodic memory at the end of the study, statistically significant differences in slopes were observed [174]. Another trial, registered as NCT01746303, evaluated the effects of long-term supplementation of blueberry powder containing *Vaccinium* species, a source of anthocyanins, including malvidin [162,163], on older adults with cognitive complaints. The trial included 94 participants who received 25 g of blueberry powder daily for 24 weeks. The tested group reported fewer cognitive symptoms and showed improved memory discrimination, indicating improved cognition. The cognitive benefit was associated with the presence of urinary anthocyanins [175]. These findings demonstrate the potential beneficial effects of anthocyanin supplementation in individuals with dementia.

In conclusion, the abovementioned studies provide evidence of the significant neuroprotective effects of malvidin and its glycosides that may be attributed to their antioxidative and anti-inflammatory mechanisms of action. The inclusion of diet-induced anthocyanins, such as malvidin and its glycosides, may present a natural alternative for prevention against the development of neurological disorders. However, further research is necessary to fully comprehend the specific ways in which anthocyanins, including malvidin and its glycosides, can contribute to the protection of brain injury.

7. Potential Applications of Malvidin and Its Glycosides

The color of a food product plays a significant role in its overall appeal. Anthocyanins, including malvidin and its glycosides, offer appealing colors of red, purple, and blue. They are naturally abundant and pose no harm to consumers, making them desirable as natural color additives. However, their limited stability, especially in comparison to artificial dyes, has restricted their widespread utilization [6,56]. The chemical stabilization of anthocyanins is currently the primary focus of recent studies due to their abundant potential applications, beneficial effects, and use as an alternative to artificial colorants [20,176]. Pazmiño-Durán et al. conducted a study to assess the anthocyanin pigment content and profile found in *Oxalis triangularis* leaves, with the aim of exploring its potential as a natural food coloring agent. The results revealed a monomeric anthocyanin content of 195 mg per 100 g of leaves, based on malvidin-3,5-diglucoside. These findings indicate that *Oxalis triangularis* possesses an appealing hue and high anthocyanin content and is safe for consumption, making it a promising candidate as a natural colorant source [177]. In addition, Mojica et al. proposed that the utilization of anthocyanins extracted from black bean coats, particularly malvidin-3-glucoside, hold promise as a natural alternative for food coloring purposes. The study revealed the presence of 32 mg of anthocyanins per gram of dry extract. These findings indicate that black-bean coats also have the potential to serve as a viable source of natural food colorants [178].

Besides the color attributes, anthocyanins, including malvidin and its glycosides, have been reported to be beneficial to health (Figure 4), possessing potential physiological activities such as antioxidant and anti-inflammatory ones. Given the notable bioactivities of malvidin and its glycosides, there is significant promise for their utilization as beneficial dietary supplements. Nevertheless, limited research has specifically examined the potential applications of isolated or purified malvidin and its glycosides. Instead, the majority of studies has focused on investigating the properties of malvidin and its glycosides within plants or derived products. Cristian Del Bo′ et al. has suggested that anthocyanin-rich foods may improve cell antioxidant defense against DNA damage. In one study, ten young volunteers received one portion of blueberries or one portion of a control jelly. One portion (300 g) of the blueberries provided about 27 g of sugars (fructose and glucose) and 348 mg of anthocyanins (malvidin-3-galactoside, delphinidin-3-galactoside, and malvidin-3-arabinoside, making up more than 50% of the total anthocyanins content) [179]. The study conducted by Garcia-Alonso et al. explored the viability of employing an anthocyanidin-rich extract derived from *Vitis vinifera* grape peel. In the test meal, the consumed extract amount of 12 g contained a total of 183.8 mg of anthocyanin monoglucosides, with malvidin-3-glucoside being the most prevalent anthocyanin. The study aimed to evaluate the impact of red-wine-extract consumption on plasma antioxidant status and the production of monocyte chemoattractant protein 1 (MCP-1) in healthy individuals. Blood and urine samples were obtained from seven volunteers after the administration of the anthocyanin extract. The intact form of anthocyanins was detected in both plasma and urine samples, alongside other metabolites of anthocyanins. Furthermore, an enhancement in antioxidant capacity and a reduction in circulating levels of MCP-1 were observed in plasma [180]. Bakuradze conducted a 9-week study with 57 healthy male volunteers to investigate the biological effects of anthocyanin-rich fruit juice. The study included an initial 1-week washout period, followed by an 8-week intervention period where participants consumed either anthocyanin-rich fruit juice or a placebo. The red fruit juice had a total anthocyanin content of 274 mg/L, and among the detected anthocyanins were malvidin-3-glucoside and malvidin-3-galactoside. The anthocyanin-rich fruit juice demonstrated DNA-protective and antioxidant effects, leading to a significant reduction in background and total DNA strand breaks. Consumption of the anthocyanin-rich fruit juice also resulted in a significant decrease in body fat and an increase in fat-free mass. Furthermore, the activity of SOD was significantly elevated after consuming the anthocyanin-rich fruit juice, and the tested group showed decreased levels of LDL and total cholesterol. In conclusion, anthocyanin-rich fruit juice has the potential to improve DNA integrity and may have an impact on lipid metabolism in humans [181]. According to the studies of Mojica et al., black bean anthocyanin-rich extracts that contain malvidin-3-glucoside, via their ability to inhibit α-glucosidase, α-amylase, dipeptidyl peptidase-IV, and ROS and decrease glucose uptake, may have antidiabetic potential [178]. Kuntz et al. conducted a study to examine the effects of consuming beverages rich in anthocyanins on oxidation-related parameters in thirty healthy female volunteers. The participants were given 330 mL of beverages daily for 14 days, including a placebo, juice, and smoothie containing 8.9, 983.7, and 840.9 mg/L of anthocyanins, respectively. The juice and smoothie were made from a mixture of red grapes and bilberries in an 80:20 ratio. The main anthocyanin present in the juice and smoothie was malvidin-3-glucoside, with concentrations of 273 and 274 mg/mL, respectively. Blood and urine samples were collected before and after each intervention. Following the ingestion of anthocyanin-rich beverages, the subjects experienced a significant increase in plasma SOD and CAT activities, indicating enhanced antioxidant defense. The antioxidant capacity was also observed after consuming the juice and smoothie. Moreover, the concentrations of malondialdehyde, a marker of oxidative stress, decreased in both plasma and urine after consuming the anthocyanin-rich beverages. These findings suggest that anthocyanin-rich beverages may protect the body against oxidative stress, which is a key factor in the development of atherosclerosis [182].

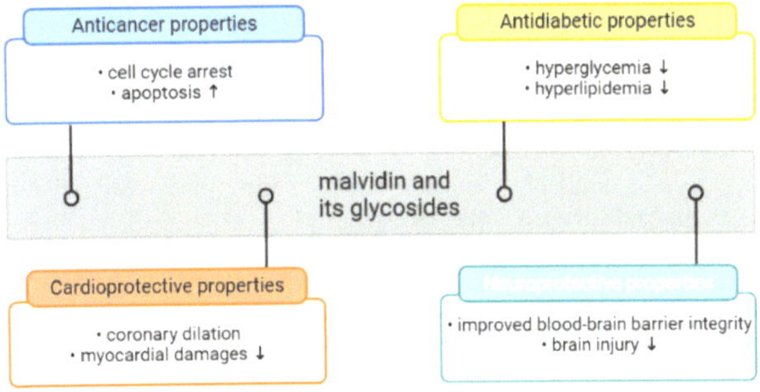

Figure 4. Health-benefit potential of malvidin and its glycosides (created by BioRender: https://www.biorender.com/ (accessed on 8 May 2023)). Downward arrow (↓) indicating decrease, upward arrow (↑) indicating increase.

Furthermore, several studies have investigated the effects of anthocyanins, including malvidin and its glycosides, on the human intestinal microbiota. Hidalgo et al. have suggested that malvidin-3-glucoside and its metabolites have the potential to influence the composition of the intestinal bacterial population by promoting the growth of *Bifidobacterium* spp. and *Lactobacillus–Enterococcus* spp. This positive modulation of the intestinal bacterial population is observed [183]. Vendrame et al. conducted a study on human volunteers to investigate the impact of consuming a wild blueberry drink for six weeks on the modulation of the intestinal microbiota. Each serving of the wild blueberry drink contained 375 mg of anthocyanins, including 49.5 mg of malvidin-galactose. The relative abundance of *Bifidobacterium* spp. significantly increased after the blueberry treatment, while *Lactobacillus acidophilus* levels also increased following the treatment. These findings suggest that regular consumption of a wild blueberry drink has a positive effect on the composition of the intestinal microbiota [184]. Zhou et al. conducted a study to assess the influence of blueberry anthocyanins on the human intestinal microbiota. Malvidin-3-glucoside was identified as the primary anthocyanin species, followed by malvidin-3-galactoside. The presence of anthocyanins was found to enhance the relative abundances of specific microbial communities, notably *Bifidobacterium* spp. These findings indicate that anthocyanins can significantly impact microbial diversity within the gut [185].

The multifaceted benefits of malvidin and its glycosides highlight their potential as nutrition enhancers. Not only can they provide appealing colors to food products, but they may also contribute to overall health and well-being. These findings suggest a potential market role for malvidin and its glycosides as functional food ingredients, offering both aesthetic and nutritional advantages. Further research and exploration of the properties and applications of malvidin and its glycosides are crucial in uncovering their full potential and maximizing their benefits for both the food industry and consumer health.

8. Conclusions

It can be concluded that malvidin and its glycosides possess significant anticancer, cardioprotective, antidiabetic, and neuroprotective properties due to their antioxidant and anti-inflammatory mechanisms of action. Various studies, including in vitro and in vivo, suggest that these molecules have the potential to counteract the onset and progression of several disease pathologies, in particular, with pathogenesis related to oxidative stress. Therefore, besides their colorant capacity, malvidin and its glycosides may have a wide range of health-promoting properties. However, further research is needed to fully understand the molecular mechanisms responsible for these effects and to explore potential new applications for these compounds. In addition, more detailed assessments of the efficacy of anthocyanin-rich products are required to support the development of new functional foods, dietary supplements, and pharmaceuticals. Overall, continued investigation into the biological activities of anthocyanins is essential for the development of novel therapeutic approaches and the promotion of public health.

Author Contributions: Conceptualization, A.M.-S., P.S. and T.K.; formal analysis, R.Z., M.J. and P.N.; data curation, K.Z.; writing—original draft preparation, A.M.-S.; writing—review and editing, P.S., T.K., M.J. and P.N.; visualization, K.Z.; supervision, R.Z. All authors have read and agreed to the published version of the manuscript.

Funding: This research received no external funding.

Institutional Review Board Statement: Not applicable.

Informed Consent Statement: Not applicable.

Data Availability Statement: Not applicable.

Conflicts of Interest: The authors declare no conflict of interest.

References

1. Kumar, S.; Pandey, A.K. Chemistry and biological activities of flavonoids: An overview. *Sci. World J.* **2013**, *2013*, 162750. [CrossRef]
2. Cook, N.S.; Samman, S. Flavonoids—Chemistry, metabolism, cardioprotective effects, and dietary sources. *J. Nutr. Biochem.* **1996**, *7*, 66–76. [CrossRef]
3. He, J.; Giusti, M.M. Anthocyanins: Natural colorants with health-promoting properties. *Annu. Rev. Food Sci. Technol.* **2010**, *1*, 163–187. [CrossRef] [PubMed]
4. Mattioli, R.; Francioso, A.; Mosca, L.; Silva, P. Anthocyanins: A Comprehensive Review of Their Chemical Properties and Health Effects on Cardiovascular and Neurodegenerative Diseases. *Molecules* **2020**, *25*, 3809. [CrossRef]
5. Sigurdson, G.T.; Tang, P.; Giusti, M.M. Cis-Trans Configuration of Coumaric Acid Acylation Affects the Spectral and Colorimetric Properties of Anthocyanins. *Molecules* **2018**, *23*, 598. [CrossRef]
6. Khoo, H.E.; Azlan, A.; Tang, S.T.; Lim, S.M. Anthocyanidins and anthocyanins: Colored pigments as food, pharmaceutical ingredients, and the potential health benefits. *Food Nutr. Res.* **2017**, *61*, 1361779. [CrossRef]
7. Kim, I.; Lee, J. Variations in Anthocyanin Profiles and Antioxidant Activity of 12 Genotypes of Mulberry (*Morus* spp.) Fruits and Their Changes during Processing. *Antioxidants* **2020**, *9*, 242. [CrossRef]
8. Hinojosa-Gómez, J.; San Martín-Hernández, C.; Heredia, J.B.; León-Félix, J.; Osuna-Enciso, T.; Muy-Rangel, M.D. Anthocyanin Induction by Drought Stress in the Calyx of Roselle Cultivars. *Molecules* **2020**, *25*, 1555. [CrossRef]
9. Sun, Y.; Zhang, Y.; Xu, W.; Zheng, X. Analysis of the Anthocyanin Degradation in Blue Honeysuckle Berry under Microwave Assisted Foam-Mat Drying. *Foods* **2020**, *9*, 397. [CrossRef] [PubMed]
10. Di Gioia, F.; Tzortzakis, N.; Rouphael, Y.; Kyriacou, M.C.; Sampaio, S.L.; Ferreira, I.C.F.R.; Petropoulos, S.A. Grown to be Blue-Antioxidant Properties and Health Effects of Colored Vegetables. Part II: Leafy, Fruit, and Other Vegetables. *Antioxidants* **2020**, *9*, 97. [CrossRef]
11. Fernandes, I.; Faria, A.; Calhau, C.; de Freitas, V.; Mateus, N. Bioavailability of anthocyanins and derivatives. *J. Funct. Foods* **2014**, *7*, 54–66. [CrossRef]
12. Alam, M.A.; Islam, P.; Subhan, N.; Rahman, M.M.; Khan, F.; Burrows, G.E.; Nahar, L.; Sarker, S.D. Potential health benefits of anthocyanins in oxidative stress related disorders. *Phytochem. Rev.* **2021**, *20*, 705–749. [CrossRef]
13. Willems, M.E.T.; Blacker, S.D. Anthocyanin-Rich Supplementation: Emerging Evidence of Strong Potential for Sport and Exercise Nutrition. *Front. Nutr.* **2022**, *9*, 864323. [CrossRef]
14. Husain, A.; Chanana, H.; Khan, S.A.; Dhanalekshmi, U.M.; Ali, M.; Alghamdi, A.A.; Ahmad, A. Chemistry and Pharmacological Actions of Delphinidin, a Dietary Purple Pigment in Anthocyanidin and Anthocyanin Forms. *Front. Nutr.* **2022**, *9*, 746881. [CrossRef] [PubMed]
15. Kong, J.M.; Chia, L.S.; Goh, N.K.; Chia, T.F.; Brouillard, R. Analysis and biological activities of anthocyanins. *Phytochemistry* **2003**, *64*, 923–933. [CrossRef] [PubMed]
16. Miller, R.; Owens, S.J.; Rørslett, B. Plants and colour: Flowers and pollination. *Opt. Laser Technol.* **2011**, *43*, 282–294. [CrossRef]
17. Kaur, S.; Tiwari, V.; Kumari, A.; Chaudhary, E.; Sharma, A.; Ali, U.; Garg, M. Protective and defensive role of anthocyanins under plant abiotic and biotic stresses: An emerging application in sustainable agriculture. *J. Biotechnol.* **2023**, *361*, 12–29. [CrossRef]
18. Wallace, T.C.; Giusti, M.M. Anthocyanins. *Adv. Nutr.* **2015**, *6*, 620–622. [CrossRef] [PubMed]
19. Castañeda-Ovando, A.; de Lourdes Pacheco-Hernández, M.; Páez-Hernández, M.E.; Rodríguez, J.A.; Galán-Vidal, C.A. Chemical studies of anthocyanins: A review. *Food Chem.* **2009**, *113*, 859–871. [CrossRef]
20. Tan, C.; Dadmohammadi, Y.; Lee, M.C.; Abbaspourrad, A. Combination of copigmentation and encapsulation strategies for the synergistic stabilization of anthocyanins. *Compr. Rev. Food Sci. Food Saf.* **2021**, *20*, 3164–3191. [CrossRef] [PubMed]
21. Konczak, I.; Zhang, W. Anthocyanins-More Than Nature's Colours. *J. Biomed. Biotechnol.* **2004**, *2004*, 239–240. [CrossRef]
22. Cappellini, F.; Marinelli, A.; Toccaceli, M.; Tonelli, C.; Petroni, K. Anthocyanins: From Mechanisms of Regulation in Plants to Health Benefits in Foods. *Front. Plant Sci.* **2021**, *12*, 748049. [CrossRef] [PubMed]

23. Horbowicz, M.; Kosson, R.; Grzesiuk, A.; Debski, H. Anthocyanins of Fruits and Vegetables—Their Occurrence, Analysis and Role in Human Nutrition. *Veg. Crop. Res. Bull.* **2008**, *68*, 5–22. [CrossRef]
24. Pojer, E.; Mattivi, F.; Johnson, D.; Stockley, C.S. The Case for Anthocyanin Consumption to Promote Human Health: A Review. *Compr. Rev. Food Sci. Food Saf.* **2013**, *12*, 483–508. [CrossRef] [PubMed]
25. Hornedo-Ortega, R.; Rasines-Perea, Z.B.; Cerezo, A.; Teissedre, P.L.; Jourdes, M. Anthocyanins: Dietary Sources, Bioavailability, Human Metabolic Pathways, and Potential Anti-Neuroinflammatory Activity. In *Phenolic Compounds—Chemistry, Synthesis, Diversity, Non-Conventional Industrial, Pharmaceutical and Therapeutic Applications*; IntechOpen: London, UK, 2022. [CrossRef]
26. Diaconeasa, Z.; Știrbu, I.; Xiao, J.; Leopold, N.; Ayvaz, Z.; Danciu, C.; Ayvaz, H.; Stănilă, A.; Nistor, M.; Socaciu, C. Anthocyanins, Vibrant Color Pigments, and Their Role in Skin Cancer Prevention. *Biomedicines* **2020**, *8*, 336. [CrossRef] [PubMed]
27. Wallace, T.C.; Giusti, M.M. Anthocyanins-Nature's Bold, Beautiful, and Health-Promoting Colors. *Foods* **2019**, *8*, 550. [CrossRef]
28. Iosub, I.; Kajzar, F.; Makowska-Janusik, M.; Meghea, A.; Geana, I.; Rau, I. Electronic Structure, Optical and Electrochemical Properties of Malvidin Molecule Extracted from Grapes. *Disp. Imaging* **2014**, *1*, 175–193.
29. Barnard, H.; Dooley, A.N.; Areshian, G.; Gasparyan, B.; Faull, K.M. Chemical evidence for wine production around 4000 BCE in the Late Chalcolithic Near Eastern highlands. *J. Archaeol. Sci.* **2011**, *38*, 977–984. [CrossRef]
30. Eker, M.E.; Aaby, K.; Budic-Leto, I.; Brnčić, S.R.; El, S.N.; Karakaya, S.; Simsek, S.; Manach, C.; Wiczkowski, W.; Pascual-Teresa, S. A Review of Factors Affecting Anthocyanin Bioavailability: Possible Implications for the Inter-Individual Variability. *Foods* **2019**, *9*, 2. [CrossRef]
31. Srinivas, K.; King, J.W.; Howard, L.R.; Monrad, J.K. Binary diffusion coefficients of phenolic compounds in subcritical water using a chromatographic peak broadening technique. *Fluid Phase Equilib.* **2011**, *301*, 234–243. [CrossRef]
32. Tanaka, Y.; Tsuda, S.; Kusumi, T. Metabolic Engineering to Modify Flower Color. *Plant Cell Physiol.* **1998**, *39*, 1119–1126. [CrossRef]
33. Mazza, G. Anthocyanins in grapes and grape products. *Crit. Rev. Food Sci. Nutr.* **1995**, *35*, 341–371. [CrossRef]
34. Tena, N.; Martín, J.; Asuero, A.G. State of the Art of Anthocyanins: Antioxidant Activity, Sources, Bioavailability, and Therapeutic Effect in Human Health. *Antioxidants* **2020**, *9*, 451. [CrossRef] [PubMed]
35. Rahman, M.M.; Ichiyanagi, T.; Komiyama, T.; Hatano, Y.; Konishi, T. Superoxide radical- and peroxynitrite-scavenging activity of anthocyanins; structure-activity relationship and their synergism. *Free Radic. Res.* **2006**, *40*, 993–1002. [CrossRef] [PubMed]
36. Pradhan, P.C.; Saha, S. Anthocyanin profiling of Berberis lycium Royle berry and its bioactivity evaluation for its nutraceutical potential. *J. Food Sci. Technol.* **2016**, *53*, 1205–1213. [CrossRef]
37. Müller, D.; Schantz, M.; Richling, E. High Performance Liquid Chromatography Analysis of Anthocyanins in Bilberries (*Vaccinium myrtillus* L), Blueberries (*Vaccinium corymbosum* L), and Corresponding Juices. *J. Food Sci.* **2012**, *77*, C340–C345. [CrossRef] [PubMed]
38. Huang, W.Y.; Zhang, H.C.; Liu, W.X.; Li, C.Y. Survey of antioxidant capacity and phenolic composition of blueberry, blackberry, and strawberry in Nanjing. *J. Zhejiang Univ. Sci. B* **2012**, *13*, 94–102. [CrossRef]
39. Fan-Chiang, H.J.; Wrolstad, R.E. Anthocyanin Pigment Composition of Blackberries. *J. Food Sci.* **2006**, *70*, 198–202. [CrossRef]
40. Lee, S.G.; Vance, T.M.; Nam, T.G.; Kim, D.O.; Koo, S.I.; Chun, O.K. Evaluation of pH differential and HPLC methods expressed as cyanidin-3-glucoside equivalent for measuring the total anthocyanin contents of berries. *Food Meas.* **2016**, *10*, 562–568. [CrossRef]
41. Mazza, G.; Kay, C.D.; Cottrell, T.; Holub, B.J. Absorption of anthocyanins from blueberries and serum antioxidant status in human subjects. *J. Agric. Food Chem.* **2002**, *50*, 7731–7737. [CrossRef]
42. Chew, L.Y.; Khoo, H.E.; Amin, I.; Azrina, A.; Lau, C.Y. Analysis of Phenolic Compounds of Dabai (*Canarium odontophyllum* Miq.) Fruits by High-Performance Liquid Chromatography. *Food Anal. Methods* **2012**, *5*, 126–137. [CrossRef]
43. García-Beneytez, E.; Revilla, E.; Cabello, F. Anthocyanin pattern of several red grape cultivars and wines made from them. *Eur. Food Res. Technol.* **2012**, *215*, 32–37. [CrossRef]
44. Kharadze, M.; Japaridze, I.; Kalandia, A.; Vanidze, M. Anthocyanins and antioxidant activity of red wines made from endemic grape varieties. *Ann. Agrar. Sci.* **2018**, *16*, 181–184. [CrossRef]
45. Lapidot, T.; Harel, S.; Granit, R.; Kanner, J. Bioavailability of Red Wine Anthocyanins as Detected in Human Urine. *J. Agric. Food Chem.* **1998**, *46*, 4297–4302. [CrossRef]
46. Gonçalves, A.C.; Campos, G.; Alves, G.; Garcia-Viguera, C.; Moreno, D.A.; Silva, L.R. Physical and phytochemical composition of 23 Portuguese sweet cherries as conditioned by variety (or genotype). *Food Chem.* **2020**, *335*, 127637. [CrossRef]
47. Gonçalves, A.C.; Bento, C.; Silva, B.M.; Silva, L.R. Sweet cherries from Fundão possess antidiabetic potential and protect human erythrocytes against oxidative damage. *Food Res. Int.* **2017**, *95*, 91–100. [CrossRef]
48. Martini, S.; Conte, A.; Tagliazucchi, D. Phenolic compounds profile and antioxidant properties of six sweet cherry (*Prunus avium*) cultivars. *Food Res. Int.* **2017**, *97*, 15–26. [CrossRef]
49. Borghesi, E.; González-Miret, M.L.; Escudero-Gilete, M.L.; Malorgio, F.; Heredia, F.J.; Meléndez-Martínez, A.J. Effects of salinity stress on carotenoids, anthocyanins, and color of diverse tomato genotypes. *J. Agric. Food Chem.* **2011**, *59*, 11676–11682. [CrossRef]
50. Blando, F.; Berland, H.; Maiorano, G.; Durante, M.; Mazzucato, A.; Picarella, M.E.; Nicoletti, I.; Gerardi, C.; Mita, G.; Andersen, Ø.M. Nutraceutical Characterization of Anthocyanin-Rich Fruits Produced by "Sun Black" Tomato Line. *Front. Nutr.* **2019**, *6*, 133. [CrossRef]
51. Martín, J.; Kuskoski, E.M.; Navas, M.J.; Asuero, A.G. Antioxidant Capacity of Anthocyanin Pigments. In *Flavonoids—From Biosynthesis to Human Health*; IntechOpen: London, UK, 2017. [CrossRef]

52. Lobo, V.; Patil, A.; Phatak, A.; Chandra, N. Free radicals, antioxidants and functional foods: Impact on human health. *Pharmacogn. Rev.* **2010**, *4*, 118–126. [CrossRef]
53. Nimse, S.B.; Pal, D. Free radicals, natural antioxidants, and their reaction mechanisms. *RSC Adv.* **2015**, *5*, 27986–28006. [CrossRef]
54. Ali, H.M.; Almagribi, W.; Al-Rashidi, M.N. Antiradical and reductant activities of anthocyanidins and anthocyanins, structure-activity relationship and synthesis. *Food Chem.* **2016**, *194*, 1275–1282. [CrossRef]
55. Schlesier, K.; Harwat, M.; Böhm, V.; Bitsch, R. Assessment of antioxidant activity by using different in vitro methods. *Free Radic. Res.* **2002**, *36*, 177–187. [CrossRef]
56. Enaru, B.; Dreţcanu, G.; Pop, T.D.; Stănilă, A.; Diaconeasa, Z. Anthocyanins: Factors Affecting Their Stability and Degradation. *Antioxidants* **2021**, *10*, 1967. [CrossRef] [PubMed]
57. Kähkönen, M.P.; Heinonen, M. Antioxidant activity of anthocyanins and their aglycons. *J. Agric. Food Chem.* **2003**, *51*, 628–633. [CrossRef] [PubMed]
58. Dudek, A.; Spiegel, M.; Strugała-Danak, P.; Gabrielska, J. Analytical and Theoretical Studies of Antioxidant Properties of Chosen Anthocyanins; A Structure-Dependent Relationships. *Int. J. Mol. Sci.* **2022**, *23*, 5432. [CrossRef]
59. Huang, W.; Zhu, Y.; Li, C.; Sui, Z.; Min, W. Effect of blueberry anthocyanins malvidin and glycosides on the antioxidant properties in endothelial cells. *Oxid. Med. Cell Longev.* **2016**, *2016*, 1591803. [CrossRef] [PubMed]
60. Toaldo, I.M.; Grootaert, C.; Bordignon-luiz, M.T. Bioactivity of malvidin 3, 5-diglucoside on endothelial cells as evidence of the antioxidant potential of V. In labrusca L. grape juices. In Proceedings of the 5th Annual Belgium Nutrition Society Congress—Adding Value to Nutrition Research, Brussels, Belgium, April 2015.
61. Paixão, J.; Dinis, T.C.; Almeida, L.M. Protective role of malvidin-3-glucoside on peroxynitrite-induced damage in endothelial cells by counteracting reactive species formation and apoptotic mitochondrial pathway. *Oxid. Med. Cell Longev.* **2012**, *2012*, 428538. [CrossRef]
62. Huang, W.Y.; Wu, H.; Li, D.J.; Song, J.F.; Xiao, Y.D.; Liu, C.Q.; Zhou, J.Z.; Sui, Z.Q. Protective Effects of Blueberry Anthocyanins against H_2O_2-Induced Oxidative Injuries in Human Retinal Pigment Epithelial Cells. *J. Agric. Food Chem.* **2018**, *66*, 1638–1648. [CrossRef] [PubMed]
63. Li, Y.; Xu, Y.; Xie, J.; Chen, W. Malvidin-3-: O-arabinoside ameliorates ethyl carbamate-induced oxidative damage by stimulating AMPK-mediated autophagy. *Food Funct.* **2020**, *11*, 10317–10328. [CrossRef]
64. Seo, H.R.; Choi, M.J.; Choi, J.M.; Ko, J.C.; Ko, J.Y.; Cho, E.J. Malvidin Protects WI-38 Human Fibroblast Cells Against Stress-induced Premature Senescence. *J. Cancer Prev.* **2016**, *21*, 32–40. [CrossRef]
65. Emadi, S.; Sadeghi, I.; Khastar, H. Malvadin Prevents Kidney from Renal Ischemia-Induced Oxidative Damage in Rats. *J. Pharm. Sci.* **2020**, *13*, 273–281.
66. Ranneh, Y.; Ali, F.; Akim, A.M.; Hamid, H.A.; Khazaai, H.; Fadel, A. Crosstalk between reactive oxygen species and pro-inflammatory markers in developing various chronic diseases: A review. *Appl. Biol. Chem.* **2017**, *60*, 327–338. [CrossRef]
67. Juan, C.A.; Pérez de la Lastra, J.M.; Plou, F.J.; Pérez-Lebeña, E. The Chemistry of Reactive Oxygen Species (ROS) Revisited: Outlining Their Role in Biological Macromolecules (DNA, Lipids and Proteins) and Induced Pathologies. *Int. J. Mol. Sci.* **2021**, *22*, 4642. [CrossRef]
68. Lugrin, J.; Rosenblatt-Velin, N.; Parapanov, R.; Liaudet, L. The role of oxidative stress during inflammatory processes. *Biol. Chem.* **2014**, *395*, 203–230. [CrossRef] [PubMed]
69. Chelombitko, M.A. Role of Reactive Oxygen Species in Inflammation: A Minireview. *Moscow Univ. Biol. Sci. Bull.* **2018**, *73*, 199–202. [CrossRef]
70. Ma, Z.; Du, B.; Li, J.; Yang, Y.; Zhu, F. An Insight into Anti-Inflammatory Activities and Inflammation Related Diseases of Anthocyanins: A Review of Both In Vivo and In Vitro Investigations. *Int. J. Mol. Sci.* **2021**, *22*, 11076. [CrossRef]
71. Kozłowska, A.; Dzierżanowski, T. Targeting Inflammation by Anthocyanins as the Novel Therapeutic Potential for Chronic Diseases: An Update. *Molecules* **2021**, *26*, 4380. [CrossRef]
72. Huang, W.; Wang, J.; Liu, Y.; Zheng, Q.; Li, C. Inhibitory effect of Malvidin on TNF-α-induced inflammatory response in endothelial cells. *Eur. J. Pharmacol.* **2014**, *723*, 67–72. [CrossRef] [PubMed]
73. Huang, W.Y.; Wang, X.N.; Wang, J.; Sui, Z.Q. Malvidin and its glycosides from vaccinium ashei improve endothelial function by anti-inflammatory and angiotensin I-converting enzyme inhibitory effects. *Nat. Prod. Commun.* **2018**, *13*, 49–52. [CrossRef]
74. Huang, W.Y.; Liu, Y.M.; Wang, J.; Wang, X.N.; Li, C.Y. Anti-inflammatory effect of the blueberry anthocyanins malvidin-3-glucoside and malvidin-3-galactoside in endothelial cells. *Molecules* **2014**, *19*, 12827–12841. [CrossRef] [PubMed]
75. Paixão, J.; Dinis, T.C.; Almeida, L.M. Malvidin-3-glucoside protects endothelial cells up-regulating endothelial NO synthase and inhibiting peroxynitrite-induced NF-kB activation. *Chem. Biol. Interact.* **2012**, *199*, 192–200. [CrossRef] [PubMed]
76. Bognar, E.; Sarszegi, Z.; Szabo, A.; Debreceni, B.; Kalman, N.; Tucsek, Z.; Sumegi, B.; Gallyas, F., Jr. Antioxidant and anti-inflammatory effects in RAW264.7 macrophages of malvidin, a major red wine polyphenol. *PLoS ONE* **2013**, *8*, e65355. [CrossRef] [PubMed]
77. Decendit, A.; Mamani-Matsuda, M.; Aumont, V.; Waffo-Teguo, P.; Moynet, D.; Boniface, K.; Richard, E.; Krisa, S.; Rambert, J.; Mérillon, J.M.; et al. Malvidin-3-O-β glucoside, major grape anthocyanin, inhibits human macrophage-derived inflammatory mediators and decreases clinical scores in arthritic rats. *Biochem. Pharmacol.* **2013**, *86*, 1461–1467. [CrossRef]
78. Bastin, A.; Sadeghi, A. The effects of malvidin on oxidative stress parameters and inflammatory cytokines in LPS-induced human THP-1 cells. *J. Cell. Physiol.* **2020**, *236*, 2790–2799. [CrossRef]

79. Bastin, A.R.; Sadeghi, A.; Abolhassani, M.; Doustimotlagh, A.H.; Mohammadi, A. Malvidin prevents lipopolysaccharide-induced oxidative stress and inflammation in human peripheral blood mononuclear cells. *IUBMB Life* **2020**, *72*, 1504–1514. [CrossRef]
80. Iban-arias, R.; Sebastian-valverde, M.; Wu, H.; Lyu, W.; Wu, Q.; Simon, J.; Pasinetti, G.M. Role of Polyphenol-Derived Phenolic Acid in Mitigation of Inflammasome-Mediated Anxiety and Depression. *Biomedicines* **2022**, *10*, 1264. [CrossRef]
81. Zhao, P.; Li, X.; Yang, Q.; Lu, Y.; Wang, G.; Yang, H.; Dong, J.; Zhang, H. Malvidin alleviates mitochondrial dysfunction and ROS accumulation through activating AMPK-α/UCP2 axis, thereby resisting inflammation and apoptosis in SAE mice. *Front. Pharmacol.* **2023**, *13*, 1038802. [CrossRef]
82. Dai, T.; Shi, K.; Chen, G.; Shen, Y.; Pan, T. Malvidin attenuates pain and inflammation in rats with osteoarthritis by suppressing NF-j B signaling pathway. *Inflamm. Res.* **2017**, *66*, 1075–1084. [CrossRef]
83. Zuo, L.; Prather, E.R.; Stetskiv, M.; Garrison, D.E.; Meade, J.R.; Peace, T.I.; Zhou, T. Inflammaging and oxidative stress in human diseases: From molecular mechanisms to novel treatments. *J. Mol. Sci.* **2019**, *20*, 4472. [CrossRef]
84. Soomro, S. Oxidative Stress and Inflammation. *Open J. Immunol.* **2019**, *9*, 1–20. [CrossRef]
85. Dominic, A.; Le, N.T.; Takahashi, M. Loop between NLRP3 Inflammasome and Reactive Oxygen Species. *Antioxid. Redox Signal.* **2022**, *36*, 784–796. [CrossRef]
86. Vandanmagsar, B.; Youm, Y.H.; Ravussin, A.; Galgani, J.E.; Stadler, K.; Mynatt, R.L.; Ravussin, E.; Stephens, J.M.; Dixit, V.D. The NLRP3 inflammasome instigates obesity-induced inflammation and insulin resistance. *Nat. Med.* **2011**, *17*, 179–188. [CrossRef] [PubMed]
87. Ma, Q. Role of Nrf2 in oxidative stress and toxicity. *Annu. Rev. Pharmacol. Toxicol.* **2013**, *53*, 401–426. [CrossRef]
88. Saha, S.; Buttari, B.; Panieri, E.; Profumo, E.; Saso, L. An Overview of Nrf2 Signaling Pathway and Its Role in Inflammation. *Molecules* **2020**, *25*, 5474. [CrossRef] [PubMed]
89. Tribble, D.L.; Jones, D.P. Oxygen dependence of oxidative stress. Rate of nadph supply for maintaining the GSH pool during hypoxia. *Biochem. Pharmacol.* **1990**, *39*, 729–736. [CrossRef]
90. Wang, B.; Wu, L.; Chen, J.; Dong, L.; Chen, C.; Wen, Z.; Hu, J.; Fleming, I.; Wang, D.W. Metabolism pathways of arachidonic acids: Mechanisms and potential therapeutic targets. *Signal Transduct. Target. Ther.* **2021**, *6*, 94. [CrossRef] [PubMed]
91. Wójcik, P.; Gęgotek, A.; Žarković, N.; Skrzydlewska, E. Oxidative stress and lipid mediators modulate immune cell functions in autoimmune diseases. *Int. J. Mol. Sci.* **2021**, *2*, 723. [CrossRef]
92. Reddy, K.K.; Vidya Rajan, V.K.; Gupta, A.; Aparoy, P.; Reddanna, P. Exploration of binding site pattern in arachidonic acid metabolizing enzymes, Cyclooxygenases and Lipoxygenases. *BMC Res. Notes* **2015**, *8*, 152. [CrossRef]
93. Liu, T.; Zhang, L.; Joo, D.; Sun, S.C. NF-κB signaling in inflammation. *Signal Transduct. Target. Ther.* **2017**, *2*, 17023. [CrossRef]
94. Lawrence, T. The nuclear factor NF-kappaB pathway in inflammation. *Cold Spring Harb. Perspect. Biol.* **2009**, *1*, a001651. [CrossRef] [PubMed]
95. Wei, Z.; Liu, H.T. MAPK signal pathways in the regulation of cell proliferation in mammalian cells. *Cell Res.* **2002**, *12*, 9–18. [CrossRef]
96. Whitmarsh, A.J.; Davis, R.J. Transcription factor AP-1 regulation by mitogen-activated protein kinase signal transduction pathways. *J. Mol. Med.* **1996**, *74*, 589–607. [CrossRef] [PubMed]
97. WHO. WHO | Cancer—Fact Sheet 2022. Available online: https://www.who.int/news-room/fact-sheets/detail/cancer (accessed on 18 April 2023).
98. Hayes, J.D.; Dinkova-Kostova, A.T.; Tew, K.D. Oxidative Stress in Cancer. *Cancer Cell* **2020**, *38*, 167–197. [CrossRef] [PubMed]
99. Schwartz, G.K.; Shah, M.A. Targeting the cell cycle: A new approach to cancer therapy. *J. Clin. Oncol.* **2005**, *23*, 9408–9421. [CrossRef]
100. Reuter, S.; Gupta, S.C.; Chaturvedi, M.M.; Aggarwal, B.B. Oxidative stress, inflammation, and cancer: How are they linked? *Free Radic. Biol. Med.* **2010**, *49*, 1603–1616. [CrossRef]
101. Vaidya, F.U.; Chhipa, A.S.; Sagar, N.; Pathak, C. Oxidative Stress and Inflammation Can Fuel Cancer. In *Role of Oxidative Stress in Pathophysiology of Diseases*; Springer: Singapore, 2020; pp. 229–258. ISBN 9789811515682.
102. Narendhirakannan, R.T.; Hannah, M.A.C. Oxidative stress and skin cancer: An overview. *Indian J. Clin. Biochem.* **2013**, *28*, 110–115. [CrossRef]
103. Nakamura, H.; Takada, K. Reactive oxygen species in cancer: Current findings and future directions. *Cancer Sci.* **2021**, *112*, 3945–3952. [CrossRef]
104. Schieber, M.; Chandel, N.S. ROS function in redox signaling and oxidative stress. *Curr. Biol.* **2014**, *24*, R453–R462. [CrossRef]
105. Braicu, C.; Buse, M.; Busuioc, C.; Drula, R.; Gulei, D.; Raduly, L.; Rusu, A.; Irimie, A.; Atanasov, A.G.; Slaby, O.; et al. A comprehensive review on MAPK: A promising therapeutic target in cancer. *Cancers* **2019**, *11*, 1618. [CrossRef]
106. Rascio, F.; Spadaccino, F.; Rocchetti, M.T.; Castellano, G.; Stallone, G.; Netti, G.S.; Ranieri, E. The pathogenic role of PI3K/AKT pathway in cancer onset and drug resistance: An updated review. *Cancers* **2021**, *13*, 3949. [CrossRef] [PubMed]
107. Park, M.T.; Kim, M.J.; Suh, Y.; Kim, R.K.; Kim, H.; Lim, E.J.; Yoo, K.C.; Lee, G.H.; Kim, Y.H.; Hwang, S.G.; et al. Novel signaling axis for ROS generation during k-ras-induced cellular transformation. *Cell Death Differ.* **2014**, *21*, 1185–1197. [CrossRef] [PubMed]
108. Perillo, B.; Di Donato, M.; Pezone, A.; Di Zazzo, E.; Giovannelli, P.; Galasso, G.; Castoria, G.; Migliaccio, A. ROS in cancer therapy: The bright side of the moon. *Exp. Mol. Med.* **2020**, *52*, 192–203. [CrossRef]
109. Liou, G.-Y.; Storz, P. Reactive oxygen species in cancer. *Free Radic. Res.* **2010**, *44*, 479–496. [CrossRef] [PubMed]

110. Singh, N.; Baby, D.; Rajguru, J.; Patil, P.; Thakkannavar, S.; Pujari, V. Inflammation and cancer. *Ann. Afr. Med.* **2019**, *18*, 121–126. [CrossRef] [PubMed]
111. Xia, L.; Tan, S.; Zhou, Y.; Lin, J.; Wang, H.; Oyang, L.; Tian, Y.; Liu, L.; Su, M.; Wang, H.; et al. Role of the NFκB-signaling pathway in cancer. *OncoTargets Ther.* **2018**, *11*, 2063–2073. [CrossRef]
112. Xia, Y.; Shen, S.; Verma, I.M. NF-κB, an active player in human cancers. *Cancer Immunol. Res.* **2014**, *2*, 823–830. [CrossRef] [PubMed]
113. Tolomeo, M.; Cascio, A. The multifaced role of stat3 in cancer and its implication for anticancer therapy. *Int. J. Mol. Sci.* **2021**, *22*, 603. [CrossRef] [PubMed]
114. Chen, X.; Andresen, B.; Hill, M.; Zhang, J.; Booth, F.; Zhang, C. Role of Reactive Oxygen Species in Tumor Necrosis Factor-alpha Induced Endothelial Dysfunction. *Curr. Hypertens. Rev.* **2008**, *4*, 245–255. [CrossRef]
115. Mantovani, F.; Collavin, L.; Del Sal, G. Mutant p53 as a guardian of the cancer cell. *Cell Death Differ.* **2019**, *26*, 199–212. [CrossRef]
116. Daniluk, J.; Liu, Y.; Deng, D.; Chu, J.; Huang, H.; Gaiser, S.; Cruz-Monserrate, Z.; Wang, H.; Ji, B.; Logsdon, C.D. An NF-κB pathway-mediated positive feedback loop amplifies Ras activity to pathological levels in mice. *J. Clin. Investig.* **2012**, *122*, 1519–1528. [CrossRef] [PubMed]
117. Diaconeasa, Z.M.; Frond, A.D.; Ştirbu, I.; Rugina, D.; Socaciu, C. Anthocyanins-Smart Molecules for Cancer Prevention. In *Phytochemicals—Source of Antioxidants and Role in Disease Prevention*; IntechOpen: London, UK, 2018.
118. Wang, L.S.; Stoner, G.D. Anthocyanins and their role in cancer prevention. *Cancer Lett.* **2008**, *269*, 281–290. [CrossRef]
119. Lin, J.; Tian, J.; Shu, C.; Cheng, Z.; Liu, Y.; Wang, W.; Liu, R.; Li, B.; Wang, Y. Malvidin-3-galactoside from blueberry suppresses the growth and metastasis potential of hepatocellular carcinoma cell Huh-7 by regulating apoptosis and metastases pathways. *Food Sci. Hum. Wellness* **2020**, *9*, 136–145. [CrossRef]
120. Wang, Y.; Lin, J.; Tian, J.; Si, X.; Jiao, X.; Zhang, W.; Gong, E.; Li, B. Blueberry Malvidin-3-galactoside Suppresses Hepatocellular Carcinoma by Regulating Apoptosis, Proliferation, and Metastasis Pathways in Vivo and in Vitro. *J. Agric. Food Chem.* **2019**, *67*, 625–636. [CrossRef]
121. Xu, H.; Zhang, J.; Huang, H.; Liu, L.; Sun, Y. Malvidin induced anticancer activity in human colorectal HCT-116 cancer cells involves apoptosis, G2/M cell cycle arrest and upregulation of p21WAFI. *Int. J. Clin. Exp. Med.* **2018**, *11*, 1734–1741.
122. Baba, A.B.; Nivetha, R.; Indranil, C.; Nagini, S. Blueberry and malvidin inhibit cell cycle progression and induce mitochondrial-mediated apoptosis by abrogating the JAK/STAT-3 signalling pathway. *Food Chem. Toxicol.* **2017**, *109*, 534–543. [CrossRef] [PubMed]
123. Dahlawi, H. Effect of Malvidin on Induction of Apoptosis and Inhibition of Cell Proliferation on Myeloid and Lymphoid Leukemia. *Sch. J. Appl. Med. Sci.* **2022**, *10*, 150–156. [CrossRef]
124. Hyun, J.W.; Chung, H.S. Cyanidin and Malvidin from Oryza sativa cv. Heugjinjubyeo Mediate Cytotoxicity against Human Monocytic Leukemia Cells by Arrest of G 2/M Phase and Induction of Apoptosis. *J. Agric. Food Chem.* **2004**, *52*, 2213–2217. [CrossRef]
125. Ouanouki, A.; Lamy, S.; Annabi, B. Anthocyanidins inhibit epithelial–mesenchymal transition through a TGFβ/Smad2 signaling pathway in glioblastoma cells. *Mol. Carcinog.* **2017**, *56*, 1088–1099. [CrossRef]
126. Sakthivel, K.M.; Kokilavani, K.; Kathirvelan, C.; Brindha, D. Malvidin abrogates oxidative stress and inflammatory mediators to inhibit solid and ascititumor development in mice. *J. Environ. Pathol. Toxicol. Oncol.* **2020**, *39*, 247–260. [CrossRef]
127. Cheng, Z.; Wang, Y.; Li, B. Illumina MiSeq reveals the influence of blueberry malvidin-3-galactoside on fecal microbial community structure and metabolizes of liver cancer mice. *PeerJ* **2018**, *6*, e27429v1. [CrossRef]
128. Pouwer, F.; Mezuk, B.; Tabák, A.G. Diabetes mellitus. In *The Routledge International Handbook of Psychosocial Epidemiology*; Kivimäki, M., Batty, D., Kawachi, I., Steptoe, A., Eds.; Routledge: Oxfordshire, UK, 2017.
129. WHO. *WHO Fact Sheet Diabetes*; WHO: Geneva, Switzerland, 2010; pp. 1–16.
130. American Diabetes Association. Diagnosis and classification of diabetes mellitus. *Diabetes Care* **2014**, *37* (Suppl. 1), S81–S90. [CrossRef] [PubMed]
131. Monnier, L.; Mas, E.; Ginet, C.; Michel, F.; Villon, L.; Cristol, J.P.; Colette, C. Activation of oxidative stress by acute glucose fluctuations compared with sustained chronic hyperglycemia in patients with type 2 diabetes. *JAMA* **2006**, *295*, 1681–1687. [CrossRef] [PubMed]
132. Evans, J.L.; Goldfine, I.D.; Maddux, B.A.; Grodsky, G.M. Oxidative stress and stress-activated signaling pathways: A unifying hypothesis of type 2 diabetes. *Endocr. Rev.* **2002**, *23*, 599–622. [CrossRef]
133. Karunakaran, U.; Park, K.G. A systematic review of oxidative stress and safety of antioxidants in diabetes: Focus on islets and their defense. *Diabetes Metab. J.* **2013**, *37*, 106–112. [CrossRef]
134. Nieto-Vazquez, I.; Fernández-Veledo, S.; Krämer, D.K.; Vila-Bedmar, R.; Garcia-Guerra, L.; Lorenzo, M. Insulin resistance associated to obesity: The link TNF-alpha. *Arch. Physiol. Biochem.* **2008**, *114*, 183–194. [CrossRef]
135. Prattichizzo, F.; De Nigris, V.; La Sala, L.; Procopio, A.D.; Olivieri, F.; Ceriello, A. "Inflammaging" as a Druggable Target: A Senescence-Associated Secretory Phenotype—Centered View of Type 2 Diabetes. *Oxid. Med. Cell Longev.* **2016**, *2016*, 1810327. [CrossRef]
136. Goto, M. Inflammaging (inflammation + aging): A driving force for human aging based on an evolutionarily antagonistic pleiotropy theory? *Biosci. Trends* **2008**, *2*, 218–230.

137. Belwal, T.; Nabavi, S.F.; Nabavi, S.M.; Habtemariam, S. Dietary anthocyanins and insulin resistance: When food becomes a medicine. *Nutrients* **2017**, *9*, 1111. [CrossRef]
138. Rózańska, D.; Regulska-Ilow, B. The significance of anthocyanins in the prevention and treatment of type 2 diabetes. *Adv. Clin. Exp. Med.* **2018**, *27*, 135–142. [CrossRef]
139. Herrera-Balandrano, D.D.; Chai, Z.; Hutabarat, R.P.; Beta, T.; Feng, J.; Ma, K.; Li, D.; Huang, W. Hypoglycemic and hypolipidemic effects of blueberry anthocyanins by AMPK activation: In vitro and in vivo studies. *Redox Biol.* **2021**, *46*, 102100. [CrossRef] [PubMed]
140. Huang, W.; Yan, Z.; Li, D.; Ma, Y.; Zhou, J.; Sui, Z. Antioxidant and anti-inflammatory effects of blueberry anthocyanins on high glucose-induced human retinal capillary endothelial cells. *Oxid. Med. Cell. Longev.* **2018**, *2018*, 1862462. [CrossRef]
141. Huang, W.; Hutabarat, R.P.; Chai, Z.; Zheng, T.; Zhang, W.; Li, D. Antioxidant blueberry anthocyanins induce vasodilation via PI3K/Akt signaling pathway in high-glucose-induced human umbilical vein endothelial cells. *Int. J. Mol. Sci.* **2020**, *21*, 1575. [CrossRef]
142. Li, D.; Zhang, Y.; Liu, Y.; Sun, R.; Xia, M. Purified anthocyanin supplementation reduces dyslipidemia, enhances antioxidant capacity, and prevents insulin resistance in diabetic patients. *J. Nutr.* **2015**, *145*, 742–748. [CrossRef]
143. Yang, L.; Qiu, Y.; Ling, W.; Liu, Z.; Yang, L.; Wang, C.; Peng, X.; Wang, L.; Chen, J. Anthocyanins regulate serum adipsin and visfatin in patients with prediabetes or newly diagnosed diabetes: A randomized controlled trial. *Eur. J. Nutr.* **2021**, *60*, 1935–1944. [CrossRef] [PubMed]
144. Hoggard, N.; Cruickshank, M.; Moar, K.M.; Bestwick, C.; Holst, J.J.; Russell, W.; Horgan, G. A single supplement of a standardised bilberry (*Vaccinium myrtillus* L.) extract (36% wet weight anthocyanins) modifies glycaemic response in individuals with type 2 diabetes controlled by diet and lifestyle. *J. Nutr. Sci.* **2013**, *2*, e22. [CrossRef] [PubMed]
145. Yazdanyar, A.; Newman, A.B. The Burden of Cardiovascular Disease in the Elderly: Morbidity, Mortality, and Costs. *Clin. Geriatr. Med.* **2009**, *25*, 563–577. [CrossRef]
146. WHO. *WHO Fact-Sheets Cardiovascular Diseases (CVDs)*; WHO: Geneva, Switzerland, 2021; pp. 1–5.
147. Pirillo, A.; Norata, G.D.; Catapano, A.L. LOX-1, OxLDL, and atherosclerosis. *Mediators Inflamm.* **2013**, *2013*, 152786. [CrossRef]
148. Gracia, K.C.; Llanas-Cornejo, D.; Husi, H. CVD and oxidative stress. *J. Clin. Med.* **2017**, *6*, 22. [CrossRef]
149. Li, J.; Stouffs, M.; Serrander, L.; Banfi, B.; Bettiol, E.; Charnay, Y.; Steger, K.; Krause, K.H.; Jaconi, M.E. The NADPH oxidase NOX4 drives cardiac differentiation: Role in regulating cardiac transcription factors and MAP kinase activation. *Mol. Biol. Cell* **2006**, *17*, 3978–3988. [CrossRef]
150. Gray, S.P.; Shah, A.M.; Smyrnias, I. NADPH oxidase 4 and its role in the cardiovascular system. *Vasc. Biol.* **2019**, *1*, H59–H66. [CrossRef] [PubMed]
151. Shekher, A.; Singh, M. Role of eicosanoid inhibition in ischemia reperfusion injury: Intact and isolated rat heart studies. *Methods Find. Exp. Clin. Pharmacol.* **1997**, *19*, 223–229.
152. Fattman, C.L.; Schaefer, L.M.; Oury, T.D. Extracellular superoxide dismutase in biology and medicine. *Free Radic. Biol. Med.* **2003**, *35*, 236–256. [CrossRef]
153. Fukai, T.; Ushio-Fukai, M. Superoxide dismutases: Role in redox signaling, vascular function, and diseases. *Antioxid. Redox Signal.* **2011**, *15*, 1583–1606. [CrossRef] [PubMed]
154. Wickremasinghe, D.; Peiris, H.; Chandrasena, L.G.; Senaratne, V.; Perera, R. Case control feasibility study assessing the association between severity of coronary artery disease with Glutathione Peroxidase-1 (GPX-1) and GPX-1 polymorphism (Pro198Leu). *BMC Cardiovasc. Disord.* **2016**, *16*, 111. [CrossRef] [PubMed]
155. Lubos, E.; Loscalzo, J.; Handy, D.E. Glutathione peroxidase-1 in health and disease: From molecular mechanisms to therapeutic opportunities. *Antioxid. Redox Signal.* **2011**, *15*, 1957–1997. [CrossRef]
156. Moss, J.W.E.; Ramji, D.P. Cytokines: Roles in atherosclerosis disease progression and potential therapeutic targets. *Future Med. Chem.* **2016**, *8*, 1317–1330. [CrossRef]
157. Wallace, T.C. Anthocyanins in cardiovascular disease. *Adv. Nutr.* **2011**, *2*, 1–7. [CrossRef]
158. Reis, J.F.; Monteiro, V.V.S.; Souza Gomes, R.; Carmo, M.M.; Costa, G.V.; Ribera, P.C.; Monteiro, M.C. Action mechanism and cardiovascular effect of anthocyanins: A systematic review of animal and human studies. *J. Transl. Med.* **2016**, *14*, 315. [CrossRef]
159. Del Bo, C.; Marino, M.; Riso, P.; Møller, P.; Porrini, M. Anthocyanins and metabolites resolve TNF-α-mediated production of E-selectin and adhesion of monocytes to endothelial cells. *Chem. Biol. Interact.* **2019**, *300*, 49–55. [CrossRef]
160. Quintieri, A.M.; Baldino, N.; Filice, E.; Seta, L.; Vitetti, A.; Tota, B.; De Cindio, B.; Cerra, M.C.; Angelone, T. Malvidin, a red wine polyphenol, modulates mammalian myocardial and coronary performance and protects the heart against ischemia/reperfusion injury. *J. Nutr. Biochem.* **2013**, *24*, 1221–1231. [CrossRef]
161. Study, A. Cardioprotective Effects of Malvidin Against Isoproterenol-Induced Myocardial Infarction in Rats: A Mechanistic Study. *Med. Sci. Monit.* **2017**, *23*, 2007–2016. [CrossRef]
162. Lee, J.; Finn, C.E.; Wrolstad, R.E. Anthocyanin pigment and total phenolic content of three Vaccinium species native to the Pacific Northwest of North America. *HortScience* **2004**, *39*, 959–964. [CrossRef]
163. Kang, H.J.; Ko, M.J.; Chung, M.S. Anthocyanin structure and ph dependent extraction characteristics from blueberries (*Vaccinium corymbosum*) and chokeberries (*Aronia melanocarpa*) in subcritical water state. *Foods* **2021**, *10*, 527. [CrossRef] [PubMed]
164. Johnson, S.A.; Figueroa, A.; Navaei, N.; Wong, A.; Kalfon, R.; Ormsbee, L.T.; Feresin, R.G.; Elam, M.L.; Hooshmand, S.; Payton, M.E.; et al. Daily blueberry consumption improves blood pressure and arterial stiffness in postmenopausal women with pre-

and stage 1-hypertension: A randomized, double-blind, placebo-controlled clinical trial. *J. Acad. Nutr. Diet* **2015**, *115*, 369–377. [CrossRef] [PubMed]
165. WHO. *Dementia Fact Sheet*; WHO: Geneva, Switzerland, 2022.
166. Liu, Z.; Zhou, T.; Ziegler, A.C.; Dimitrion, P.; Zuo, L. Oxidative Stress in Neurodegenerative Diseases: From Molecular Mechanism to Clinical Applications. *Oxid. Med. Cell Longev.* **2017**, *2017*, 2525967. [CrossRef] [PubMed]
167. Büttner, R.; Schulz, A.; Reuter, M.; Akula, A.K.; Mindos, T.; Carlstedt, A.; Riecken, L.B.; Baader, S.L.; Bauer, R.; Morrison, H. Inflammaging impairs peripheral nerve maintenance and regeneration. *Aging Cell* **2018**, *17*, e12833. [CrossRef] [PubMed]
168. Hu, W.T.; Howell, J.C.; Ozturk, T.; Gangishetti, U.; Kollhoff, A.L.; Hatcher-Martin, J.M.; Anderson, A.M.; Tyor, W.R. CSF cytokines in aging, multiple sclerosis, and dementia. *Front. Immunol.* **2019**, *10*, 480. [CrossRef]
169. Henriques, J.F.; Serra, D.; Dinis, T.C.P.; Almeida, L.M. The anti-neuroinflammatory role of anthocyanins and their metabolites for the prevention and treatment of brain disorders. *Int. J. Mol. Sci.* **2020**, *21*, 8653. [CrossRef]
170. Zhang, J.; Wu, J.; Liu, F.; Tong, L.; Chen, Z.; Chen, J.; He, H.; Xu, R.; Ma, Y.; Huang, C. Neuroprotective effects of anthocyanins and its major component cyanidin-3-*O*-glucoside (C3G) in the central nervous system: An outlined review. *Eur. J. Pharmacol.* **2019**, *858*, 172500. [CrossRef]
171. Lin, Y.C.; Tsai, P.F.; Wu, J.S.B. Protective effect of anthocyanidins against sodium dithionite-induced hypoxia injury in C6 glial cells. *J. Agric. Food Chem.* **2014**, *62*, 5603–5608. [CrossRef]
172. Gilani, S.J.; Bin-Jumah, M.N.; Al-Abbasi, F.A.; Imam, S.S.; Alshehri, S.; Ghoneim, M.M.; Shahid Nadeem, M.; Afzal, M.; Alzarea, S.I.; Sayyed, N.; et al. Antiamnesic Potential of Malvidin on Aluminum Chloride Activated by the Free Radical Scavenging Property. *ACS Omega* **2022**, *7*, 24231–24240. [CrossRef]
173. Lapi, D.; Chiurazzi, M.; Di Maro, M.; Mastantuono, T.; Battiloro, L.; Sabatino, L.; Ricci, S.; Di Carlo, A.; Starita, N.; Guida, B.; et al. Malvidin's effects on rat pial microvascular permeability changes due to hypoperfusion and reperfusion injury. *Front. Cell Neurosci.* **2016**, *10*, 153. [CrossRef] [PubMed]
174. Aarsland, D.; Khalifa, K.; Bergland, A.K.; Soennesyn, H.; Oppedal, K.; Holteng, L.B.A.; Oesterhus, R.; Nakling, A.; Jarholm, J.A.; de Lucia, C.; et al. A Randomised Placebo-Controlled Study of Purified Anthocyanins on Cognition in Individuals at Increased Risk for Dementia. *Am. J. Geriatr. Psychiatry* **2023**, *31*, 141–151. [CrossRef]
175. McNamara, R.K.; Kalt, W.; Shidler, M.D.; McDonald, J.; Summer, S.S.; Stein, A.L.; Stover, A.N.; Krikorian, R. Cognitive response to fish oil, blueberry, and combined supplementation in older adults with subjective cognitive impairment. *Neurobiol. Aging* **2018**, *64*, 147–156. [CrossRef] [PubMed]
176. Cortez, R.; Luna-Vital, D.A.; Margulis, D.; Gonzalez de Mejia, E. Natural Pigments: Stabilization Methods of Anthocyanins for Food Applications. *Food Sci. Food Saf.* **2016**, *16*, 180–198. [CrossRef]
177. Pazmiño-Durán, E.A.; Giusti, M.M.; Wrolstad, R.E.; Glória, M.B.A. Anthocyanins from Oxalis triangularis as potential food colorants. *Food Chem.* **2001**, *75*, 211–216. [CrossRef]
178. Mojica, L.; Berhow, M.; Gonzalez de Mejia, E. Black bean anthocyanin-rich extracts as food colorants: Physicochemical stability and antidiabetes potential. *Food Chem.* **2017**, *229*, 628–639. [CrossRef]
179. Del Bó, C.; Riso, P.; Campolo, J.; Møller, P.; Loft, S.; Klimis-Zacas, D.; Brambilla, A.; Rizzolo, A.; Porrini, M. A single portion of blueberry (*Vaccinium corymbosum* L.) improves protection against DNA damage but not vascular function in healthy male volunteers. *Nutr Res.* **2013**, 33, 220–227. [CrossRef]
180. Garcia-Alonso, M.; Minihane, A.M.; Rimbach, G.; Rivas-Gonzalo, J.C.; de Pascual-Teresa, S. Red wine anthocyanins are rapidly absorbed in humans and affect monocyte chemoattractant protein 1 levels and antioxidant capacity of plasma. *J Nutr. Biochem.* **2009**, *20*, 521–529. [CrossRef]
181. Bakuradze, T.; Tausend, A.; Galan, J.; Groh, I.A.M.; Berry, D.; Tur, J.A.; Marko, D.; Richling, E. Antioxidative activity and health benefits of anthocyanin-rich fruit juice in healthy volunteers. *Free Radic. Res.* **2019**, *53*, 1045–1055. [CrossRef]
182. Kuntz, S.; Kunz, C.; Herrmann, J.; Borsch, C.H.; Abel, G.; Fröhling, B.; Dietrich, H.; Rudloff, S. An-thocyanins from fruit juices improve the antioxidant status of healthy young female vol-unteers without affecting anti-inflammatory parameters: Results from the randomized, double-blind, placebo-controlled, cross-over Anthonia (Anthocyanins in Nutrition Investigation Alliance) study. *Br. J. Nutr.* **2014**, *112*, 925–936. [CrossRef] [PubMed]
183. Hidalgo, M.; Oruna-Concha, M.J.; Kolida, S.; Walton, G.E.; Kallithraka, S.; Spencer, J.P.; de Pascu-al-Teresa, S. Metabolism of anthocyanins by human gut microflora and their influence on gut bacterial growth. *J. Agric. Food Chem.* **2012**, *60*, 3882–3890. [CrossRef] [PubMed]
184. Vendrame, S.; Guglielmett, S.; Riso, P.; Arioli, S.; Klimis-Zacas, D.; Porrini, M. Six-week con-sumption of a wild blueberry powder drink increases bifidobacteria in the human gut. *J. Agric. Food Chem.* **2011**, *59*, 12815–12820. [CrossRef] [PubMed]
185. Zhou, L.; Xie, M.; Yang, F.; Liu, J. Antioxidant activity of high purity blueberry anthocyanins and the effects on human intestinal microbiota. *LWT* **2020**, *117*, 108621. [CrossRef]

MDPI AG
Grosspeteranlage 5
4052 Basel
Switzerland
Tel.: +41 61 683 77 34

Nutrients Editorial Office
E-mail: nutrients@mdpi.com
www.mdpi.com/journal/nutrients

www.ingramcontent.com/pod-product-compliance
Lightning Source LLC
Chambersburg PA
CBHW041408210926
51850CB00013B/3214
* 9 7 8 3 7 2 5 8 3 4 1 7 4 *